Metal-Catalysis in Industrial Organic Processes

"Chi ama la pratica senza la teoria è come il marinaio che s'imbarca senza bussola e sestante e non sa mai dove viene portato"
("Without theory to guide him the experimenter is as lost as a sailor setting out without compass or rudder")

Leonardo da Vinci

Metal-Catalysis in Industrial Organic Processes

Edited by

Gian Paolo Chiusoli
Department of Organic and Industrial Chemistry, University of Parma, Parma, Italy

Peter M. Maitlis
Department of Chemistry, The University of Sheffield, Sheffield, UK

RSCPublishing

Front cover design by Ting Chou Hu,
Department of Chemistry, The University of Sheffield, UK

ISBN-10: 0-85404-862-6
ISBN-13: 978-0-85404-862-5

A catalogue record for this book is available from the British Library

Published by The Royal Society of Chemistry,
Thomas Graham House, Science Park, Milton Road,
Cambridge CB4 0WF, UK

Registered Charity Number 207890

For further information see our web site at www.rsc.org

Typeset by Macmillan India Ltd, Bangalore, India
Printed and bound by Henry Ling Ltd, Dorchester, Dorset, UK

Preface

Catalysis is the driving force of the modern chemical industry. Although the chemical industry today is disliked by the media who use it as a convenient peg on which to hang many evils of modern society, it accounts for a very substantial proportion of the manufacturing wealth creation of a country. It has been estimated that some 20% of the total Gross Domestic Product is directly or indirectly due to the chemical industry.

Within the industry it is hard to overstate the importance of catalysis which allows the replacement of traditional stoichiometric reactions that are frequently polluting and wasteful of energy by milder, less energy consuming and more environmentally friendly, catalytic procedures. Many of these new industrial processes rely on catalysis by metals, and modern catalysts resemble Nature's enzymes more and more in their efficiency and selectivity: polymerization catalysts offer a good example. Thus we need to understand the way in which metal catalysts act and the logic that underlies their utilization. This is the philosophy that has motivated the present book. In seven chapters on the most important specific processes, plus two appendices on basic chemistry related to homogeneous and heterogeneous catalyses, it covers metal-catalyzed processes used in a wide variety of reactions from oxidation to polymerization, from modern processes for refining fuel and making fibres to synthesizing pharmaceuticals such as the profens.

The large elements of organic, organometallic, inorganic, and physical chemistry together with information on a variety of modern industrial processes, makes it ideally suited as an advanced general textbook for chemistry students and their teachers; it will also be welcomed by researchers in industrial and Government laboratories. We consider new as well as well-established industrial processes, emphasizing the improvements that modern catalyzed reactions offer over traditional routes. A typical example is making acetic acid where the last hundred years has seen the progressive replacement of (biological) fermentation first by hydrocarbon oxidation and then by metal-catalyzed methanol carbonylation. The last has indeed gone through three further changes of process catalyst involving first cobalt, then rhodium, and now iridium plus ruthenium, each bringing further improvements.

In each case we discuss the underlying mechanisms, which can then be used as a source of new ideas to overcome drawbacks of existing processes in order

v

to improve them or to invent new ones. We believe that students do not need too much mnemonic information since that is easy to find on the web today. They require instead to be shown the path of reasoning that will lead to improvements and to the discovery of new processes. Thus in addition to the descriptive text and illustrative Figures, we are including Discussion Points in each Chapter, which are intended to spark inquiry and debate with colleagues and teachers.

Our authors are researchers in industry and academia, from Germany, Italy, South Africa, Spain, and the UK and have worked all over the world in their fields, in teaching and in the chemical industry. We thank them warmly for their work and for sharing their knowledge with us. We also thank Marta Catellani, Cathy Dwyer, José Fraile, and Glenn Sunley for reading parts of the manuscript and for their comments and suggestions.

Gian Paolo Chiusoli (Parma, Italy) and Peter Maitlis (Sheffield, UK)
December 2005

Contents

Chapter 2 Formation of C–O Bonds by Oxidation
Mario G. Clerici, Marco Ricci and Giorgio Strukul

Chapter 4 **Syntheses Based on Carbon Monoxide**
Peter Maitlis and Anthony Haynes

Chapter 5   Carbon–Carbon Bond Formation

*Fausto Calderazzo, Marta Catellani and
Gian Paolo Chiusoli*

Chapter 7   Polymerization Reactions
Gerhard Fink and Hans-Herbert Brintzinger

Appendix 1   Basic Organometallic Chemistry Related to Catalytic Cycles
Peter Maitlis and Gian Paolo Chiusoli

Glossary

Glossary of abbreviations, acronyms and special terms used in *Metal-Catalysis in Industrial Organic Processes*

Term	Definition or meaning
ABS	copolymer of acrylonitrile/styrene grafted onto acrylonitrile/butadiene rubber.
acac	acetylacetonate (MeCOCHCOMe).
agostic bond	a C–H bond that interacts with a metal centre.
ammoxidation	oxidation of ammonia and propylene or propane with O_2 to give acrylonitrile.
ammoximation	the reaction of carbonyl compounds with ammonia, under oxidative conditions, to yield the corresponding oxime.
API	Active Pharmaceutical Ingredient.
atropoisomers	conformers which can give rise to enantiomers due to restricted rotation about a single bond.
autoxidation	the air-oxidation of organic compounds.
bite angle	the angle subtended by the two donor atoms at the metal in a bidentate ligand, for example 71.7° for $Ph_2PCH_2PPh_2$ (dppm), and 97.7° $Ph_2P(CH_2)_4PPh_2$ (dppb).
chiral switch	enantiopure preparation and commercialization of a compound that has been previously employed as racemate.
C_n symmetry	rotation of an object about this axis by an angle of 360/n gives a superposable object.
cone angle	a measure of the bulk of a ligand when attached to a metal centre: literally the solid angle of the ligand and its substituents subtended at the metal, introduced by Tolman. Typical values are 107°, $(P(OMe)_3$; 118°, PMe_3; 145°, PPh_3; 170°, PCy_3. There are now many variations, especially where allowances are made for "empty spaces" between substituents; see also *bite angle*.

(*continued*)

Term	Definition or meaning
cyclone	device using centrifugal force to separate powders.
dba	dibenzylideneacetone, $PhCH=CHCOCH=CHPh$; an easily synthesized ligand useful for stabilizing Pd(0), as $Pd_2(dba)_3$
diastereoisomers	stereoisomers not related as mirror images.
dispersion	the number of active sites per mass of a heterogeneous catalyst, related to the size and distribution of the primary particles of the active phase. Techniques of selective adsorption, e. g. CO for precious metals, are used to measure it.
E, Z-, trans-, cis-	E, Z describe the stereochemistry about double bonds; *trans* and *cis* denote configurations about single bonds in organic chemistry and the relative position of ligands in metal complexes.
emulsion polymerization	a system where the initiator is dissolved in an aqueous phase and the monomer is in the form of oil droplets; polymerization occurs in colloidal particles and leads to high molecular weights.
enantiomer	one of a pair of molecular entities which are mirror images of each other and non-superposable.
enantiomeric excess	the percentage of an enantiomer (R or S) over the racemate (R + S) abbreviated as $ee = (R-S/R+S)$ %.
enantioselective synthesis	preferential formation in a chemical reaction of one enantiomer over the other.
EPD	atactic, amorphous, rubberlike ethylene-propylene-diene terpolymer.
ethylene	common (older) name for ethene ($CH_2=CH_2$), but still widely used, especially in industry.
gasoline	common term (US) for petrol.
HDPE	crystalline, linear High-Density Polyethylene.
isomorphous substitution	the substitution of, for example a T-atom in a zeolite, by another having similar cationic radius and coordination.
kinetic isotope effect	the change in rate of a chemical reaction connected with the replacement of a C–H with a C–D bond.
LDPE	Low Density Polyethylene, particularly soft but tough arising from long-chain branches.

(*continued*)

Term	Definition or meaning
LLDPE	Linear Low-Density Polyethylene with soft, short-chain branches arising from 5–10% of C_3–C_8 olefins.
MFI	acronym defining the framework type of important zeolites, e.g., silicalite-1 (S-1), ZSM-5, titanium silicalite-1 (TS-1), according to the rules of International Zeolite Association (IZA).
mordenite	zeolite of hydrated calcium sodium potassium aluminium silicate.
MTBE; ETBE	methyl *tert*-butyl ether; (also ETBE, ethyl *tert*-butyl ether) octane enhancers for petrol (gasoline)
NADPH	Nicotinamide Adenine Dinucleotide Phosphate, reduced form
nanoparticles	particles of controlled size with diameters of the order of nanometers
naphtha	comprises the mixture of liquid *n*-alkanes, C_5H_{12} to C_9H_{20}
naphthenates	salts of naphthenic acids, cheap mixtures of carboxylic acids containing saturated carbocycles such as cyclopentyl and cyclohexyl rings.
Nylon-6,6	-[NH(CH$_2$)$_6$NHOC(CH$_2$)$_4$CO]$_n$-, a synthetic polyamide prepared by the two step polycondensation of hexa-methylenediamine and adipic acid.
Nylon-6	a synthetic polyamide prepared by the ring opening condensation of caprolactam
PB	atactic, amorphous poly(1-butene)
PET	-(CO-p-C$_6$H$_4$-COOCH$_2$CH$_2$O)$_n$-, a synthetic polyester prepared by esterification of terephthalic acid with methanol and subsequent polycondensation of the ester with ethylene glycol.
PIB	rubberlike polyisobutene
PMP	low-density, stereoregular poly(4-methyl-1-pentene)
polyether polyols	polymers of propylene oxide and polyhydric alcohols (eg., glycerol) used for the production of rigid (MW > 3000) and flexible (MW < 3000) polyurethanes
PP	crystalline, isotactic polypropene

(*continued*)

Term	Definition or meaning
prochirality	the geometric property of an achiral arrangement of atoms which is capable of becoming chiral in a single de-symmetrization step.
propylene	common (older) name for propene ($CH_3CH{=}CH_2$), but still widely used, especially in industry.
PS	polystyrene, oldest synthetic polymer, atactic, amorphous, clear and brittle.
Re-face, Si-face	a face of a trigonal atom is designated *Re* if the ligands of the trigonal atom appear in a clockwise sense in order of CIP (Cahn-Ingold-Prelog) priority when viewed from that side of the face. The opposite arrangement is termed *Si*.
stereogenic centre	a centre within a molecular entity that may be considered a focus of stereoisomerism.
stereospecific reaction	when starting materials differing only in their configuration are converted into distinct stereoisomeric products. Such a reaction also is stereoselective; if however the same stereo-isomer is formed the reaction is stereoselective but not stereospecific.
surface area	of a heterogeneous catalyst is calculated from the mono-layer capacity, which is defined as the amount of adsorbate needed to cover the surface with a complete monolayer of molecules. The experimental determination is through the adsorption-desorption isotherm of a chosen adsorbate of known dimensions, usually N_2 at 77 K.
T-atom	atom at a tetrahedral position in the lattice of a zeolite.
TDI	toluene diisocyanate, used to make polyurethanes
telechelic	polymer containing terminal functional groups
TS-1: Titanium silicalite-1	material obtained by the isomorphous substitution of Ti for Si in the lattice of the pure silica Silicalite-1 (S-1).
zeolites	aluminosilicates with framework structures enclosing cav-ities that can be occupied by water or large ions, and which can readily be exchanged; for details see *http://www.iza-structure.org/databases/*

CHAPTER 1

Introduction: Catalysis in the Chemical Industry

PHILIP HOWARD,[a] GEORGE MORRIS[b] AND
GLENN SUNLEY[b]
[a] BP Lubricants UK Limited, Pangbourne
[b] BP Chemicals Limited, Hull

1.1 Catalysis in the Chemical Industry

1.1.1 The Importance of Catalysis

In 1985 the National Academy of Sciences of the United States published a landmark study "Opportunities in Chemistry" which mapped out some of the important discoveries in the field over the preceding 20 years. The very first point made in the Pimentel Report (named after the chairman of the study group) was that the successful competitiveness of the chemical industry depends critically on the constant improvements of existing processes and the introduction of new ones. Thus advances in chemical catalysis and synthesis hold the key to a successful chemical industry. Indeed they estimated that a large proportion (*ca.* 20%) of the entire US Gross National Product is generated through the use of catalytic processes.

More than 20 years have elapsed since the Pimentel Report and all mankind has benefitted enormously from the improvements to our lives that catalysis has brought. We now have access to cheaper and more effective fuels, to new drugs and medications, new polymers and other materials with useful properties, and new routes to a whole host of commodity and fine chemicals. Especially significant are the new, energy-saving, and environmentally more friendly ("greener") methodologies that chemists have devised to make the chemicals. These changes have largely been brought about by better catalysts. And metal catalysts, developed jointly by industrial and academic chemists, form one of the main classes of present-day industrial catalysts. We now understand how many catalysts work and are beginning to tune them to high degrees of selectivity and activity; in some cases such catalysts now begin to rival enzymes, the catalysts of Nature.

1

This book is primarily a textbook that aims to help students see chemistry from the perspective of the industrial or academic scientist who wants to find new processes for making compounds. To do this it examines and classifies the transformations that organic compounds undergo when catalyzed by metals. Many new and profitable processes based on metal catalyses have been developed by industry. The intellectual stimulus that the study of catalytic reactions has given to chemistry is also reflected in the award of Nobel Prizes to the many who have made significant contributions to the science of catalysis and the role of metal catalysts, for example, Ostwald (1909), Haber (1918), Bergius and Bosch (1931), Natta and Ziegler (1963), Fischer and Wilkinson (1973), Knowles, Noyori, and Sharpless (2001) and, most recently, Chauvin, Grubbs and Schrock (2005). Their contributions range from the development of basic kinetic principles, to high pressure processes, and to new catalysts for hydrogenation, stereospecific polymerization, enantioselective reactions and olefin metathesis.

Ostwald recognized that catalysis was about the interplay of reaction rates, i.e. the fundamental role played by a catalyst to accelerate one kinetic pathway against several other different thermodynamically feasible pathways. Since the catalyst does not appear in the reaction product, catalysis has been to some extent the Cinderella of chemistry: vital, hard-working, but unrecognized. Although in a few cases consumers buy a product that contains a catalyst: the enzymes in detergents, the cerium oxide coating on the walls of ovens, the car exhaust catalyst, or the yeast to make bread, wine or beer – even there the consumers are not buying a catalyst – they are buying cleaner clothes, washed with less electricity, a self-cleaning oven, a car that pollutes less, and the means to make food and drink. Generally, the value of catalysis lies not in the catalyst itself but in the products or effects they produce.

Catalysts and catalysis are fundamental to being able to produce the fuels, polymers, medicines, plant growth regulators and herbicides, paints, lubricants, fibres, adhesives and a vast array of other consumer products. As well as catalysis contributing some 20% to the Gross Domestic Product of the USA, it is also estimated that 80% of all chemicals processes, with a total value of more than US$1800 billion, involve a catalyst at some point.

In 2001 it was estimated that the world merchant market for catalysts was worth ca. US$25 billion, divided roughly equally between refining, petrochemicals, polymers, environmental (20–25% each) and with about 11% being used in fine chemicals. Refining is about the production of fuels (Chapter 3, Box 2), petrochemicals cover many of the basic commodity chemicals and the monomers required for the polymer industries; fine chemicals include pharmaceuticals and agrochemicals, as well as flavours and fragrances; and environmental is about exhaust gas and waste product clean-up. Vehicle catalytic converters use catalysts, as does the production of the main tonnage polymers: polyethylene, polypropylene, polystyrene, polyvinyl chloride and polyethylene terephthalate.

But these catalyst sales figures do not reflect the number of tonnes or the value added by the catalysts in each sector. For example even though the use of catalysts (by volume) are similar for refining and polymers, the annual world production of gasoline in 2004 was about 1 billion tonnes, that of polyethylene was around 50 million tonnes.

Catalysts add value in many ways, ranging from reducing the cost of manufacture to increasing the quality of a chemical product, to the production of novel chemical compounds and to the reduction in environmental emissions. To take one example, the catalytic cracking of crude oil was started by Houdry in 1936 using simple silica/alumina catalysts in a fixed bed; further development of refining technology has not only enabled huge increases in volumes of gasoline fuel obtained from a barrel of oil (now about 50%), but has also led to dramatic quality improvements by increasing the octane rating, while reducing sulfur and aromatics. Processes used to bring this about include fluid catalytic cracking (FCC), isomerization, catalytic reforming, hydrotreating, and hydro-cracking (see Section 3.2).

Catalytic converters (containing precious metal catalysts dispersed on ceramic honeycomb structures that oxidize carbon monoxide and unburnt hydrocarbons to carbon dioxide and water, and reduce nitrogen oxides to nitrogen) are now fitted on more than 85% of new cars, and achieve emission reductions of over 90%.

Catalysis has also resulted in the creation of novel chemical structures, especially in the area of polymers, which give rise to new products and applications, as is discussed in Chapter 7. In 1933 the first polyethylene (PE), a highly branched flexible and soft polymer known as Low Density Polyethylene (LDPE), was made by the free radical polymerization of ethylene in a very high pressure (2000 bar) process. This was followed in the 1950s by catalytic routes to new higher density polyethylene (HDPE), and polypropylene (PP), Sections 7.3.1 and 7.3.2. One of the first contributions to society of the new HDPE was in providing a stiff, light and easy to mould material ideal for the 1958 Hula Hoops craze. Today we may complain about plastic bags littering the streets and plastic waste but the advantages of these light, strong, low cost, easy to manufacture, chemically resistant polymers are numerous – from HDPE pipes which do not rot or corrode underground for our water and gas pipes, to sterile, disposable medical equipment and to lightweight packaging keeping our food fresh whilst reducing freight costs.

Another example of how catalysis plays a key role in enabling our lives is in the synthesis of pharmaceuticals. Knowles's development, at Monsanto in the early 1970s, of the enantioselective hydrogenation of the enamide precursor to *L*-DOPA (used to treat Parkinson's disease), using a Rh-chiral phosphine catalyst (Section 3.5), led to a share in the Nobel prize. His co-laureates, Noyori and Sharpless, have done much to inspire new methods in chiral synthesis based on metal catalysis. Indeed, the dramatic rise in the demand for chiral pharmaceutical products also fuelled an intense interest in alternative methodologies, which led to a new one-pot, enzymatic route to *L*-DOPA, using a tyrosine phenol lyase, that has been commercialized by Ajinomoto.

1.1.2 Chemical Processes

Whether the end product is a pharmaceutical or a plastic bag, each industry is part of a value chain. The pharmaceutical company that markets the drug depends on fine chemicals or speciality producers to make the active pharmaceutical or its precursors. In turn, the commodity chemical companies supply the building blocks for the basic fine chemicals, whilst the refiners produce the feedstock for the commodity chemical companies.

If we consider a simplified, four-fold chain involving, 1) refining, 2) commodity (including petrochemicals and polymers), 3) fine (or speciality) chemicals, and 4) pharmaceuticals, the progression from refining to pharmaceuticals, is one of scale and product value – the latter being a consequence of the former. Refining and commodity chemicals are about very large plants, dedicated to one process. They are said to be *process intensive*: a lot of material is processed in one particular unit and so benefits from economies of scale. As the size of a reactor, for example, increases, the cost of manufacture increases roughly with its surface area. However, its throughput increases with its volume, hence the cost of the reactor per unit of feed is proportional to area/volume i.e. the capital cost only increases to the power of 2/3. In fact the actual value used is to the power 0.6 since manpower and other services do not increase with size. So as the process scale increases, so the product becomes cheaper. A single refinery unit, such as an FCC, can process up to 8 Mt/a. The units are run within narrow limits of feedstock quality and operating conditions. The crude oil feedstock comes literally straight out of the ground with little pre-treatment other than a simple distillation. As a consequence refining processes have huge economies of scale enabling profits to be made on what are sometimes very narrow margins. In addition carbon efficiencies are high, while manpower needs are relatively low, with continuous 24-hour operations. Since the products are rarely pure compounds, but often complex mixtures of hydrocarbons characterized by their boiling point and some simple chemical and physical properties, product separation is usually straightforward.

Commodity chemicals share the same benefits of economies of scale. Typical process units are 1–2 orders of magnitude smaller than refinery units, although large methanol synthesis plants can produce up to several Mt/a (Section 4.7.1). Since a chemically pure material is being produced, often in a stoichiometric reaction, the catalyst system now becomes more specialized, the reactor may require special metallurgy, and product purification starts to be an issue.

The products are not necessarily single compounds. Polymers for example are a range of compounds of varying molecular weights but which are chemically identical. For commodity chemicals the expense of feed purification, a more exotic catalyst, the corrosion resistant metallurgy and the need for a product purification train, start to add cost. But like refinery processes these are dedicated to one product, run 24 hours a day and are part of an integrated complex benefiting from heat and waste recovery. As the cost to build these refinery and petrochemical complexes is many billions of dollars, time is needed to pay back the investment. Once built, both commodity chemical and refinery processes will operate for more than 25 years producing the same product since either the product is well developed serving a large entrenched market, such as gasoline, or because the product has a multitude of uses, such as acetic acid, so that demand is not overly sensitive to the vicissitudes of any particular application. European production capacities of some major chemicals are estimated at, ethylene (24 Mt/a); propylene (17 Mt/a); ethylene dichloride (11 Mt/a); polypropylene, benzene (both 10 Mt/a); HDPE, ethylbenzene, and vinyl chloride monomer (all *ca.* 7 Mt/a); styrene, urea, and LDPE (all *ca.* 6 Mt/a); p-terephthalic acid and ester, methanol, and LLDPE (*ca.* 4 Mt/a). Other production capacities are given in the appropriate chapters; a useful rule of thumb is that world capacity of many chemicals is often around 3 times the European tonnage. Many important commodity chemicals (eg., adipic acid, caprolactam, glycols, acrylates, vinyl acetate) are produced by routes involving oxidation: the impacts of economics on these processes are also discussed in Chapter 2, Sections 2.2, 2.4, and 2.8–2.15.

The fine and speciality chemicals industries span a wide range from the manufacture of well defined, characterizable chemical compounds such as are used in pharmaceuticals, to making compounds which have a critical performance in a defined end-use application for a specific customer, such as inks for ink-jet printers where the design of the printer is closely matched to the performance of the ink. Indeed the manufacture of a catalyst is an example of a speciality chemical in itself.

The scale of production of a fine chemical can range from a few t/a up to tens of kt/a. Two of the final intermediates in the synthesis of the world's top selling drug, the anticholesterol Lipitor®, or atorvastin, are only produced on a scale of 500 t/a each. The consequence of the smaller scale and more specialist nature of these industries, is that the processes are far more likely to be batch or discontinuous. Thus process units will be flexible to produce more than one product and the product's lifetime in the market may be comparatively short, either for economic or performance reasons. However, product added value is high.

The final sector is the pharmaceutical (which can also extend to crop protection and agrochemicals) sector that produces a formulated product, the drug, which contains the Active Pharmaceutical Ingredient, API. Increasingly pharmaceutical companies are contracting out the synthesis of the API to the fine chemicals sector whilst focussing themselves on the drug discovery phase, the drug formulation and clinical trials. The drug formulation takes place using general-purpose equipment of low capital cost, where the emphasis is on quality control, product purity and avoidance of cross-contamination. The cost of production is much smaller compared to the price than in the fine chemicals sector. For the Pfizer drug Lipitor® one estimate puts the price of the drug at 20 times the cost of production of the API. But the R&D and clinical trial costs for the pharmaceutical industry are high and its R&D investment can be compared to the capital investments required in refining and commodity chemicals.

So how do the different characteristics of the various industrial sectors affect what they look for in developing and understanding catalysts, catalysis and catalytic processes, and how do they look for it? The most obvious difference is that as you go from refining to pharmaceuticals you need to spend a lot more money on R&D – both in absolute and relative terms. On average an oil and gas company spends about 1% of its sales on R&D, a purely commodity chemical company about 2.5–5%, a fine chemicals producer around 5–7% whilst the average for the pharmaceutical sector is about 15%. In 2004 among the world's top 700 companies, pharmaceutical/bio-industry companies spent $67 billion on R&D, the commodity chemical companies around $19 billion and the oil and gas industry about $5 billion. Pfizer alone spent $7 billion on R&D, being the second highest of all US industries, only behind Ford.

The reason for these large differences lies in the nature of the sectors. In refining and commodity chemicals more focus goes into improving existing processes. Since they are large scale and often replicated around the world, small improvements in selectivity, yield, heat recovery, catalyst cost or other process efficiencies, reap big rewards. There is relatively little emphasis on products since these are well defined and industry accepted. Due to the high cost of the feedstock as a percentage of the overall cost of production, new lower cost feedstocks are often a source for "step-out" processes. But small increases in process efficiencies can still result from detailed understanding of the kinetics, by-product formation or the catalytic cycles. The long life cycle of the product means that time can be invested to obtain this understanding, and one will get the full lifetime of a patent to protect from the competition. There is more emphasis on purely process improvements in refining, and more on the catalyst and mechanisms in commodity chemicals. This reflects the maturity of the refining processes where sufficient mechanistic understanding has been obtained and catalyst developments are still possible but ever harder to justify.

In the fine chemicals and pharmaceuticals sectors the emphasis is on the product, and getting new products to market as quickly as possible. Once a product is launched, depending on its projected life cycle, there may be little point in optimizing a process which is a smaller fraction of the product value and which may even become redundant within the time needed for further R&D. Fundamental understanding is more applied to new processes to give new products, to reduce the cost of a product to make a competitor's process uneconomic, or to the development of new applications for the product – both to increase its demand and prolong its product life. In the case of pharmaceuticals where quality control is very important, it may be that once a drug is approved, this approval is not only against a chemical structure but also against the process for its manufacture. Thus, any changes in process could require the lengthy and costly process of re-approval.

One of the consequences of this different emphasis is that whereas in the commodity chemicals sector one patent is produced for every $4.5 million of R&D, in pharmaceuticals the equivalent figure is $18 million. It is easier to make incremental inventions around one's own established process than make step-out discoveries in drug syntheses.

1.1.3 Evolution of the Catalysis Based Industries

The shape of the industries employing catalysis is continually changing, driven by economic, geographic, demographic, regulatory, consumer and environmental factors.

Although refining can be considered to be a mature industrial sector with no new refinery having been built in the USA for nearly 30 years, pressure is growing on both the volume and quality of transport fuels. The increased demand for diesel in Europe, due to its higher fuel efficiencies, the rapidly burgeoning car ownership in countries like Russia, India and China, and environmental pressures (such as reduction of diesel particulate emissions, reducing sulfur in all fuels and the need for high octane, clean gasoline; see Section 3.2), all require rethinking of fuel quality. So it is inevitable that new and more efficient refineries, producing even cleaner fuels will have to be built or existing complexes drastically improved. This will bring new challenges for catalysis.

In the chemicals sector, from commodity to pharmaceuticals, many changes are happening. Europe, the USA and Japan still have the largest chemical based manufacturing industries but over the period 1999–2004, the average *annual growth* in chemical production in China was 13.4%, 6.7% in India and 4.8% in the Middle East compared to 2.2% in the EU, 1.4% in the US and 1.3% in Japan. In 2004 China's chemical industry was one quarter the size of that of the EU, but is expanding rapidly, driven by a GDP in near-double-digit growth, with a dramatic rise in demand for all types of consumer goods.

The obvious driver for the expansion of the chemical industry in the Middle East is the local availability of the basic feedstocks – methane, ethane and naphtha. Qatar with one of the world's largest gas fields is attracting multi-billion dollar investments in gas conversion technology (GTL, Chapter 4) such as Fischer-Tropsch to produce high quality, sulfur free fuels (Section 4.7.2). It has been predicted that over the next 10 years 40% of the world's new ethylene production will be built in the Middle East. SABIC, the Saudi Arabian State Petrochemical Company was rated no.34 in the Chemical & Engineering News global chemical company survey in 1990, while in 2004 it was at no.12. Industry observers predict that Iran will soon have the second largest chemicals production in the region. The commodity products of the Middle East will be exported to the fast growing Asian markets, and to the USA and Europe. The latter could then become net importers as their local industries will not be able to compete with the cost of production in the Middle East.

In the fine chemicals and pharmaceutical sectors lower labour rates, increased local demand and acceptance of international standards in quality management are also taking manufacturing away from Europe and the US. India has become the no.1 manufacturer of the APIs for generic drugs; 70% of all such APIs used in Europe are now manufactured in India and China.

The manufacturing growth in the Middle East and Asia is not all "bad" news for the companies in the developed markets. The huge cost of establishing the infrastructure means that these markets are increasingly being opened up to foreign investors in order to access the capital, construction and operating expertise, and the latest technology.

Although the trends in the location of the manufacturing processes are clear, the ultimate key to profitability is largely in technology. The owner of a proprietary technology, whether that is a chemical process or a chemical compound and its application, can influence where a process is operated, how the products are marketed, or the cost of a licence to operate that process. International companies are building R&D centres in the growing Asian markets to access an increasingly educated local staff and to show their commitment to the country in which they are investing. But equally, national companies are now looking outwards; for example SABIC manufactures outside Saudi Arabia, invests in R&D facilities in the USA, Netherlands and India, and sponsors academic work around the world. Their drivers are the same – access to proprietary technology.

1.1.4 Applying Catalysis

For the chemical and pharmaceutical sectors R&D expenditure has been estimated to be split into 14% for basic, 27% applied and 59% development. These are very broad categories that mean different things to each industry sector. In commodity chemicals the activities range from catalyst discovery, mechanistic understanding and catalyst characterization to the construction and operation of pilot plants, customer trials, plant design and optimization, catalyst scale-up and validation. Whilst the role of the engineer – reaction, process or chemical – is well appreciated during the development, implementation, operation and optimization of the catalytic process, engineering is also an inherent part of the basic phase. Reactor and process design can overcome some deficiencies of the catalyst by designing regeneration, recovery and recycle systems to make an apparently unattractive system economic. And cost engineers can also direct the early stages of R&D towards the right goals in terms of catalyst and process performance. Increasingly today extensive computer modelling takes place before any plant is built. Many areas are involved: the suppliers of the raw materials, the manufacturers of catalysts, promoters, additives, and equipment, and collaborative ventures with Universities and customers. The situation is similar in fine chemicals but with greater emphasis on the product and its applications in the applied and development stages. For pharmaceuticals, the corresponding activities range from the identification of active families of chemicals, to drug discovery, process scale-up through to pre-clinical and clinical trials. Additionally R&D is also carried out post-drug approval in order to monitor long term patient safety and the drug effectiveness in broader patient populations.

The costs of these latter development activities and the possible failure of processes, formulations or APIs at these stages have serious financial implications for any organization. Whilst the expenditure on developing new catalysts and new polymers can cost many hundreds of millions of dollars, in pharmaceuticals the total R&D cost for a single drug is now approaching US$1 billion.

Catalysts are often divided into three groups: heterogeneous, homogeneous and biocatalytic. Each of the industrial sectors uses these classes to differing extents. Thus 90% of all refinery processes are heterogeneous whereas homogeneous catalysis is used in commodity and fine chemicals for highly selective conversions to products requiring little or no purification. Biocatalysis and homogeneous metal catalysis are increasingly used in the fine, pharmaceuticals and food industries. While the most widely used industrial reactions are still acid/base catalyzed, increasingly catalysts are more likely to be metal oxides, on economic and environmental grounds (see Section 5.2 for examples). Metal catalysis is becoming ever more important as the demands on selectivity, activity, and product purity increase.

Some 75% of all catalytic processes are heterogeneous but the detailed understanding often comes from homogeneous processes. It has been said that the first response of an industrialist is to heterogenize a catalyst since this leads to a potentially simpler process but that the academic often makes the case for the homogeneous since it is potentially easier to study. The actual situation is more complex, since even in refining several processes use homogeneous catalysis, while in commodity chemicals there are many examples of heterogeneous catalysis. Some of these are described in Chapter 5.

The complexity of many heterogeneous systems used in multi-phase reactions, the use of a solid support, the difficulty in analyzing highly dispersed active sites and the bifunctional nature of many solid supported metal catalysts, make a detailed and complete study challenging. The simpler homogeneous systems teach many of the principles of catalysis: active sites, reaction mechanisms, reaction kinetics and catalytic cycles, which can often be applied elsewhere.

1.2 Selection of a Chemical Process: What Does the Catalyst Do?

1.2.1 Feedstocks: Availability and Cost

In developing a new route to a chemical one of the first things a chemist needs to consider is feedstock availability and cost. This is the area where catalysis often has its biggest impact as it can facilitate the move from a higher- to a lower-cost feedstock. Indeed the discovery of a more efficient catalytic route to a product can result in the gradual or complete substitution of one type of feedstock for another. In extreme cases a new process may be so economically advantageous that older processes become uncompetitive and are shut down. The new producer can undercut the existing producer in price, take market share and still make a decent profit whilst the old producer can be left nursing losses, even when using a fully depreciated plant. Even if shutdown economics are not feasible most, if not all, new manufacturing capacity will tend to be based upon the most economically advantageous feedstock route.

A dramatic example of how metal catalysis can impact on the feedstock used to make an organic chemical can be seen in the evolution of the manufacture of acetic acid over the last 100 years or so, as summarized in Section 4.2. The methanol carbonylation based route has such an economic advantage over the other routes that currently more than 85% of acetic acid manufacturing capacity is based upon methanol carbonylation. The cost of making acetic acid via methanol carbonylation is approximately half of that of the two-step ethylene based Wacker route. A large part of this cost reduction can be attributed to the fact that methanol is approximately one third of the price of ethylene on a per tonne basis, though this ratio varies with time and market conditions.

The raw materials available for the manufacture of organic chemicals are crude oil, natural gas, coal and biomass. It is estimated that currently more than 95% of organic chemicals are derived from crude oil and natural gas, with only small contributions from coal and biomass. These raw materials are converted into the main building blocks of the modern organic chemicals industry, which are alkanes, alkenes, aromatics and synthesis gas. These materials can be considered to be the primary intermediates from which, in volume terms, most organic chemicals are made. There are of course exceptions such as the naturally occurring organic materials used to make relatively low volume products such as pharmaceuticals. An example of a pharmaceutical made from a natural product is Bristol-Myers Squibb's blockbuster cancer treatment drug Taxol$^{®}$, which is made from 10-deacetylbaccatin, a compound which is extracted from the needles of the English yew tree.

The dominance of crude oil and natural gas as sources of primary feedstock for organic chemicals stems from their ready availability in large volume at relatively low price, ease of handling and transportation, and the flexibility with which they can be used to generate most of the primary intermediates on which the organic chemicals industry is based. The availability and price of crude oil, natural gas and coal are very much tied to their primary use as fuels. Variations in supply and demand for the use of crude oil, natural gas and coal in the generation of energy thus have a strong impact on the cost of making an organic chemical. As the chemical industry is also a big energy user there is also a secondary effect on costs caused by variations in raw materials prices.

Today the main chemical use of coal is in the manufacture of synthesis gas and certain aromatic compounds. However coal is likely to play an increasingly important role in the manufacture of organic chemicals, particularly in the rapidly developing economies of Asia, some of which have relatively little crude oil and natural gas but large reserves of coal.

In volume terms the main chemical made from biomass is ethanol, more than 90% of which is made via fermentation, with so-called *synthetic ethanol*, made via ethylene hydration, accounting for the remainder. An increasingly important class of materials made from biomass are long chain methyl esters produced via the transesterification of glycerides. These materials can be blended into conventional diesel to produce what is known as biodiesel. A particularly interesting primary intermediate is *syngas* (synthesis gas; Section 4.1.2), which can be generated from virtually any carbon source, including biomass and waste materials. As such it has the potential to become a universal intermediate for a wide variety of organic chemicals.

1.2.2 Feedstocks: Thermodynamic and Kinetic Feasibility

When comparing a number of routes to a chemical a key consideration is the thermodynamic feasibility of the proposed reaction, the two thermodynamic parameters of interest being the Gibbs energy of reaction, ΔG_r, and the enthalpy of reaction, ΔH_r. The ΔG_r gives a view of the equilibrium conversion possible for a proposed reaction. To use it, we first write an idealized fully balanced equation for the conversion in question, treating the reaction as an equilibrium. The ΔG_r can then be calculated across a range of likely reaction temperatures using thermodynamic values from the literature.

As a general rule a reaction is not considered feasible industrially if the ΔG_r > *ca.* 40 kJ/mol, as the equilibrium conversion for the reaction lies too much in favour of the reactants to be practical. (It should be noted that the relevant number is the ΔG_r found at the temperature of the process; at different temperatures the ΔG_r will of course be different). There may be exceptions to this general rule of thumb when special circumstances prevail, such as the continuous removal of a product as it is formed, for example by the condensation, precipitation or use of a selective membrane. A further strategy is to use a large excess of one reactant over another, thus driving the conversion of the lower concentration reactant. The ΔG_r should only be used as a preliminary guide to reaction feasibility. A more thorough analysis, to focus down on the most promising routes, involves calculating the equilibrium constants from ΔG_r across a range of temperatures and subsequently the equilibrium mole fractions of reactants and products as a function of temperature and pressure. For reactions with gases which result in a change in molar reaction volume, pressure may be used to influence the position of the equilibrium conversion and in this way a reaction with a seemingly unfavourable ΔG_r can be made industrially viable. Thus for methanol synthesis, Equation 1, the equilibrium conversion can be moved to the right by increasing the reaction pressure. Conversely for ethylbenzene dehydrogenation, Equation 2, the equilibrium conversion to styrene is favoured by reducing the reaction pressure.

$$2H_2(g) + CO(g) \rightleftarrows MeOH(g) \tag{1}$$

$$PhCH_2CH_3(g) \rightleftarrows PhCH{=}CH_2(g) + H_2(g) \tag{2}$$

The interplay between temperature and equilibrium conversion provides a useful guide as to the appropriate temperature window under which to operate a chemical process. Clearly there is little point in operating a process at temperatures where the equilibrium conversion is negligible, as the higher the equilibrium conversion the better from a practical and cost point of view. In cases where the equilibrium conversion reduces with increasing temperature a compromise has to be struck between the temperatures required for a decent equilibrium conversion of reactants and that for a commercially viable rate of reaction.

A clear illustration of this occurs in the manufacture of methanol from CO and hydrogen (Section 4.7.1), where the first industrial process (commercialized by BASF in 1923) used a ZnO–Cr_2O_3 based catalyst. The equilibrium conversion of CO and hydrogen to methanol becomes increasingly unfavourable as the temperature of the reaction is increased; the ΔG_r is + 21 kJ/mol at 227°C and + 45 kJ/mol at 327°C. However the ZnO–Cr_2O_3 catalyst was of low activity and hence required high temperatures, 320–380°C, to achieve commercially viable rates. To achieve acceptable conversions per pass the process consequently ran at very high pressure, 340 bar. As a result the capital and energy costs for the process were high. In the 1960's ICI discovered an improved catalyst based upon Cu-Zn-Al-oxide, which operated under much milder conditions (50–100 bar and 240–260°C) and with higher selectivity. The capital and energy costs of this low pressure methanol process are consequently substantially less.

The enthalpy of reaction, ΔH_r, is the other important thermodynamic parameter to consider. On its own, whether a reaction is exothermic or endothermic will not determine if a reaction is industrially feasible or not. Both exothermic and endothermic processes are known in industry, methanol carbonylation to acetic acid (Equation 3; ΔH_r −123 kJ/mol at 200°C), being an example of the former and the steam reforming of methane to synthesis gas, (Equation 4; ΔH_r + 227 kJ/mol at 800°C), being an example of the latter.

$$MeOH + CO \rightarrow MeCOOH \qquad (3)$$

$$CH_4 + H_2O \rightarrow CO + 3H_2 \qquad (4)$$

The ΔH_r does however have an important bearing on plant design and costs. An endothermic reaction requires energy to drive it, which is an additional cost. The energy has to be supplied efficiently to the reaction section of the plant, which has a significant impact on reactor design and cost.

In the case of an exothermic reaction the energy generated needs to be efficiently removed otherwise a runaway reaction could occur with potentially disastrous consequences. The catalyst itself may also be sensitive to even small increases in reaction temperature which may shorten its lifetime or result in a loss in reaction selectivity. The energy generated in an exothermic reaction may be recovered, often as steam, and used as a cost credit; that energy has to be efficiently removed from the reaction section of the plant, again with impact on reactor design and cost.

Lastly a thermodynamically feasible reaction is not necessarily a commercially viable one, even if the feedstock costs are low. A second factor then comes into play, that of reaction kinetics. If a reaction is unfeasibly slow it will not be commercially viable. For example a very slow reaction may require a reactor so large it may not be economically practical. This is, of course, the role of catalysis, to speed up the rate of formation of a desired product, with a more selective catalyst speeding up the rate of formation of a desired product more than that of unwanted by-products. (We note however, that catalysis cannot change the equilibrium conversion for a reaction, as it is purely a kinetic phenomenon.)

1.2.3 Economics: The Costs of Making a Chemical

The fully built-up costs of manufacturing and delivering a chemical to market comprise several elements: *a*) *variable costs*: feedstock, utilities (cooling and process water, energy in the form of electricity, steam, and fuel), catalyst and other chemicals, including waste disposal; *b*) *fixed costs*: maintenance, labour, and overheads; *c*) *costs relating to depreciation and return on capital*: building the plant and associated equipment; and lastly *d*) *transportation costs*, including packaging. Catalysis can impact on nearly all of these elements, for example via the reaction rate and selectivity, reaction conditions, catalyst lifetime, effluent treatment and separation costs, plant design and materials of construction. To help illustrate how the fully built-up cost of a typical commodity organic chemical might be composed let us imagine we have developed a remarkable new process for the direct oxidation of methane to methanol, Equation 5, based upon a newly discovered selective precious metal catalyst. The main competing reaction is the over-oxidation of methane to CO_2 and water, Equation 6, which results in a significant loss of process yield on both methane and oxygen. The overall process yield to methanol for this hypothetical process is 85% based upon methane and 58% based upon oxygen. The fully built-up costs of such a process might break-down per tonne of product produced as follows: methane+oxygen, 54%; other chemicals and catalyst, 3%; utilities, 7%; depreciation, 5%; transportation, 5%; fixed costs, 12%; and lastly a contribution for return on investment of 14%. It has been estimated that such a direct single-step process has the potential to reduce the cost of making methanol by 25% compared with the established two-step route via syngas.

$$CH_4(g) + 0.5O_2(g) \rightarrow MeOH(g) \tag{5}$$

$$CH_4(g) + 2O_2(g) \rightarrow CO_2(g) + 2H_2O(g) \tag{6}$$

It should be noted that the exact cost breakdown will vary from case to case and will depend on a large variety of circumstances, even for the same process operated in different locations. The costs of making a speciality chemical, such as a pharmaceutical or an agrochemical, are more complex, as such products are generally speaking formulations with one or more active ingredients. In the direct methanol synthesis process the fully built-up cost is dominated by the raw materials, methane and oxygen. Indeed if one considers the cash cost of production, that is the cost before depreciation and return on capital are factored in, then feedstock makes up approximately two-thirds of the cost of production of a commodity chemical. For a refinery process, such as a fluid catalytic cracker, the contribution of feedstock to the cash cost can be as high as 80%. For fine chemicals the contribution of feedstock to costs is somewhat less, around 30 to 40%, and other costs, such as labour are correspondingly higher. Hence the reason why the efficient use of feedstock in terms of overall process yields is so important, particularly in the manufacture of refinery products and commodity chemicals. Catalysts have a crucial role to play here as they directly affect the reaction selectivity and conversion through their impact on reaction rates. In a once-through-process the yield is simply a function of conversion and selectivity. However many, particularly large scale, chemical processes are continuous and recycle unreacted material to ensure high overall process yields. Hence losses of unreacted material related to the efficiency of the recycling and other plant operations, such as process purges, also play a role in determining the overall process yield. Process purges are often necessary in continuous recycling processes to prevent a build-up of unwanted material, typically one or more of the by-products, in the reactor, which can eventually have a detrimental effect on the reaction kinetics. Thus reducing the formation of by-products by using a more selective catalyst can improve the overall process yield both directly and indirectly.

The costs relating to depreciation and return on investment relate directly to the overall capital cost of the plant and associated equipment. The higher the cost of the plant the higher the return has to be in absolute terms to justify the investment. A return of 10 to 20% of the capital employed is typically built into the fully-built up cost to reflect this. Again catalysis can have a significant impact here, for example a more selective catalyst can simplify the plant design and cost by reducing or eliminating the need to remove unwanted by-products. A more active catalyst may allow a smaller and less costly reactor to be used. A catalyst which enables a process to operate at lower pressure can also significantly reduce the cost of a plant. The choice of catalyst can also influence the materials of construction of a plant, since corrosive catalyst systems (containing for example strong acids or halides) will require expensive materials. Other chemicals are sometimes required in product and effluent treatment steps to remove unwanted by-products or residues and the costs of these can be reduced by catalyst improvements.

Utility costs can also be influenced by the catalyst used in a process. For example the energy used to compress gases can be decreased by a catalyst that allows the reaction pressure to be significantly reduced whilst retaining productivity. Examples here include low pressure processes for the manufacture of methanol from synthesis gas (Section 4.7.1) and the manufacture of linear low density polyethylene, LLDPE, by a metal catalysed polymerization compared to the high pressure radical catalyzed polymerization which gives low density polyethylene, LDPE (Section 7.2.1). Similarly a catalyst that reduces the amount of by-products made in a process can significantly reduce the energy required, for example in distillation, to purify the product. An example here is the highly selective production of acetic acid via methanol carbonylation compared to the unselective oxidation of naphtha (Section 4.2.2), which gives a slate of products (formic acid, acetic acid, propionic acid, and acetone) which require extensive distillation to separate. Improved catalyst life can also have a significant impact on catalyst cost, as the catalyst has to be replaced less often. This also reduces plant downtime and in effect increases plant capacity in terms of how much material can be made in a given time period. Substitution of an expensive catalyst for a less expensive one can also affect costs, for example the substitution of rhodium by lower cost iridium in methanol carbonylation.

Though catalyst costs only make up a of a small percentage of making a chemical it should be noted that, where margins are small and volumes are large, for example in commodity chemicals, a reduction in the cost of catalyst per tonne of product made can be very significant. In the case of a fine chemical if the cost of the catalyst is prohibitively high it may even make a catalytic route uncompetitive with a non-catalytic route. Currently in fine chemicals production there is increased emphasis on the efficient recovery of metal catalysts from batch processes via techniques such as immobilization and the use of novel reaction media such as supercritical fluids and ionic liquids [see, for example, *Chemical & Engineering News*, September 5, 2005, p 55].

The choice of catalyst may affect fixed costs via its effect on the overall plant design; a complicated catalytic process may require more maintenance than a relatively simple one for example. A catalyst improvement may also allow the throughput of a plant to be significantly increased, so in effect less labour and overhead is employed per unit of product manufactured. In the manufacture of much lower volume chemicals, such as fine chemicals used as intermediates to make pharmaceuticals, a selective catalytic conversion may allow one or more reaction steps to be eliminated and product work-up and purification to be significantly streamlined, which can reduce labour costs.

Catalysis has little direct impact on transportation costs for a given product. However catalysis can be used to convert gaseous and solid materials that are expensive to transport into cheaper to transport liquids. Examples of this include the conversion of methane in remote locations, via synthesis gas, into methanol or liquid hydrocarbons [Sections 4.1.2 and 4.7].

1.2.4 Safety and Environmental Impact

A further important criterion for the success of a chemical process is its environmental impact. Legislation, for example on emissions, can have an effect here. However many of the economic drivers discussed in Section 1.2.2 are also related to minimizing the impact of a chemical process on the environment; for example, the more efficient use of feedstock, catalyst or other chemicals results in less waste, reduced processing uses less energy and water, and increased plant throughput with minimum additional equipment means less energy and materials are required to make new plant items. Eliminating one or more processing steps can impact on all of these factors and more. Thus reducing costs and minimizing the impact on the environment are not mutually exclusive, and catalysis has an increasingly important role to play. Similarly, safety considerations are also crucial to the development of a successful chemical process.

1.2.5 Product Properties and Value

The price a chemical commands in the marketplace is a key factor in determining whether a chemical process for making it will be successful or not. Pricing is a complex issue but in simple terms the price of a chemical is determined by the balance of supply and demand. For a given producer the price of a chemical has to be sufficient to justify the cost incurred in making it, plus a reasonable return; if the price of a chemical falls below the cash cost of production for a significant time the producer will shut down capacity, either temporarily or permanently. Organic chemicals are sold as intermediates to make other chemicals and for their properties as materials, polymers, pharmaceuticals, fuels, herbicides, etc. The route and the catalysis chosen can affect the final nature and hence the value of a product in the marketplace. Catalysts which reduce or eliminate impurities can allow a price premium to be placed on a product or allow it to enter a new market. An example of the use of metal catalysis in improving product quality can be seen in the BP process for the manufacture of purified terephthalic acid (PTA), via the Co/Mn/bromide catalyzed oxidation of *p*-xylene. The reaction produces 4-carboxybenzaldehyde as a minor by-product. However the main use for terephthalic acid is in the manufacture of polyethylene terephthalate, PET, via a condensation reaction with ethylene glycol. The 4-carboxybenzaldehyde can interfere in this polymerization reaction by causing premature chain termination. PTA is obtained by hydrogenating the 4-carboxybenzaldehyde to 4-methylbenzoic acid using a Pd catalyst. The terephthalic acid can then be separated by fractional crystallization, giving a fibre grade product suitable for the manufacture of PET.

The impact of metal catalysts on product properties is dramatically illustrated by their widespread use in the manufacture of polyolefins with widely differing properties (Chapter 7). Using Ziegler-Natta catalysts ethylene can be polymerized at low pressure to make high density polyethylene, HDPE, which has high crystallinity. The crystallinity can be reduced by incorporating branches into the polymer structure via the incorporation of alpha-olefins such as 1-hexene, to make linear low density polyethylene, LLDPE. Metallocene based catalysts are particularly effective at making LLDPE and give products with improved tear and puncture resistance. Also because of the improved strength less material is required per application. For these reasons metallocene based polyethylenes can find niche applications and are often referred to as specialities rather than as commodities, as a premium can be charged for them based upon performance.

Another area where metal catalysis can have a profound influence on product value and properties is in enantioselective synthesis, through control of the environment around the co-ordination sphere of the metal via ligand modification. An example is in the manufacture of S-metolachlor, the active ingredient in Syngenta's Dual® Gold® herbicide. The use of a material greatly enriched in the S,S-diastereoisomer gives a much more active herbicide requiring less material to be used per application. The key step in the synthesis of S-metolachlor involves the enantioselective hydrogenation of an aryl-imine intermediate using an iridium/chiral phosphine based catalyst (Section 3.6).

1.2.6 What Makes a Successful Process?

There is no single answer as it very much depends on the competitive landscape. The nature of the industry determines criteria for success: a good commercial catalysis technology for a fine chemical or pharmaceutical may not be so for refining or a commodity chemical. The availability of a feedstock at a significantly reduced price, perhaps because of a catalyst breakthrough elsewhere, may make a new route to a chemical economic, and an established route uncompetitive. On the other hand the costs of existing processes are driven down a learning curve making them more competitive with time.

Market timing is also important. There may be only a limited amount of time to fill a gap in the marketplace before a competitor comes in and takes market share, therefore speed to market may be more important than developing a fully optimized synthetic route to a product. This is especially true for a pharmaceutical, as the product life cycle is short because clinical trials and product registration eat into the length of time a proprietary product has before it comes off patent protection and generic versions can enter the marketplace. Although it has been suggested that the development times for a new metal-catalyzed process are too long and costly to be suitable for the manufacture of pharmaceuticals, there are now many examples, some of which are discussed in Chapters 3 and 5, and metal catalysts are increasingly used also there [see also *Chemical & Engineering News* September 5, 2005, p.40].

To summarize: many factors (thermodynamics, kinetics, catalysis and other reaction chemistry, chemical analysis, engineering, instrumentation and process control, materials science and metallurgy, economics, environmental and safety issues, and market awareness) have to be considered in developing a successful new catalytic process. For these reasons it is a very team oriented activity, bringing together a whole host of disciplines. This makes it a challenging but rewarding experience.

1.3 Developing Metal-Catalysis – the Role of Fundamental Understanding

Let us suppose that, using the principles described in the previous sections, we have been through an exercise and identified a reaction that might be the basis for a chemical product. For a commodity chemical involving a relatively simple transformation we have suitable feedstocks or reactants and we are confident that the thermodynamics are adequate. For a fine chemical we will probably wish to carry out a reaction at one reactive centre in the molecule, in the presence of others. We then need to find answers to several questions. How do we go about designing or discovering a new or better catalyst to bring about the desired reaction? What is the basis of our understanding of catalysts, and how do we apply these fundamental ideas? And, finally, how well can we predict the outcome of our experiments now, and what are the prospects for the future?

1.3.1 Catalytic Cycles

A feature common to all catalysts is the catalytic cycle, which is a main theme of this book. Such a cycle is a series of reaction steps involving the catalyst species and the reactants which leads to the products. So far as we are concerned here, the catalysts will largely be metals, either as liganded complexes (soluble or supported), or as aggregates (nanoparticles) on a surface, which is typically the metal, an oxide, or other support. While the catalysis is taking place the catalyst is continually changing from one state to another at the molecular or atomic level; however after the reaction is over it may be recovered in its original form.

Steps in the catalytic cycle typically involve *a*) the activation of the metal centre into a reactive form; *b*) the movement of the reactants to the active centre, and the binding and activation of the reactants; *c*) the making and breaking of the reactant bonds to generate new species; which *d*) are then liberated as products, with the return of the catalyst to a state where it is ready to begin the cycle again and bind further reactants once more.

In an ideal situation all these steps follow sequentially at rates which allow an effective cycle to be maintained. However many things can go wrong: we may have chosen an inappropriate solvent so that the substrates did not bind, or one substrate might bind too strongly and not react, the coupling reaction(s) might not take place, or the product might be too firmly bound and hence not liberated. In such cases nothing happens and there is no catalysis. An alternative situation occurs when the desired steps do occur but only in part, or too slowly. How can we improve the steps so that the cycle runs better (faster)?

1.3.2   How we Study What a Catalyst Does

Our understanding of a catalytic cycle is based on a variety of data. At the macroscopic level we make observations about which substrates do or do not react, the products they give, and the kinetics of the reaction. At the microscopic level we can study the structure around the metal and how it interacts with individual molecules, and we can study single steps with model reactions. Even a single reaction step can prove subtle or complex, and the whole catalytic cycle for a reaction may be the result of years of meticulous study or no more than a plausible speculation.

1.3.3   Reaction Kinetics and the Catalytic Cycle

We can measure the rates of the overall catalyzed reaction as a function of variables such as catalyst concentration, substrate concentration, temperature, solvent (for a liquid phase reaction), gas flow rate (over a heterogeneous catalyst) and so on. Such a kinetic analysis gives a useful guide, but it must then be interpreted in terms of a chemically meaningful model.

In a catalytic cycle where the reaction passes through a number of steps there is the concept of the *rate controlling step*, just as in any other reaction. It is defined as the step which, if it is speeded up or slowed down by changing its rate constant (as opposed to the concentration terms), has the greatest relative effect on the overall reaction rate. Under those conditions it is the key step in the reaction, and if we can speed up that step we speed up the overall process, or if we can speed it up relative to a side reaction we can improve the overall selectivity. As concentrations are varied the rate controlling step of the catalytic cycle may change. A good example is in the Rh catalysed carbonylation of MeOAc to Ac_2O (Section 4.2.8) where intermediate MeCOI can build up at high Rh concentrations and MeOAc concentration rather than Rh concentration becomes rate controlling. Similar kinetic effects are seen in heterogeneous catalysis, for example when a surface becomes saturated with a reactant or reactants, and in enzyme reactions due to formation of intermediate enzyme substrate complexes.

The rates of the overall reactions can be related to the rate law expressions of the individual steps by using the *steady state approximation*. However simple kinetic data alone may not distinguish a mechanism where, for example, a metal and an olefin form a small amount of complex at equilibrium that then goes on to react, from one in which the initial complex undergoes dissociation of a ligand and then reacts with the olefin. As a reaction scheme becomes more complex such steady state approximations become more complicated, but numerical methods are now available which can simulate these even for complex mixtures of reactants.

In looking at reaction kinetics we are generally considering the relationship between reaction rate and concentrations in at least one of the bulk phases present – typically either the molar concentration in a liquid phase or the partial pressure in the gas phase. The kinetics at the catalyst site, whether it is homogeneous or heterogeneous, may not be reflected in the rates of the overall reaction because of mass transfer effects. These may involve a gas dissolving in a liquid to reach a homogeneous catalyst site, or the diffusion of a gas or liquid through the pores of a solid to reach a heterogeneous catalyst site. Such effects may become apparent if certain parameters of the system are varied, such as the rate of mixing in a gas/liquid reaction or the catalyst particle size in a gas/solid process.

As well as looking at reaction rates as a function of concentration and factors which might affect mass transfer we can also investigate the influence of temperature on reaction rate. For any single reaction step the rate constant of the reaction can be expressed either in the form of the Arrhenius equation or, more usually, for the purposes of interpreting the data mechanistically, the Eyring equation. The enthalpy and entropy of activation of the reaction, $\Delta H^{\ddagger}$ and $\Delta S^{\ddagger}$ are the parameters that inform us about the transition state on the reaction pathway of the rate determining step. The enthalpy of activation is the energy required to go from reactants to the transition state, while the entropy of activation is interpreted as being related to the relative order in the reactants and the transition state. A negative (decreasing) entropy of activation may indicate greater order and a positive (increasing) entropy of activation less order. Thus a positive entropy of activation can indicate a dissociative reaction, a bound species leaving a metal site for example, whereas a negative entropy of activation may indicate an associative process of a catalyst and a reactant species.

In an industrial context the reaction kinetics are often studied anyway not only for insight into a catalyst mechanism, but to determine, for example, the required reactor size and to predict how the reaction is to be controlled.

1.3.4   Model Studies – Structures and Reactions

We also need to understand the nature of the catalyst itself and to identify the metal species present. In a cycle with several steps several species will be present. Characterizing the metal species present in a catalyst system and relating them to the mechanistic cycle is generally easier for homogeneous systems because a range of spectroscopic methods may be used to study them at the molecular level. The most important of these are NMR and IR spectroscopy which, under favourable conditions, may even be used to observe metal species while catalysis is taking place. Structural information from X-ray crystallography of metal complexes is also useful to define metal species isolated from catalyst systems. However a single species or form of the catalyst may predominate and secondary species may be difficult to detect and quantify; it may even be that the major form of the metal complex is not directly involved in the catalytic cycle, as found in the homogeneous hydrogenation of prochiral enamides catalysed by chiral rhodium phosphine complexes where it turns out that the minor, less stable isomer is the more reactive and leads to most of the product (see Section 3.5.2).

Sometimes it is not the catalyst metal itself which is used but a neighbouring metal with similar properties. For example, the methylrhodium complex $[Rh(Me)(CO)_2I_3]^-$ was widely accepted as an intermediate in the carbonylation of MeOH to AcOH involving $[Rh(CO)_2I_2]^-$ and MeI even though it had never been observed. This was in part because the corresponding Ir complex was stable and well characterized. The rhodium species was subsequently observed by carrying out the reaction in a non polar solvent at a very high [MeI] which increased the rate of its formation and slowed down further reaction (see Section 4.2.4).

For heterogeneous catalysts it is more difficult to study the metal sites and their interaction with reactants under catalytic conditions. Many solid catalysts consist of metal crystallites or oxide particles in a variety of sizes and forms which themselves contain a number of different environments for the metal. Much of the routine characterization of these catalysts takes place before and after, rather than during, the catalytic reaction. Other studies probing surface bound species may have to be carried out under high vacuum rather than under the typical working conditions of the catalyst (see Appendix B for a summary of some relevant aspects of surface science).

Thus even if we do not observe the intermediate species in a metal catalysed reaction under working conditions, we can make good models or manipulate the conditions to enable us to observe these intermediates, or in some way modify the metal complexes to make the structural types we believe are important.

1.3.5   How to Apply Understanding to Discovering and Improving Catalysts

Since a considerable body of knowledge has now been built up on all the aspects of metal catalyzed reactions, in order to select a catalyst for a new reaction or to look for a new catalyst for a known reaction we can call on the information available about the structure and activity of catalyst systems. In some systems we can systematically vary the steric and electronic properties of the ligand system. Another possibility might be to mimic unusual biocatalysis mechanisms, or to use structured supports with novel properties. All of this work is about controlling the activity of the metal centres by manipulating the metal environment. Sometimes the idea and understanding may not be quite right, but can nevertheless lead to interesting results.

The workers at Monsanto started their search for a catalyst for MeOH carbonylation starting from a rhodium phosphine hydroformylation complex. When this did not work they had the insight to add MeI, knowing the propensity of MeI to react with metals such as Rh and it worked. We now know that the hydroformylation catalyst had been destroyed by the reaction and that the MeOH carbonylation catalyst would be a poor catalyst for hydroformylation. The time from deciding to attempt the reaction to discovering the catalyst was about twelve days. In contrast, the time taken by the group at Ciba-Geigy to perfect the catalyst for the key enantioselective hydrogenation step in the synthesis of S-metolachlor was thirteen years during which time they had to learn by testing a great number of options for catalyst and ligand.

As the target of synthesis becomes increasingly difficult for a catalyzed reaction molecular modelling has been widely used. The efficiency of modelling algorithms coupled with increasing computer power and in particular the widespread acceptance of density functional theory (DFT) methods has made calculations of structures and transition states for quite complicated metal ligand species widely accessible. However even for a single reaction the accuracy of these methods for transition metal species cannot yet reliably predict relative rates when it is considered that a few kJ/mol difference in activation energies between two competing pathways results in an order of magnitude difference in relative rates. Nevertheless, approximations will probably continue to become more accurate and methods for exploring wider ranges of potential pathways in a reasonable time will develop.

Just as computing power has increased in recent years so has the capability of analytical and robotic techniques. The methods of small scale high throughput and combinatorial screening, initially developed for testing pharmaceuticals, are now being applied to catalysis research. These methods are enabling the rapid testing of large numbers of component combinations such as metals and supports. This is often done within a framework of knowledge that certain metal or ligand combinations are expected to have activity but also includes the search for novel formulations.

References

Application of thermodynamics to industrial problems, including the use of ΔG_r in assessing reaction feasibility, is discussed in detail with worked examples in Chapter 7 of D. R. Stull, E. F. Westrum and G. C. Sinke,

The Chemical Thermodynamics of Organic Compounds, Krieger, Malabar, 1987. Thermodynamic values for common inorganic and organic compounds are given in Chapters 8 and 9. The use of ΔG_r in the calculation of equilibrium constants is discussed in Chapter 5.

Techniques. The methods of small scale high throughput and combinatorial screening, initially developed for testing pharmaceuticals, are now being applied to catalysis research. These methods are enabling the rapid testing of large numbers of component combinations such as metals and supports. This is often done within a framework of knowledge that certain metal or ligand combinations are expected to have activity but also includes the search for novel formulations.

Comprehensive overviews of the organic chemicals industry:

H. A. Wittcoff, B. G. Reuben and J. S. Plotkin, *Industrial Organic Chemicals*, 2nd edition, Wiley-Interscience, New Jersey, 2004.

K. Weissermel and H.-J. Arpe, *Industrial Organic Chemistry*, 4th edition, Wiley-VCH, Weinheim, 2003.

Kirk-Othmer Encyclopedia of Chemical Technology, Wiley-Interscience, 2004.

Statistical information on reserves, production, consumption and prices of oil, coal and natural gas *The BP Statistical Review of World Energy*, BP p.l.c., London, published annually; also at *www.bp.com/statisticalreview*

Prices of major commodity chemicals, market reports and useful industry updates and perspectives, are in *ICIS Chemical Business*, and *Chemical & Engineering News*, published weekly.

Industry perspectives on the use of catalysts in asymmetric synthesis are *Chemical and Engineering News*, June 14 2004, pages 47–62, and September 5, 2005, pages 40–60.

Information on the EU chemical industry, general information on petrochemicals and comparisons to the worldwide chemical industry is available from the European Chemical Industry Council, CEFIC at *www.cefic.be*

Data on the world pharmaceutical industries can be obtained from IMS at *www.imshealth.com*

Data on the R&D expenditure of various industrial sectors is published by the UK's Department of Trade and Industry at *www.innovation.gov.uk*

CHAPTER 2

Formation of C–O Bonds by Oxidation

MARIO G. CLERICI,[a] MARCO RICCI[b] AND
GIORGIO STRUKUL[c]

[a] Via Europa 34, 20097, San Donato Milanese, Italy

[b] Polimeri Europa S.p.A- Istituto Guido Donegani, Via Fauser, 4, 28100, Novara (NO), Italy

[c] Dipartimento di Chimica, Università di Venezia, Dorsoduro 2137, 30123, Venezia, Italy

2.1 Review: The Basic Chemistry of Oxygen

2.1.1 Diradical Nature of the Dioxygen Molecule

By contrast with many other simple diatomic molecules such as N_2 or F_2, the oxygen molecule (O_2, or dioxygen) is paramagnetic as it has two unpaired electrons in the ground state. The highest occupied molecular orbitals (HOMO's) are a pair of π^* orbitals of identical energy, so that the two highest energy electrons have no reason to spin-pair. Consequently, dioxygen may be regarded as a diradical, a formulation that is very useful in understanding its chemistry. In fact most of its reactions proceed in one-electron steps: electrons added to the O_2 molecule populate antibonding (π^*) orbitals and weaken the O–O bond (Figure 1). This effect is evident in both the O–O bond length and its dissociation energy. This makes the resulting superoxo ($O_2^{\bullet-}$) and peroxo (O_2^{2-}) species much more reactive.

As rationalized by Sheldon and Kochi, metal catalyzed oxidations can be conveniently divided into two types: *homolytic* and *heterolytic*.

$$O_2 \xrightarrow{e^-} O_2^{\bullet-} \xrightarrow{e^-} O_2^{2-}$$

bond energy (kJ/mol)	494	368	188
length (Å)	1.21	1.26	1.49

Figure 1 *Basic data concerning the O_2 molecule and the $O_2^{\bullet-}$ and O_2^{2-} anions.*

$$Co^{3+} + e^- \rightleftharpoons Co^{2+} \qquad E_0 = 1.8$$

$$Mn^{3+} + e^- \rightleftharpoons Mn^{2+} \qquad E_0 = 1.5$$

$$Cu^{2+} + e^- \rightleftharpoons Cu^+ \qquad E_0 = 0.17$$

Figure 2 *Examples of redox couples in 3d metals.*

Homolytic oxidations involve free radical intermediates, are catalyzed by first row transition metals and characterized by one-electron redox steps. Some examples are shown in Figure 2. These metals vary considerably in their redox properties. However, the redox potentials change substantially with changes in solvent and changes in the ligand bound to the metal ion.

In homolytic reactions the hydrocarbon to be oxidized is generally not coordinated to the metal and is oxidized outside the coordination sphere via a radical chain. These processes are common and constitute the basis for several very important industrial applications (Sections 2.2 and 2.3). However radical chains are difficult to control, they do not often preserve the configuration of the substrate and typically lead to the formation of a wide variety of products. Consequently, reactions involving one-electron processes with dioxygen as the oxidant generally show only moderate to low selectivities towards the desired product.

By contrast heterolytic oxidations generally require activation of the substrate by coordination to the metal and are often highly selective and stereospecific. They do not involve free radical intermediates and use second- and third-row transition metals which conserve their oxidation state or change it by a two electron redox step.

2.1.2 Metal-Oxygen Complexes

The various oxygenated species that may play roles in oxidation processes are schematically represented in Figure 3. The first step consists of the formation of the dioxygen adduct which can have either a superoxo structure (**A**) if the metal is a one-electron donor, or a peroxo structure (**B**) if the metal is a two electron donor. These superoxo or peroxo complexes can be considered as the formal, but not chemical, analogues of the superoxide and peroxide anions (Figure 1). Among the different species shown in Figure 3, only peroxo (**B**), oxo (**D**), alkylperoxo (**G**) and hydroperoxo (**H**) are directly involved as the oxidizing species in oxygen transfer processes.

Superoxo complexes (**A**) have long been studied because they are relevant models for naturally occurring oxygen carriers. However they constitute only the first step of a complex reaction sequence in which oxygen is transferred to a suitable substrate. For example, the so-called "cobalt salen" complex (Figure 4) was historically the first synthetic analogue of vitamin B_{12} (cobalamine). It gives a relatively stable O_2 complex in the solid state, where dioxygen coordination occurs in a bent end-on mode. The N_2O_2-ligand array about cobalt in the complex is essentially square planar.

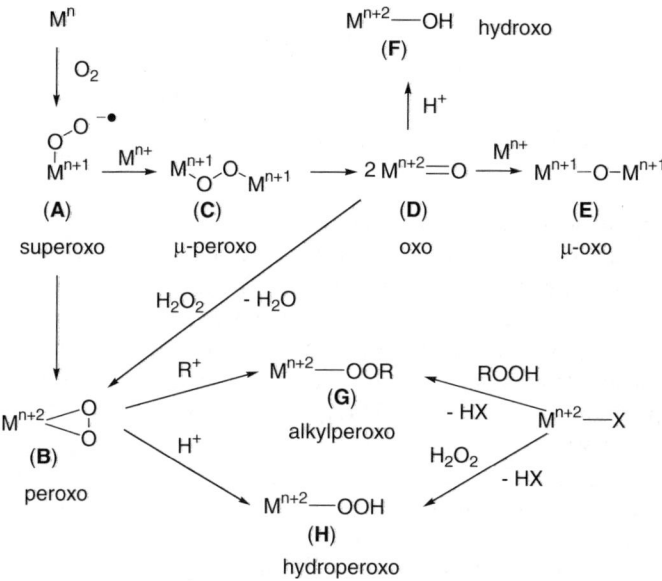

Figure 3 *Schematic representation of metal-oxo- and -peroxo-species involved in metal-promoted oxidations.*

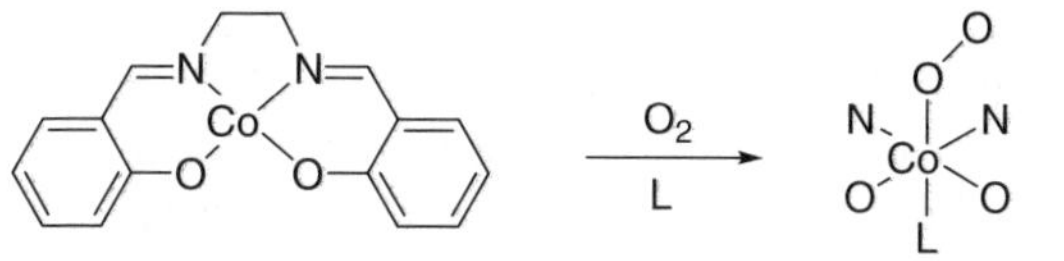

Figure 4 *Dioxygen uptake by a cobalt-salen complex.*

When O_2 coordinates to the cobalt it occupies an axial position in an octahedral structure. Coordination of O_2 is assisted by the binding of an amine to the other axial position; this precisely mimics the situation found in natural systems.

The formation of the O_2 complex usually involves some transfer of electron density from the metal to the O_2 ligand. In other words the cobalt is partially oxidized. This partial oxidation is facilitated by N-donor ligands that compensate for the electron withdrawal by O_2.

Peroxo complexes (Figure 3, (**B**)) are far more numerous than superoxo complexes and are known for early d^0 transition metals (*e.g.* Ti(IV), V(V), Nb(V), Cr(VI), Mo(VI), and W(VI)) and for the groups 8–10 metals (*e.g.* Ru(II), Ru(IV), Co(III), Rh(III), Ir(III), Pd(II), and Pt(II)). Peroxo complexes can be synthesized by two main methods (Figure 3): by reaction with hydrogen peroxide for early transition metals; and, by direct reaction of dioxygen with the reduced two-electron donor metal complex for the groups 8–10 metals.

In the early 1960's when Vaska first discovered that Ir(I) could directly bind dioxygen (Figure 5), the possibility to exploit this activation step to oxidize hydrocarbons seemed to be at hand. Unfortunately this chemistry did not lead

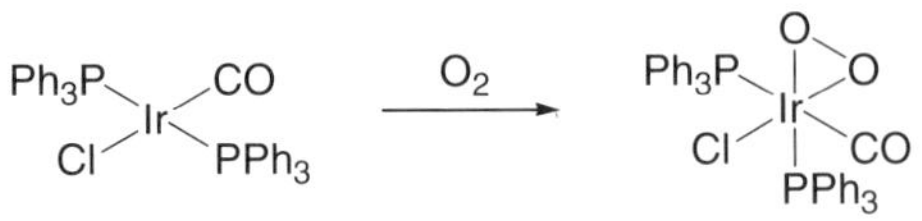

Figure 5 *Dioxygen activation by Vaska's complex.*

to oxidations of any significant practical utility with noble metals. However peroxo complexes of the early metals are now recognized to be important reactive intermediates in catalytic oxidations. The major reason is their ability to form stable oxo species (hence starting a possible catalytic cycle) after delivery of one oxygen atom to the substrate to be oxidized. This pathway is precluded for noble metals as they generally do not form stable oxo species.

Oxo metal complexes (Figure 3, (**D**)) are rather common for the early transition elements, and stable oxo compounds such as MnO_4^-, ReO_4^-, RuO_4 and OsO_4, in which the metal is in its highest oxidation state are well known and are used by organic chemists as stoichiometric hydrocarbon oxidizing reagents.

In the past 25 years, $Fe^{(V)}=O$ and $Mn^{(V)}=O$ have also emerged as reactive intermediates in the oxidation of hydrocarbons: they are not formed by interaction with dioxygen but rather by monooxygen donors (see below).

Surface oxo species are generally involved in the activation of dioxygen on a heterogeneous catalyst. Oxygen is chemisorbed on groups 8–10 metals in this way even at relatively low temperatures and its high reactivity can be exploited for the oxidation of, for example, ethylene to ethylene oxide (Section 2.4), CO to CO_2 or in catalytic combustion. In the same way, metal oxide or mixed-metal oxide catalysts generally employ their already existing surface oxo species for the high temperature selective oxidation of hydrocarbons (*e.g.* see Sections 2.8–2.11). Oxygen vacancies are subsequently replenished by dioxygen.

Alkylperoxo (**G**) and hydroperoxo (**H**) complexes are very important intermediates in industrial oxidation, but they generally form from alkylhydroperoxides and hydrogen peroxide respectively (Figure 3). They are considered in more detail in Sections 2.5–2.7, 2.12–2.14.

2.1.3 Biomimetic Oxidations

Nature has devised a wide variety of mechanisms to make dioxygen available to living organisms. For example, hemeproteins are ubiquitous and participate in oxygen transport (hemoglobin, myoglobin), electron transport (cytochromes C), and oxygen redox chemistry (cytochromes P-450 and peroxidases). Figure 6 shows the complex pathway, consisting of a series of electron and proton transfer steps, adopted by cytochrome P-450 to incorporate one oxygen atom into a hydrocarbon.

In vivo, the source of protons and electrons necessary to generate the active, formally oxoiron(V) oxidant from the iron(II) dioxygen complex is provided by the cofactor NADPH. Synthetic models have long been studied by chemists in

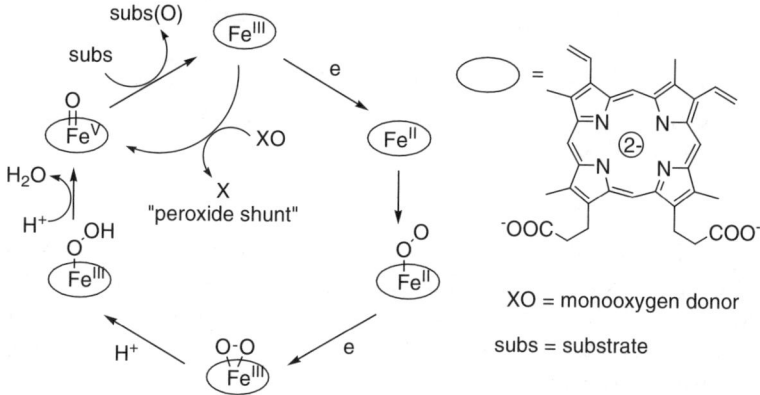

Figure 6 *The complex pathway adopted by cytochrome P-450 to incorporate one oxygen atom into a hydrocarbon.*

order to mimic (and simplify) the complex behaviour of natural systems and try to make it available for transformations of synthetic utility. In protein-free models, the 2-electron reducing equivalents (the sacrificial co-reductant) have been provided by BH_4^-, ascorbic acid, hydrogen, zinc dust, *etc.* and synthetic catalytic systems have been devised using metallo-porphyrin complexes. The need for a co-reductant was circumvented by a pathway known as the "peroxide shunt". This makes use of monooxygen donors such as peracids, alkylhydroperoxides, hypochlorite, iodosobenzene, *etc.*, that allow a direct path from Fe(III) to Fe(V). In 1979 Groves and coworkers were the first to report a model system made of Fe(III) meso-tetraphenylporphyrin (TPP) chloride in combination with iodosobenzene for the epoxidation of olefins and the hydroxylation of alkanes.

Over the past 25 years, biomimetic model systems have been extensively studied and a wide variety of interesting oxidation processes such as the epoxidation of olefins, the hydroxylation of aromatics and alkanes, the oxidation of alcohols to ketones, *etc.*, have been accomplished: some of these are also known in enantioselective versions with spectacular *ee*'s. The vast majority of these transformations were obtained using monooxygen donors such as those mentioned above as primary oxidants. The complexity of the catalysts and the practical impossibility to use dioxygen as the terminal oxidant have so far prevented the use of such systems for large industrial applications, but some small applications in the synthesis of chiral intermediates for pharmaceuticals and agrochemicals, are finding their way to market.

2.1.4 Hydrogen Peroxide and Alkylhydroperoxides

Hydroperoxides represent a reduced and easier to control form of dioxygen. As is clear from Figure 1, in peroxides the O–O bond is longer and its energy lower than in free dioxygen. Therefore it is possible to deliver one of the oxygens of hydroperoxides under mild conditions and in a controlled manner.

$$Co^{II} + ROOH \longrightarrow Co^{III} + RO\bullet + OH^-$$

$$Co^{III} + ROOH \longrightarrow Co^{II} + ROO\bullet + H^+$$

Figure 7 *Scheme demonstrating the cobalt-catalyzed decomposition of hydroperoxides (Haber-Weiss mechanism).*

Stable hydroperoxides that have been extensively used in catalytic oxidations are hydrogen peroxide (H_2O_2), *t*-butylhydroperoxide (*t*-BuOOH), cumylhydroperoxide (cumOOH) and ethylbenzenehydroperoxide. They readily react with suitable transition metal precursors (see Figure 3) to give a variety of species (some of which are stable enough to have been isolated) involved in oxygen transfer to hydrocarbons, namely peroxo (**B**), oxo (**D**), alkylperoxo (**G**) and hydroperoxo (**H**). With the exception of oxo species, where hydroperoxides operate as monooxygen donors to the metal, in the other cases the O–O moiety remains intact. In all cases only one of the peroxygens is utilized in oxygen transfer, the other is used to make H_2O or alcohols.

Their behaviour in oxidation is largely dependent on the type of metal used as catalyst. With one-electron redox systems homolytic oxidation prevails and hydroperoxides are simply decomposed in the catalytic system to generate radical species. Figure 7, also known as the Haber-Weiss mechanism, shows the intermediates formed with for example a Co(II)/Co(III) redox couple that may eventually lead to decompositon of the hydroperoxide with formation of dioxygen and alcohol (or water).

With two-electron redox systems heterolytic oxidations generally occur. The Haber-Weiss mechanism is not accessible and hence the metal can selectively transfer one oxygen atom to a suitable substrate. This is the basis for a wide variety of catalytic oxidation reactions, some of which have found applications in industry and are dealt with below. Some examples are shown in Figure 8. Regeneration of the hydroperoxo/alkylperoxo species can easily occur by reaction of hydrogen peroxide/alkylhydroperoxide with the corresponding hydroxo/alkoxo intermediate according to Figure 3. This will close the catalytic cycle.

2.2 Cyclohexane Oxidation to Cyclohexanol and Cyclohexanone and to Adipic Acid: on the Way to Nylon-6,6

Adipic acid is mainly used to produce nylon-6,6, a synthetic polyamide used in clothing, in the automobile industry, and in construction; it also finds application in polyurethanes, plasticizers, and synthetic lubricants. Manufacture of nylon accounts for 89% of adipic acid consumption in North America, and 55–65% in Western Europe and Japan.

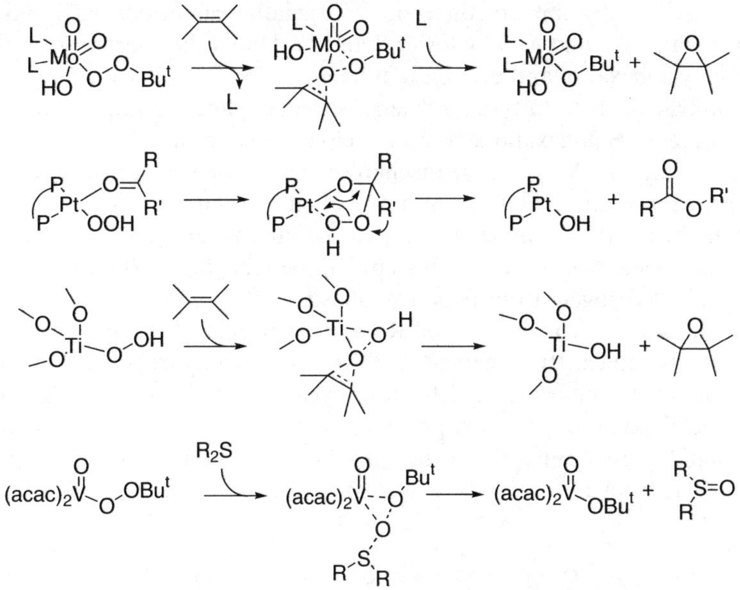

Figure 8 *Examples of oxygen transfer to different substrates using hydroperoxo or alkylperoxo species:* **A**, *the epoxidation of olefins catalyzed by Mo(VI) complexes as in the Oxirane process;* **B**, *the Baeyer-Villiger oxidation of ketones catalyzed by Pt(II) complexes;* **C**, *the epoxidation of olefins catalyzed by Ti(IV) silicates;* **D**, *the oxidation of organic sulfides catalyzed by V(V) complexes.*

Global production of adipic acid is put at around 2.9 Mt/a and its consumption grows at around 3% per year. More than 90% of the production is achieved by oxidation of cyclohexane which, in turn, is obtained by hydrogenation of benzene or, in much smaller amounts, from the naphtha fraction of crude oil. Other routes to adipic acid are based on benzene (*via* cyclohexanol, obtained by partial hydrogenation to cyclohexene and subsequent hydration) or on phenol (through its hydrogenation to cyclohexanol, cyclohexanone or a mixture of both).

Cyclohexane is initially oxidized by air in the presence of cobalt salts: a mixture of cyclohexanol and cyclohexanone (known as KA oil) is produced, which is eventually converted into adipic acid by a further oxidation with nitric acid.

2.2.1 KA Oil from Cyclohexane

The first step of the process, the liquid-phase oxidation of cyclohexane to a mixture of cyclohexanol and cyclohexanone, known as KA (Ketone + Alcohol) oil, was developed in the 1940s. Cyclohexane is oxidized with air at 150–160°C and 8–20 bar. Soluble cobalt(II) naphthenate is commonly used as catalyst. Cyclohexanol and cyclohexanone (ratio 1:1 to 3.5:1) are obtained. Since they (as well as the intermediate cyclohexylhydroperoxide) are more easily oxidized than cyclohexane itself, only a limited conversion of the latter can be achieved: this is never higher than 10% and usually lower than 6%. Under these conditions, the selectivity to cyclohexanol and cyclohexanone is around 70%. By-products are mainly carboxylic acids (n-butyric, n-valeric, succinic, glutaric, adipic), formed by ring cleavage of cyclohexanone and/or cyclohexyloxy radicals.

Although details of the reaction mechanism have not yet been elucidated there is little doubt that, at least at low cobalt concentrations, homolytic autoxidation chemistry dominates here, with the intermediacy of a number of free radical intermediates. Three main steps can be recognized: i) initiation by abstraction of a hydrogen atom from cyclohexane (RH) with formation of a cyclohexyl radical (R); ii) reaction of $R^{•}$ with molecular oxygen to form a cyclohexylperoxy radical $RO_2^{•}$ which iii) abstracts a new hydrogen atom from a second cyclohexane molecule, thus forming cyclohexylhydroperoxide ROOH and a new cyclohexyl radical which propagates the radical chain. Alcohol and ketone formation predominantly occur via self-termination of cyclohexylperoxy radicals Equation 1:

$$2 \quad \bigcirc\!\!-OO^{•} \quad \longrightarrow \quad \bigcirc\!\!=\!O \quad + \quad \bigcirc\!\!-OH \quad + O_2 \qquad (1)$$

In this regard the metal catalyst has little effect, if any, on the process selectivity but it helps to speed up the hydroperoxide decomposition by the well established Haber-Weiss mechanism (Figure 7).

The cyclohexyloxy radicals ($RO^{•}$ in Figure 7) can react with the hydroperoxide, again affording $ROO^{•}$ species and cyclohexanol Equation 2:

$$RO^{•} + ROOH \rightarrow ROO^{•} + ROH \qquad (2)$$

Alternatively cleavage of an α C–H bond can give cyclohexanone.

To improve selectivity the oxidation can be run in the presence of stoichiometric amounts of boron compounds such as boric acid (H_3BO_3), metaboric acid (HBO_2), or boric anhydride (B_2O_3). Under the reaction conditions, these boron compounds afford cyclohexyl borates which are more resistant toward further oxidation and are eventually hydrolyzed to cyclohexanol and boric acid. The latter is then recycled to the oxidation reactor. The use of boron compounds results in higher cyclohexanol:cyclohexanone ratios (around 10:1) and, more important, in an increased selectivity: almost 90%, even when cyclohexane conversion is pushed up to 15%. This boron modified technology is operated by several companies around the world, despite the higher investments and the higher operating costs arising from the need to recover and recycle boric acid.

2.2.2 Adipic Acid from KA Oil

In the second step of the process, the cyclohexanol/cyclohexanone mixture is further oxidized to adipic acid by nitric acid at 75–80°C and 1–4 bar.

Again, metal salts mainly of vanadium(V) and copper(II) are used as catalysts.

The reaction starts with the oxidation of cyclohexanol to cyclohexanone by nitric acid; several paths are then available for cyclohexanone transformation into adipic acid. The main one involves cyclohexanone nitrosation to 2-nitrosocyclohexanone which further reacts with nitric acid to afford 2-nitro-2-nitrosocyclohexanone (Figure 9). The latter gives, upon hydrolysis, 6-nitro-6-hydroximinohexanoic acid which eventually breaks down to adipic acid and nitrous oxide (N_2O).

A second pathway which also contributes significantly to the adipic acid production, involves formation of 1,2-cyclohexanedione, perhaps by hydrolysis of the ketoxime tautomer of 2-nitrosocyclohexanone. Conversion of the ketoxime and of the diketone to adipic acid requires a vanadium(V) catalyst (Figure 10). The resulting vanadium(III) species, VO^+, is eventually reoxidized by nitric acid. The copper(II) apparently helps to reduce multiple nitrosation of cyclohexanone, as that eventually affords glutaric acid.

Yields of adipic acid range around 92%. Typically for each mole of adipic acid produced, 2 moles of nitric acid are converted to N_2O which could, in principle, be used for other processes, such as the Solutia process for the direct oxidation of benzene to phenol (Section 2.14).

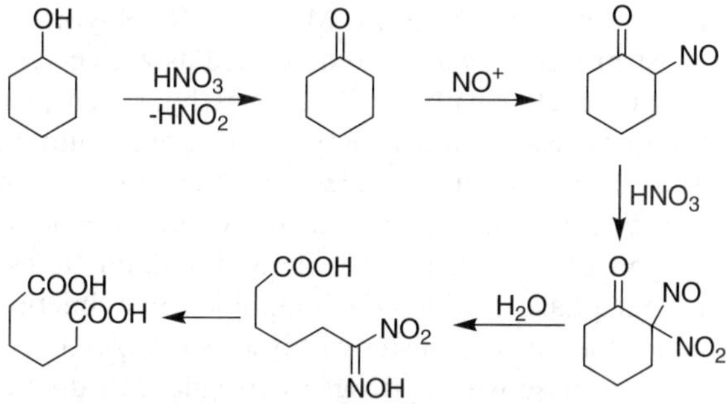

Figure 9 *The nitrosation route by which cyclohexanol is converted into adipic acid.*

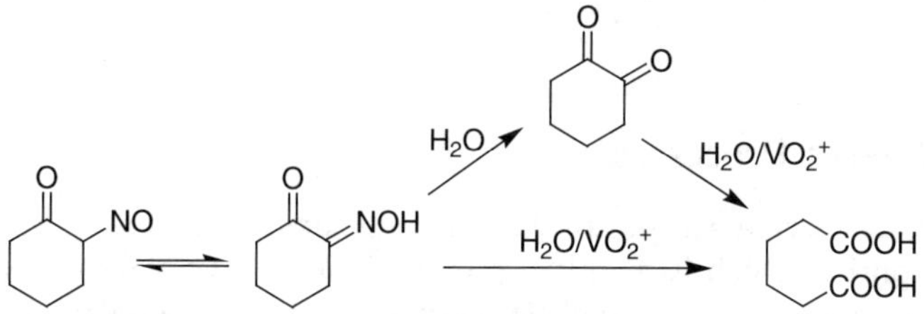

Figure 10 *The V and Cu catalyzed route to adipic acid.*

Box 1 Alternative Processes to Adipic Acid

The traditional two-step process to adipic acid has very low cyclohexane conversion and co-produces NO_x. Huge efforts have therefore been devoted to develop new more efficient processes.

Direct conversion of cyclohexane to adipic acid by air is made possible by the use of a high concentration of cobalt acetate as catalyst (Gulf process). Adipic acid is obtained with 70–75% selectivity at 80–85% conversion of cyclohexane. The main by-product is glutaric acid. It is believed that at these high concentrations, cobalt is not only a catalyst for hydroperoxide decomposition, but directly reacts with cyclohexane (Equation B1).

$$C_6H_{12} + Co^{(III)} \rightarrow C_6H_{11}{}^{\bullet} + Co^{(II)} + H^+ \qquad (B1)$$

When cyclohexanone forms via a cyclohexyl radical, a similar sequence of reactions occurs at the carbon α to the carbonyl group affording 1,2-cyclohexanedione which is eventually cleaved to adipic acid.

Cyclohexene has been also evaluated as a raw material for adipic acid production. Its oxidative cleavage by hydrogen peroxide can be catalyzed by peroxidic phospho-tungstates (C. Venturello, M. Ricci, Eur. Pat. Appl. 122804, 1984; *Chem. Abs.* 1985, **102**, 95256; see C. Venturello, R. D'Aloisio, J.C.J. Bart, M. Ricci, *J. Mol. Catal.* 1985, **32**, 221 for the catalyst structure) or even by sodium tungstate (K. Sato, M. Aoki, R. Noyori, *Science* 1998, **281**, 1646) The last procedure has been described as a green route to adipic acid: cyclohexene is reacted with 30% aqueous hydrogen peroxide in the presence of sodium tungstate and trioctylmethylammonium hydrogen sulphate as catalysts. The reaction provides a nice example of phase-transfer catalysis. The reaction medium is formed by two immiscible liquid phases: an aqueous solution of hydrogen peroxide and sodium tungstate, and an organic, cyclohexene phase in which the lipophilic trioctylmethylammonium hydrogen sulphate (the phase-transfer catalyst) is dissolved. Tungstate ions react in the aqueous phase with hydrogen peroxide affording an oxidizing, anionic peroxo-species which is extracted into the organic phase by the quaternary ammonium cation. In the organic phase, the peroxo complex reacts with cyclohexene forming the desired product and restoring the reduced form of the catalyst which is then again partitioned between the organic and the aqueous phases.

2.2.3 Related Processes

In the past, acetic acid was produced by aerobic oxidation of *n*-butane (and also of light naphtha, which is mainly a mixture of liquid *n*-alkanes up to C_9H_{20}). The process catalyzed by cobalt(II) acetate, closely resembled the one-step oxidation of cyclohexane to adipic acid. Today acetic acid is mostly produced by methanol carbonylation (see Section 4.2); however oxidation could again become important in this field should SABIC develop its innovative technology for air oxidation of ethane catalyzed by mixed oxides of molybdenum, vanadium, palladium, and niobium (Annex 1).

2.3 *p*-Xylene Oxidation to Terephthalic Acid. Polyethylene Terephthalate: on the Way to Fibres for Shirts

Terephthalic acid (1,4-benzenedicarboxylic acid) and its dimethyl ester (dimethyl terephthalate, DMT) are commonly used to produce high molecular weight, linear polyesters characterized by high melting point, high crystallinity, and excellent fiber-forming properties. Particularly important is PET (polyethylene terephthalate), the largest volume synthetic fibre, obtained from terephthalic acid (or DMT) and ethylene glycol.

The production of terephthalic acid and DMT is one of the largest industrial applications of homogeneous catalysis, and has great economic significance. Together terephthalic acid and DMT ranked 25th in volume of all chemicals produced in the USA in 2000, and 11th among the organic chemicals, with a current global production estimated to be well over 25 Mt/a, almost all prepared by oxidation of *p*-xylene which, in turn, is mainly obtained by naphtha reforming.

The syntheses of terephthalic acid and DMT are basically similar but, during the early years of the polyester industry, in the 1950s and 1960s, DMT could be made in a purer form than the acid and was the preferred raw material. In the 1960s however with the development of the Amoco Mid-Century process, terephthalic acid gained acceptance and, since then, it has become the preferred feedstock (based on cost and performance) so that today nearly all of the new PET capacity is based on it. Thus we consider only the synthesis of terephthalic acid in this section.

Autoxidation of methylbenzenes is an inefficient process, even in the presence of transition metal salts as catalysts. The oxidation of xylenes to the corresponding dicarboxylic acids is even more difficult since the methyl group of the intermediate toluic acid is deactivated towards the oxidation by the electron-withdrawing carboxyl substituent on the same aromatic ring. Nevertheless partially oxidized intermediates (the most abundant of which is usually 4-formylbenzoic acid, or 4-carboxybenzaldeyde) must be kept as low as possible in the final terephthalic acid, since they act as terminating agents in the subsequent polymerization reactions to prepare PET. At the same time, selectivity must be high, mainly to avoid the formation of coloured by-products which alter the desired optical properties of the fibres.

Despite these difficulties efficient catalysts and conditions have been developed which allow nearly quantitative oxidation, by air, of both methyl groups.

Several process variations exist: *e.g.*, bromide and manganese ions can be added to the conventional cobalt-based catalytic system. However the general outline is common to all the processes.

Soluble cobalt salts (acetate or naphthenate) are used as catalysts, most often together with manganese and bromide ions. Particularly in the presence of bromide source (as HBr, sodium bromide, or even organic bromides), the rate of the oxidation of methylbenzenes increases by up to 400 x. None of the other halogens approaches bromide in its promoting activity. The maximum effect is achieved with a 1:1 cobalt:bromine atomic ratio.

The most common practice is to run the oxidation in acetic acid at 175–225°C and 12–30 bar. *p*–Xylene is almost completely converted and typical yields are higher than 98% (96–97% after product purification). Small amounts of xylene and acetic acid are however lost due to complete oxidation to carbon oxides.

In the presence of bromide the oxidation is thought to be mediated by the very reactive bromine atoms, without any direct interaction between cobalt(III) and the methylbenzene (Equations 3–4; X is a generic anionic ligand):

$$X_2Co^{(III)}Br \rightarrow Co^{(II)}X_2 + Br^{\bullet} \tag{3}$$

$$Br^{\bullet} + ArCH_3 \rightarrow HBr + ArCH_2^{\bullet} \tag{4}$$

Under the reaction conditions, the benzyl radical is trapped by dioxygen and transformed into a benzylperoxy radical which eventually reacts with cobalt(II) affording an aromatic aldehyde and regenerating cobalt(III) (Equations 5–6):

$$ArCH_2^{\bullet} + O_2 \rightarrow ArCH_2OO^{\bullet} \tag{5}$$

$$ArCH_2OO^{\bullet} + Co^{(II)}X_2 + HBr \rightarrow ArCHO + X_2Co^{(III)}Br + H_2O \tag{6}$$

Finally the aldehyde is oxidized to the carboxylic acid through a chain mechanism involving a metal (Equations 7–10; M = Co or Mn):

$$ArCHO + M^{(III)}X_3 \rightarrow ArC(O)^{\bullet} + M^{(II)}X_2 + HX \tag{7}$$

$$ArC(O)^{\bullet} + O_2 \rightarrow ArC(O)OO^{\bullet} \tag{8}$$

$$ArC(O)OO^{\bullet} + M^{(II)}X_2 + HX \rightarrow ArCO_3H + M^{(III)}X_3 \tag{9}$$

$$ArCO_3H + ArCHO \rightarrow 2\ ArCOOH \tag{10}$$

The overall stoichiometry is simply given by Equation 11:

$$\text{(CH}_3\text{-C}_6\text{H}_4\text{-CH}_3) + 3\,O_2 \longrightarrow (\text{HOOC-C}_6\text{H}_4\text{-COOH}) + 2\,H_2O \qquad (11)$$

Manganese, if present, speeds the reaction rate by up to five times. Its role however is poorly understood. One possibility is that it intervenes in the aldehyde oxidation: manganese has a lower oxidation potential than cobalt; so the reoxidation of Mn(II) by the ArC(O)OO$^{\bullet}$ radical is easier than that of Co(II). As a result, the aldehyde oxidation is better catalyzed by manganese than by cobalt. Other co-substrates (such as acetaldehyde, paraldehyde, or methyl ethyl ketone) can also be used, but most of these variations have only limited success.

The oxidation of *p*-xylene to terephthalic acid is by far the most important process based on the oxidation of methyl aromatics. However other similar processes are also operated industrially and oxidize toluene to benzoic acid or *m*-xylene to isophthalic acid. The latter is used as comonomer with terephthalic acid in bottles for carbonated drinks, and for special polyesters, and its production is roughly 2% of that of the terephthalic derivatives.

Discussion Point DP1: *Since autoxidation processes occur merely on exposure of organic compounds to air, what roles do metals play in such reactions? What properties do the metals need to be most effective in their roles? Why are industrial processes generally carried out at low conversion under recycling conditions?*

2.4 Ethylene Oxide by Ag-catalyzed Oxidation of Ethylene: for Antifreeze and Detergents

World production capacity of ethylene oxide is currently *ca.* 17.1 Mt/a. Most of it (*ca.* 75%) is consumed for the production of glycols, namely ethylene glycol (polyesters, antifreezes), diethylene glycol (polyurethanes, unsaturated polyesters resins, antifreezes), triethylene glycol (solvent, plasticizers, gas dehydration) and higher glycols. Other uses are in the production of ethoxylates for surface active agents (*ca.* 11%), ethanolamines, glycol ethers and polyol polyethers for polyurethanes.

Early production was based on the chlorohydrin process (Box 2), characterized by high yields but also encumbered by technical and environmental problems due to the use of chlorine. The introduction, in the 1940–1950s, of the oxidation of ethylene on silver catalysts with lower costs and better environmental standards, led to its phasing out, virtually completed by the end of the 1960s.

Box 2 Chlorohydrin Processes to Ethylene and Propylene Oxides

The industrial production of ethylene and propylene oxides was historically dependent on the chlorohydrin process, a multistep procedure that proceeds via the stoichiometric reaction of propylene (or ethylene) with chlorine and water to yield a mixture of chlorohydrin isomers (only one for ethylene) and hydrochloric acid. The epoxide is formed upon reaction of the chlorohydrins with calcium or sodium hydroxide. All the chlorine used in the process eventually ends up as chlorinated organic and inorganic by-products (Equation B2).

$$XCH{=}CH_2 + H_2O + Cl_2 \longrightarrow n\ \underset{\substack{| \\ OH\ Cl}}{XCH{-}CH_2} + m\ \underset{\substack{| \\ Cl\ OH}}{XCH{-}CH_2} + HCl$$

$$n\ \underset{\substack{| \\ OH\ Cl}}{XCH{-}CH_2} + m\ \underset{\substack{| \\ Cl\ OH}}{XCH{-}CH_2} + NaOH \longrightarrow \underset{O}{XCH{-}CH_2} + NaCl + H_2O \qquad (B2)$$

$$(X = H,\ CH_3;\ n{+}m = 1)$$

Currently the chlorohydrin process is only used for the epoxidation of propylene, where it still accounts for some 48% of world installed capacity. The yields are 88–89%. In most cases, the plant is integrated with a chloro-alkali facility that supplies both the required chlorine and sodium hydroxide. The recycle to the electrolysis cells of the brine solution produced in the dehydrochlorination step has been considered but not applied, most probably, for technical and economic reasons. In general the aqueous solution of calcium or sodium chloride is disposed of.

The epoxidation reaction is apparently simple and the major side reactions consist of the total oxidation of ethylene and ethylene oxide to water and carbon dioxide. Acetaldehyde and formaldehyde are present in traces, since their further oxidation is fast under reaction conditions (Figure 11).

2.4.1   Air- and Oxygen-based Industrial Processes

The oxidation of ethylene on silver catalysts developed along two distinct lines leading to two different commercial processes, one employing air and the other pure oxygen as oxidant. Both processes adopt the fixed bed technology, with reactors consisting of bundles of several thousand tubes, each having a length of *ca.* 10 m and an internal diameter of 20–40 mm. The multitubular option is dictated by the need for efficient heat exchange between the catalyst particles and the external coolant fluid in order to keep the strongly exothermic reactions under control (Figure 11). The hot spots that can develop in the catalyst bed

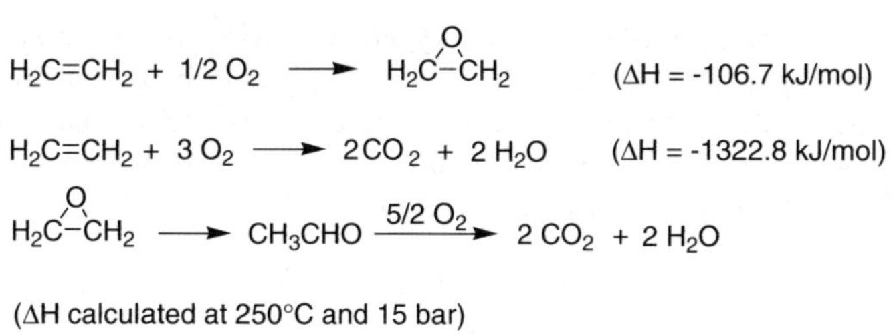

$$H_2C{=}CH_2 + 1/2\ O_2 \longrightarrow H_2\overset{O}{\overset{\triangle}{C}}CH_2 \qquad (\Delta H = -106.7\ kJ/mol)$$

$$H_2C{=}CH_2 + 3\ O_2 \longrightarrow 2\ CO_2 + 2\ H_2O \qquad (\Delta H = -1322.8\ kJ/mol)$$

$$H_2\overset{O}{\overset{\triangle}{C}}CH_2 \longrightarrow CH_3CHO \xrightarrow{5/2\ O_2} 2\ CO_2 + 2\ H_2O$$

(ΔH calculated at 250°C and 15 bar)

Figure 11 *Ethylene oxidation on Ag catalyst.*

not only decrease the selectivity but also lead to a faster ageing through the sintering of silver particles.

The temperature and the pressure are chosen in the ranges 200–300°C and 15–25 bar. Ethylene concentration at the reactor inlet varies between 5–40% while that of oxygen is between 5–9%, as set by flammability considerations. Normally, the air process operates on the lower side of the concentration ranges. A high selectivity is achieved through a low per-pass conversion of ethylene, the continuous recovery of the product from the exit stream in a water absorber and the recycle of most of the gas leaving the absorber, except for a purge stream necessary to avoid the accumulation of inerts added with the intake gases. Overall selectivities of 70–80 % are normally achieved in current oxygen processes, with selectivities of up to 83% sometimes reported at start-up. Overall yields in the air process are lower, in the range 65–75%.

The air-based process was the first to be commercialized in 1937, some years after the discovery of the reaction by Lefort in 1931. The oxygen-based process introduced in 1958, was however more competitive and has become increasingly preferred for new installations and for the revamping of earlier ones. It is now estimated that the use of air is only competitive for production capacities of 20 kt/a or less, for which an oxygen separation unit would be too expensive.

The epoxidation catalyst is finely dispersed metallic silver together with alkali or alkaline earth metal promoters, on a low surface area carrier ($<1\ m^2\ g^{-1}$), generally a high purity α-alumina. A larger surface area support would promote greater activity through a greater dispersion of silver at the expense however of a lower selectivity as the longer diffusion path of the epoxide in the pores would favour its further transformation, *e.g.*, into acetaldehyde. The addition of ppm chlorine compounds in the feed (*e.g.*, 1,2-dichloroethylene or halogenated aromatics) selectively inhibits the sites responsible for ethylene combustion. The optimization of supports and of silver deposition, and the careful selection of promoters and inhibitors have contributed to the increase of selectivities from 50% or so to sometimes above 80% in current catalysts.

2.4.2 Proposed Epoxidation Mechanisms on Ag Catalysts

The oxidation mechanism is still being debated, with the discussion centred on the nature of surface species responsible for the different reaction paths. It is generally agreed however that at least three sorts of oxygen species are chemisorbed on silver: undissociated oxygen (Equation 12), monoatomic surface oxygen (Equation 13), and monoatomic subsurface oxygen, in order of increasing adsorption strength. The last species arises from the diffusion of surface oxygen to the region below the outermost layer of Ag atoms, a process that requires a relatively high temperature ($> 100°C$).

$$O_2 + Ag \rightarrow O_2^-{}_{(ads)} + Ag^+ \tag{12}$$

$$O_2 + 4\,Ag \rightarrow 2\,O^{2-}{}_{(ads)} + 4\,Ag^+ \tag{13}$$

In early proposals the species responsible for epoxidation was identified as the adsorbed molecular oxygen, $Ag \cdot O_{2(ads)}$, while combustion was attributed to monoatomic $Ag \cdot O_{(ads)}$ (Equations 14–16). The oxidation step envisages the transfer of one atom of molecularly adsorbed oxygen to the double bond, while the other remains adsorbed on silver. The consumption of the latter by the total oxidation of ethylene restores the site vacancies necessary for the continuation of catalysis. Up to a maximum of six oxygen atoms are required for the combustion of one ethylene molecule. Thus, the combination of the reactions (Equation 14) and (Equation 15) predicts that the maximum attainable selectivity in the epoxidation of ethylene is 6/7, *i.e.*, 85.7% (Equation 16). A lower selectivity should normally be expected because some monoatomic oxygen independently formed by dissociative adsorption (Equation 13) raises the level of ethylene combustion above that predicted by Equation 16.

$$Ag \cdot O_{2(ads)} + CH_2{=}CH_2 \rightarrow C_2H_4O + Ag \cdot O_{(ads)} \tag{14}$$

$$6\,Ag \cdot O_{(ads)} + CH_2{=}CH_2 \rightarrow 2\,CO_2 + 2\,H_2O + 6\,Ag \tag{15}$$

$$6\,Ag \cdot O_{2(ads)} + 7\,CH_2{=}CH_2 \rightarrow 6\,C_2H_4O + 2\,CO_2 + 2\,H_2O + 6\,Ag \tag{16}$$

However this rather elegant model did not withstand new evidence arising from the development of improved catalysts and by progress in surface characterization techniques. In fact the selectivity limit of 85.7% has sometimes been surpassed and a growing mass of data on the interaction of silver with the oxygen, ethylene and other molecules relevant to the reaction, indicates that monoatomic oxygen is the species responsible for both the epoxidation and the combustion of ethylene. What makes the difference to the direction of the oxidation path is the nature of the local environment to the oxygen sites (*i.e.*, the nature and the surface density of co-adsorbed species), and it is found that the properties of adsorbed monoatomic oxygen vary significantly as a function of surface coverage (θ). At low θ, surface oxygen is strongly adsorbed on Ag and is nucleophilic, while at higher θ the competition of electronegative species for silver electrons is greater, with the effect of decreasing the charge on individual oxygen sites. The Ag–O bond becomes progressively weaker at increasing coverage with the species assuming a more electrophilic character. Consistent with this, in the oxidation of ethylene by a precovered surface, the

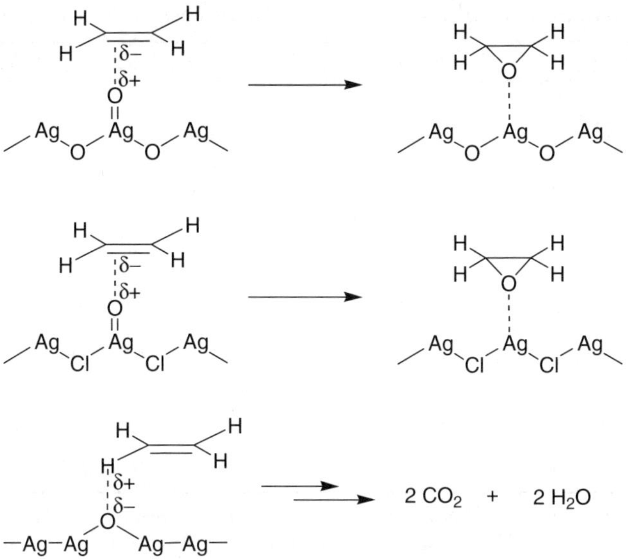

Figure 12 *Representation of the proposed mode of the oxidation on a silver surface.*
(Adapted from R.A. van Santen, H.P.C.E. Kuipers, Advances in Catalysis,
*1987, **35**, 265 (D.D. Eley, H. Pines, P.B. Weisz (Eds), Academic Press), with*
permission from Elsevier.)

epoxide selectivity is a function of oxygen coverage and increases progressively
with it to reach the highest values at near to saturation coverage.

In the currently accepted mechanism the weakly adsorbed oxygen present at
high coverage is the species responsible for the electrophilic attack on the double
bond leading to epoxidation, while the strongly adsorbed nucleophilic oxygen
initiates the combustion process by splitting one C–H bond. The co-adsorption
of chlorine has the effect of increasing the ratio of electrophilic to nucleophilic
sites. Isotope experiments indicate the attack may occur asymmetrically on the
double bond with only partial retention of configuration (Figure 12).

In industrial reactors under steady state catalytic conditions the high oxygen
coverage required for high selectivity is not easily reached. The addition of
chlorine compounds in the cycle gas is therefore necessary for best performance
in commercial processes. Carbon dioxide too improves the selectivity, probably
through a titration of nucleophilic oxygen species. However the promoter effect
is limited to a well defined range of concentration, *i.e.*, ppm amounts for
chlorine and *ca.* 6–8% for carbon dioxide; at higher levels they start behaving
as inhibitors.

2.4.3 Is the Epoxidation of Olefins Other than Ethylene Feasible on Silver Catalysts?

Despite much research the direct oxidation of propylene on silver catalysts
remains an unselective reaction yielding primarily carbon oxides and a range of

Table 1   *Epoxidation of olefins with molecular oxygen on CsCl-Ag/Al$_2$O$_3$^{a}*

	Olefin	Product	Conversion (%)	Selectivity (%)
1			12	92
3			19	95
5	H₃C—	H₃C—	100	0
6			43	92

^{a} Adapted from J.R. Monnier, The Direct Epoxidation of Higher Olefins Using Molecular Oxygen *Appl. Catal. A*, vol. 221, p.73, copyright 2001, with permission from Elsevier.

partial oxidation products. The reason lies in the ease of abstraction of an allylic C–H of the methyl group. This is reflected in the bond energy of the allylic C–H in propylene (356 kJ/mol) compared to that of the vinylic C–H in ethylene (444 kJ/mol). Thus, the use of peracids and hydroperoxides is generally preferred for the epoxidation of longer chain and cyclic olefins. By contrast this aspect of propylene reactivity forms the basis of the acrolein/acrylic acid and acrylonitrile processes.

However the epoxidation of olefins lacking allylic and other reactive C–H bonds with molecular oxygen has recently been achieved on silver catalysts (Table 1). In 1997 the Eastman Chemical Company started the manufacture of 3,4-epoxy-1-butene, the product of mono-epoxidation of butadiene, on a semiworks production scale (entry 1). Remarkably enough the presence of benzylic hydrogen, as in *p*-methyl styrene (entry 5), drives the oxidation towards combustion, while sterically hindered allylic C–H's, as in norbornene (entry 6), are inert to oxidation.

Discussion Point DP2: *Silver is unique among metal catalysts for its high selectivity in the epoxidation of ethylene. Rationalize this result on the basis of the different properties of surface metal-oxygen bonds and of the different adsorption behaviour of hydrocarbons on transition and non-transition metals.*

2.5   Propylene Oxide: to Biocompatible Propylene Glycol

The world installed capacity for propylene oxide production is *ca.* 5.9 Mt/a. It is a raw material for many end products and intermediates in the chemical industry. Polyether polyols, produced by the reaction with polyhydric alcohols and employed in the manufacture of polyurethanes, represent the largest outlet for propylene oxide (*ca.* 65%). Other applications are in the production of propylene glycols (20%), glycol ethers (5%), isopropanolamines, alkoxylates, cellulose ethers, propylene carbonate, allyl alcohol, and 1,4-butanediol. The biocompatibility and biodegradability of propylene and polyalkylene glycols is an advantage over analogous products from ethylene oxide. Their use is recommended when good compatibility of the product with living organisms and the environment is required, for example in the food, cosmetic, and pharmaceutical industries.

The direct oxidation of propylene on silver catalysts has been intensively investigated, but has failed to provide results with commercial potential. Selectivities are generally too low and the isolation of propylene oxide is complicated by the presence of many by-products. The best reported selectivities are in the range 50–60% for less than 9% propylene conversion. The relatively low selectivity arises from the high temperature necessary for the silver catalysts, the radical nature of molecular oxygen, as well as the allylic hydrogens in propylene. Thus alternative routes have been studied based on the use of oxidants able to act heterolytically under mild conditions. Hypochlorous acid (chlorine+water) and organic hydroperoxides fulfill these requirements and their use has led to the introduction of the chlorohydrin (Box 2) and the hydroperoxide processes, both currently employed commercially.

The development of the hydroperoxide processes by Halcon, ARCO and Shell was prompted by the discovery, made in the late 1960s, that certain metal oxides of Groups 4-6 are good catalysts for the epoxidation of olefins with hydroperoxides (Figure 13). These are produced by the autoxidation of hydrocarbons containing a tertiary or a benzylic C–H bond, such as isobutane ($R^1 = R^2 = CH_3$), ethylbenzene ($R^1 = H$, $R^2 = C_6H_5$), or cumene ($R^1 = CH_3$, $R^2 = C_6H_5$), carried out in one section of the plant. The use of hydroperoxides implies that one molecule of alcohol is produced per molecule of propylene oxide in the epoxidation stage. In fact much more alcohol is co-produced owing

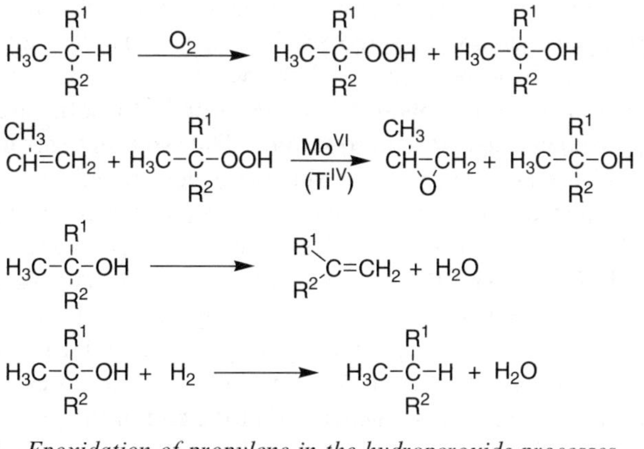

Figure 13 *Epoxidation of propylene in the hydroperoxide processes.*

to the less than 100% selectivity in both the autoxidation and epoxidation reactions (Figure 13, Equation 17,18). The oxidant is employed in stoichiometric amounts, as in the chlorohydrin process, even though the hydroperoxide processes are clearly catalytic. However profitable use can be made of the co-produced alcohol which makes the process more attractive. Generally ethylbenzene or isobutane is used, and the corresponding 1-phenylethanol and t-butanol are transformed into polymer grade styrene or isobutene for octane enhancers in gasoline (MTBE, ETBE).

Nonetheless the presence of a co-product implies that the value of propylene oxide is significantly affected by the demand/pricing of the co-product and difficulties can arise in balancing two different markets that may occasionally experience diverging dynamics of growth. On the whole however, the hydroperoxide processes have been successful, with a market share growing progressively up to the present level of 52%.

The process originally developed by Halcon/ARCO, with a current market share of *ca.* 18%, is based on the use of t-butyl hydroperoxide (TBHP) as the oxidant (Figure 13, $R^1 = R^2 = CH_3$, Equation 17,18). Isobutane is oxidized with air at *ca.* 135–140°C, under pressure, in a typical autoxidation reaction, yielding comparable quantities of TBHP and t-butyl alcohol (TBA), with conversions around 40%. The epoxidation of propylene is a liquid phase homogeneous reaction carried out at *ca.* 120°C, under pressure, in the presence of a soluble Mo(VI) catalyst. Yields on propylene are *ca.* 92%. The ratio of the co-produced TBA to propylene oxide is in the range 2.4 to 2.7. TBA is mostly dehydrated to isobutene and etherified with methanol or ethanol for the production of octane enhancers.

Ethylbenzene hydroperoxide (EBHP) is used in two other processes, developed by Halcon/ARCO and by Shell with a whole market share of *ca.* 34% (Figure 13, $R^1 = H$, $R^2 = C_6H_5$, Equation 17,18). The differences in this case are substantial and the process developed by Shell stands out since it uses fixed bed technology. EBHP is produced by the autoxidation of ethylbenzene at 140–160°C, limiting the conversion to somewhat below 15% to minimize the decomposition of the hydroperoxide. The selectivity to EBHP is in the range 80–85%, with the balance being a mixture of 1-phenylethanol and acetophenone. The epoxidation of propylene, catalyzed in the Shell process by Ti(IV) supported on silica (Ti/SiO$_2$) and by a soluble organic salt of Mo(VI) in the Halcon/ARCO process, is operated in the liquid phase at *ca.* 100–120°C. In both cases, the 1-phenylethanol co-product is dehydrated to styrene. The yields of propylene oxide are 91–92%, with a styrene to propylene oxide ratio close to 2.2.

An advanced version of the hydroperoxide process, in which the alcohol is transformed back into the starting hydrocarbon, was commercialized by Sumitomo in 2003 ($R^1 = CH_3$, $R^2 = C_6H_5$, equations 17,18,20). As the end-use of the co-product is no longer a discriminating issue for the choice of oxidant, the preference was for cumene hydroperoxide (CHP) over other hydroperoxides, on the grounds of its higher stability and superior performance in the epoxidation stage. Cumene is regenerated at the end of the process by the dehydration-hydrogenation of cumyl alcohol and recycled to the autoxidation reactor. In practice, hydrogen and oxygen are consumed to yield equimolar amounts of epoxide and water (Equation 21); in this aspect the process resembles the monooxygenase type of reactions.

$$CH_3CH=CH_2 + H_2 + O_2 \longrightarrow CH_3CH-CH_2 + H_2O \qquad (21)$$
$$\underset{O}{}$$

The use of peroxidic oxidants in the epoxidation of alkenes has been the subject of detailed mechanistic investigations. Ti, V, Mo, and W compounds in their highest oxidation state are the most active catalysts and organic hydroperoxides are the oxidants of choice. The reaction is best carried out in non-polar solvents, such as hydrocarbons, with the exclusion of protic and polar compounds, because of their competition for coordination sites on the metal (Equation 22). The alcohol co-produced in the reaction behaves as an inhibitor retarding the epoxidation of the olefin.

$$[Mo^{VI}ROH] \xrightleftharpoons{ROH} Mo^{VI} \xrightleftharpoons{RO_2H} [Mo^{VI}RO_2H] \qquad (22)$$

Hydrogen peroxide is generally ineffective because of its poor solubility in hydrocarbon media and because of the unavoidable presence of water associated with its use. There are some notable exceptions to this rule, for example olefins able to coordinate to the catalyst through a hydroxy- or carboxy-group, such as allyl alcohols and acrylic acid. These are readily epoxidized by hydrogen peroxide in protic solutions (see also Section 2.7.1). Generally speaking Group 4–6 metal catalysts have a high sensitivity to inhibition by polar compounds.

The active species is an electrophilic peroxo-metal complex containing an intact hydroperoxide ligand (Figure 14). The high-valent metal acts as a strong Lewis centre facilitating the heterolysis of peroxidic oxygen.

In the epoxidation step, the peroxidic oxygen vicinal to the metal center is transferred to the double bond forming the oxirane ring (Figure 14, bottom line, for Mo(VI) catalysts). The active species is regenerated by reaction of the alkoxymetal species with the organic hydroperoxide.

2.6 Hydrogen Peroxide Route to Propylene Oxide

The epoxidation of propylene with hydrogen peroxide has been implemented to pilot plant scale (2,000 t/a) by EniChem. The reaction is catalyzed by titanium silicalite-1 (TS-1). Although TS-1 has similar composition to the Ti/SiO$_2$ developed by Shell for the epoxidation of propylene with organic hydroperoxides (*ca.* 2.5 wt% Ti, as TiO$_2$) the reactivity is quite different. TS-1 prefers H$_2$O$_2$ and protic solvents, and is almost inactive with hydroperoxides or in apolar media. The reverse holds for Ti/SiO$_2$. The reason is found in the different

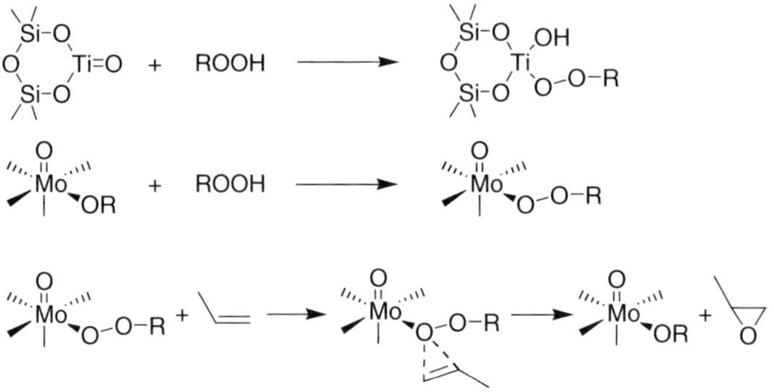

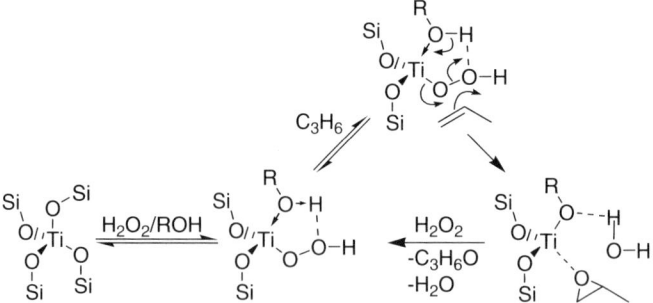

Figure 14 *Schematic representation of the probable mode of formation and of reaction of peroxy metal species in oxidation reactions.*

Figure 15 *Epoxidation of propylene on TS-1 (Si atoms belong to -O-Si-O- bonds of TS-1 lattice.*

structures and surface properties: TS-1 is crystalline, microporous and hydrophobic while Ti/SiO_2 is amorphous, mesoporous (with a broad pore size distribution) and hydrophilic. The reasons for the relationship of the properties of the surface to the reaction medium and the oxidant are discussed in the Annex 2.

The active species too is characterized by rather unusual properties and is different from soluble Ti-peroxides, known for their inertness in the epoxidation of simple olefins. As originally proposed by Clerici et al., the active sites contain Ti-OOH species, stabilized in a cyclic structure by the coadsorption of a protic molecule, generally the alcohol solvent. The oxygen transfer step to the double bond has electrophilic character thus resembling other peroxidic oxidants (Figure 15).

Using hydrogen peroxide for the manufacture of propylene oxide means that water is co-produced in the reaction and no by-product needs to be disposed of or utilized. The reaction is carried out in a dilute solution of hydrogen peroxide (< 10 wt%) in aqueous methanol, below 60°C and moderate pressure. The reaction is fast even with $\leq 1\%$ H_2O_2.

Box 3 Epoxidation of Higher Olefins with Hydrogen Peroxide

In comparison with soluble catalysts, titanium silicalite is extremely easy to recover and reuse, or to be used in continuous processes. However, it is a poor catalyst in the epoxidation of long chain or cyclic olefins, which are too large to enter the TS-1 micropores and to approach the active sites. In these cases, epoxidation by hydrogen peroxide is still possible by using a tetra-alkylammonium (or phosphonium) salt of the peroxidic phosphotungstate anion $[PW_4O_{24}]^{3-}$ as the catalyst (C. Venturello, R. D'Aloisio, J.C.J. Bart, M. Ricci, *J. Mol. Catal.* 1985, **32**, 221). The reactions are run batchwise, in biphase aqueous/organic media. Typical yields based on hydrogen peroxide are 80–95% and the selectivity on the olefins is usually higher than 95%, thus allowing the recovery of excess olefin substantially unchanged. This epoxidation procedure has been developed on a commercial scale for the production of, *e.g.*, isobutyl 3,4-epoxybutyrate, the key intermediate in the synthesis of the nootropic drug oxiracetam.

The yield based on hydrogen peroxide is almost quantitative with negligible decomposition of the oxidant and negligible oxidation of methanol. The main by-products are propylene glycol and its methyl ethers formed by the solvolysis of the oxirane ring. The selectivity however is high and can be made almost quantitative (*ca.* 98%) by the addition of ppm amounts of bases as acidity moderators. Propylene oxide is recovered by distillation from the effluent stream that is then recycled to the epoxidation reactor. Only a small purge is necessary to avoid the accumulation of water and glycols in the cycle stream.

The hydrogen peroxide process is characterized by lower environmental impact, a simpler process scheme and lower investment. However hydrogen peroxide is relatively expensive and still produced in plants with relatively low nameplate capacities, though the construction of a 300 kt/a propylene oxide plant has been announced by Dow Chemical and BASF.

2.7 Asymmetric Epoxidation, Dihydroxylation and Sulfide Oxidation: New Routes to Chiral Agrochemicals and Pharmaceuticals

2.7.1 Epoxidation of Allylic Alcohols

Some general points concerning the economic and chemical requirements needed for an enantioselective process to find commercial applications are listed in Appendix A. Given these constraints it is no coincidence that the cradle of all metal catalyzed enantioselective transformations was hydrogenation (Chapter 3, Sections 3.5 and 3.6), as it was the first reaction to be well-understood and the first to lead to a commercial process: the synthesis of L-dopa developed by Knowles at Monsanto in 1974. Epoxidation came next in the early 1980's, thanks to the seminal work of Sharpless and Katsuki on the epoxidation of allylic alcohols using a [Ti(O-*i*-Pr)$_4$]/diethyl tartrate catalyst and an organic hydroperoxide as the oxidant (Equation 23). Knowles, Sharpless and Noyori won the Nobel Prize for Chemistry in 2001.

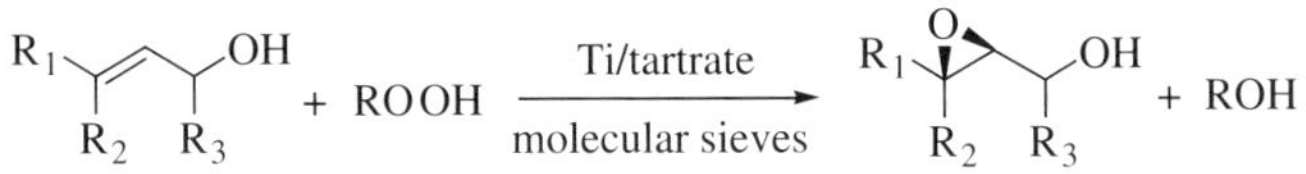

$$R_1 \underset{R_2 \quad R_3}{\diagdown} OH \; + \; ROOH \quad \xrightarrow[\text{molecular sieves}]{\text{Ti/tartrate}} \quad R_1 \underset{R_2 \quad R_3}{\diagdown} OH \; + \; ROH \qquad (23)$$

While homogeneous hydrogenations are generally carried out using platinum metal (Rh, Ru) complexes, modified with expensive and synthetically elaborate chiral diphosphines as catalysts, the Sharpless catalyst is based on an inexpensive early transition metal compound modified with an easily accessible chiral natural product (diethyl tartrate). The low cost of the catalyst can easily tolerate the moderate turnover numbers generally achieved. The method is used for the commercial synthesis of glycidol (Figure 16, **A**), used as a multi purpose chiral building block (*e.g.*, in the synthesis of lipids and phospholipids), that is operated by PPG-Sipsy on a multi tonne per year scale. The same company has developed the epoxidation of allylic alcohols (Figure 16, **B** and **C**) with excellent enantioselectivities (> 95% *ee*).

The first commercial application of Sharpless technology was in the synthesis of (+)-disparlure (Figure 17), which is operated by Upjohn in a multi-10 kg scale.

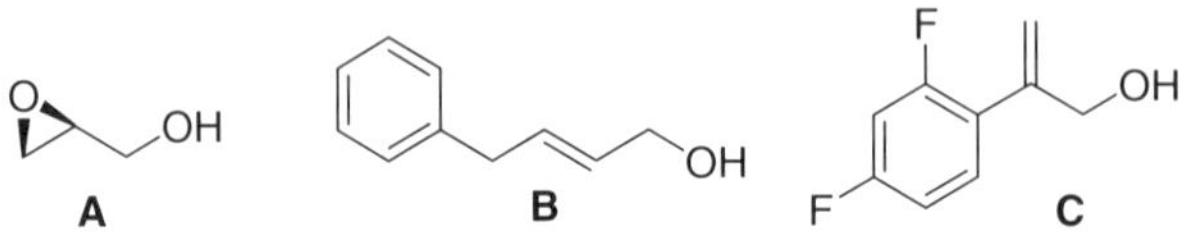

Figure 16   *Glycidol (**A**) and other allylic alcohols (**B** and **C**) giving good enantioselectivities in epoxidation.*

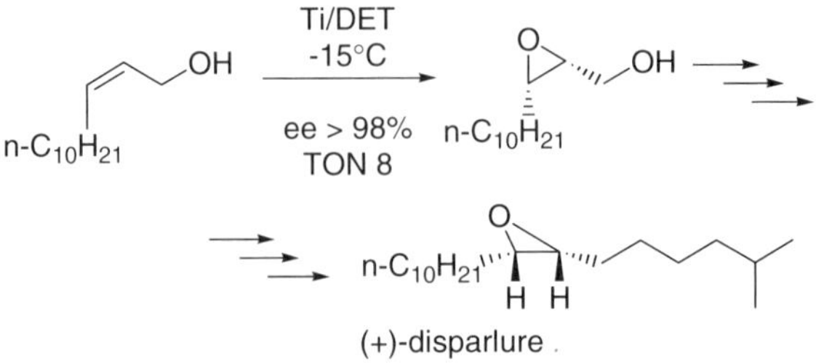

Figure 17   *Key steps in the synthesis of (+)-disparlure.*

Disparlure is a sex pheromone for the gipsy moth. This leaf-feeding insect causes much damage to trees in the Eastern United States but attempts to control its spread through conventional techniques using insecticides have had only limited success. The new strategy is to use disparlure, the sex attractant emitted by the female gipsy moth, to lure male moths into a trap. Alternatively it can be used to confuse males and prevent them from finding females, reducing mating. This has obvious advantages for the environment since it needs only a few grams of disparlure instead of spreading tons of insecticides. Success depends on the use of the (+) isomer of disparlure. As shown in Figure 17, the key step in the synthesis is the enantioselective epoxidation of Z-2-tridecenol. The further conversion to (+) disparlure requires three conventional organic steps.

The mechanism of the asymmetric epoxidation of allylic alcohols with the Sharpless-Katsuki catalyst is assumed to be very similar to the one described for the Halcon-ARCO process in Section 2.5. The key point is that the chiral tartrate creates an asymmetric environment about the titanium center (Figure 18). When the allylic alcohol and the *t*-butyl hydroperoxide bind through displacement of alkoxy groups from the metal, they are disposed in such a way as to direct oxygen transfer to a specific face of the C=C double bond. This point is crucial to maximize enantioselectivity.

The two faces of the C=C double bond are prochiral, which means that opposite enantiomers result from oxygen transfer to one face rather than to the other. The binding of the alcoholic function of the substrate to titanium imposes some steric constraints so that one face of the C=C double bond is forced over the other with respect to the metal. This effect of the alcoholic function is known as *secondary interaction*; it is not directly involved in the actual oxygen transfer, but very important in directing it. The epoxidation reaction proceeds by oxygen transfer from the coordinated alkylperoxy ligand to the olefin via the so-called "butterfly" intermediate, followed by

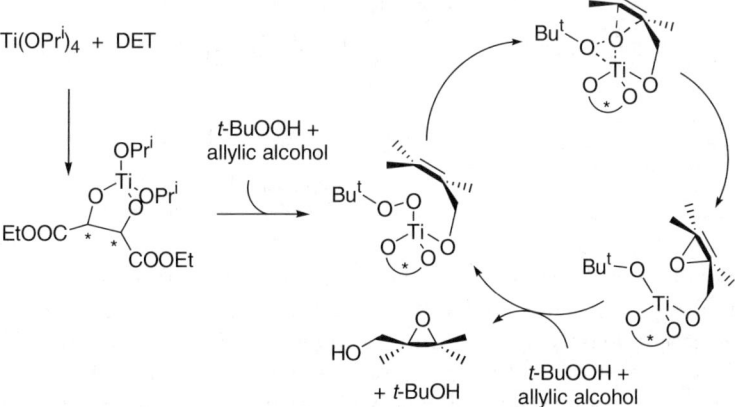

Figure 18 *Secondary interactions in the asymmetric epoxidation of allylic alcohols.*

displacement of the *t*-butoxy and allyloxy epoxide moieties by fresh *t*-butyl hydroperoxide and allylic alcohol.

A major limitation to the Sharpless-Katsuki epoxidation is that its utility is largely confined to oxidation of allylic alcohols. Homoallylic alcohols are oxidized less cleanly and the oxidation of simple olefins shows little enantio-selectivity. This is presumably because the stereochemical control depends on "anchoring" the substrate to a particular site on the metal by means of an auxiliary coordinating function.

The key for the commercial success of the enantioselective oxidation of allylic alcohols relies mainly on the low cost of both the catalyst and the chiral auxiliary. Potential problems with the Sharpless oxidation system arise from the need to handle large amounts of hydroperoxides for large scale work and the moderate to low catalytic activity.

2.7.2 Epoxidation of Simple Olefins

The asymmetric oxidation of indene to the corresponding epoxide (Equation 24) is carried out commercially by Sepracor on a small scale. Chiral indene oxide is an intermediate in the synthesis of crixivan (an HIV protease inhibitor). Reaction is carried out at 5°C with moderately high turnover numbers in the presence of an "exotic" donor ligand ("P_3NO", 3-phenylpropylpyridine N oxide) and sodium hypochlorite as the terminal oxidant. A similar epoxidation of a simple *cis* olefin (Equation 25) leads to an enantiomerically pure amino-alcohol used in the synthesis of taxol, a potent anticancer drug.

$$(24)$$

$$(25)$$

The epoxidation of simple olefins which cannot benefit from secondary inter-actions brings some formidable problems that were solved by sophisticated catalyst design, mainly by the groups of Jacobsen and Katsuki in the 1990's. A class of square planar salen complexes was chosen (Figure 19, for example) capable of giving a metal-oxo derivative by reaction with monooxygen donors such as iodosobenzene or sodium hypochlorite (the preferred oxidant). A series

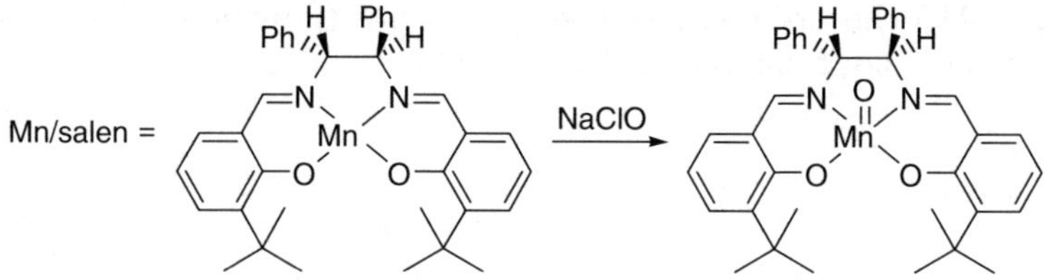

Figure 19 *A bulky Mn salen complex used for enantioselective oxidation reactions and its oxo adduct formed by reaction with sodium hypochlorite.*

of different bulky groups were placed at the periphery of the macrocyclic ligand, with the aim of directing the approach of the substrate in a "locked" rigid fashion towards the oxo ligand in the coordination plane by purely steric effects. The chiral information is generally placed in the 5-membered ring containing the two nitrogens. Altogether several hundred complexes were synthesized, changing the metal and the substituents on the ligand. The structure had to be optimized for each reaction, but the end result was very successful.

A so far still unsolved problem is the direct enantioselective epoxidation of simple terminal olefins. For example the epoxidation of propylene that was achieved with a 41% *ee* almost twenty years ago by Strukul and his coworkers using Pt/diphosphine complexes is still unsurpassed. Unfortunately such low *ee*'s are of no practical interest. The problem was circumvented by Jacobsen using hydrolytic kinetic resolution of racemic epoxides (Equation 26) and is practised on a multi 100 kg scale at Chirex. The strategy used is to stereoselectively open the oxirane ring of a racemic chiral epoxide leaving the other enantiomer intact. Reactions are carried out to a 50% maximum conversion. The catalyst belongs to the metal-salen class described above and can be recycled. The products are separated by fractional distillation.

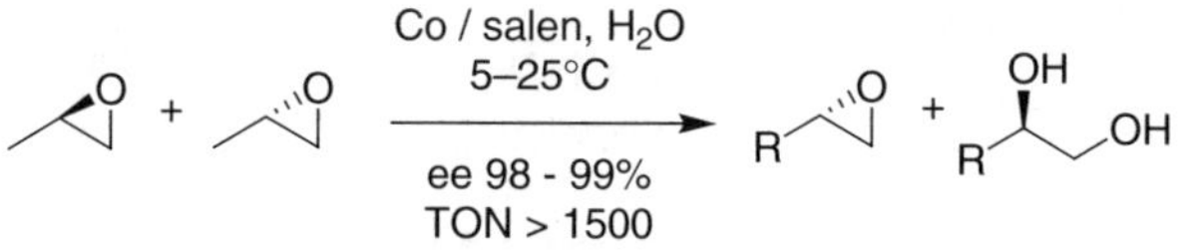

$$\tag{26}$$

2.7.3 Vicinal Dihydroxylation of Olefins

It has been known for decades that osmium tetroxide catalyzes the H_2O_2 oxidation of olefins to *cis*-1,2-diols but the cost, toxicity and volatility of OsO_4 have limited its use to the organic research laboratory. Industrial interest was aroused in the 1990's by *i*) the invention of a system for vicinal hydroxylation using an electrochemical device as the ultimate oxidant; and *ii*) the discovery of conditions to carry out the reaction enantioselectively, mainly by Sharpless and his group. The reaction is currently carried out by Chirex to produce (R)-2-chorophenyl-1,2-dihydroxyethane (Equation 27), used as a chiral building block, on a multi 10 kg scale.

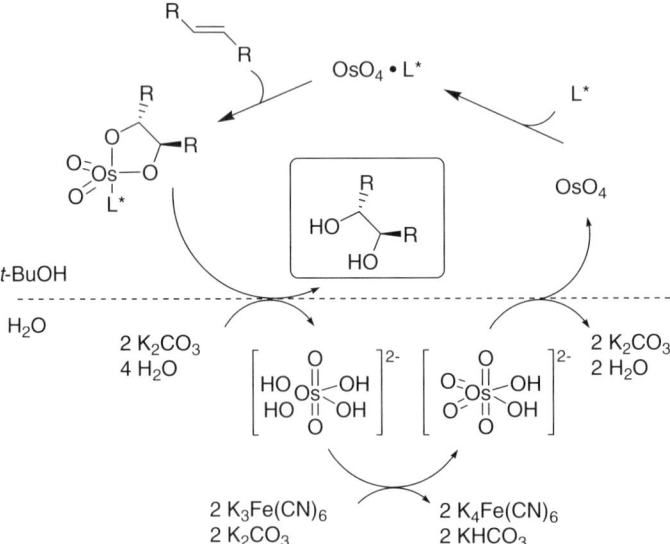

$$(27)$$

The chiral ligand used is based on a phthalazine (PHAL) modified by two dihydroquinidine (DHQD) substituents. Other asymmetric dihydroxylation reactions for the synthesis of pharmaceuticals have been developed at Chirex and Pharmacia/Upjohn.

The catalytic cycle for the asymmetric dihydroxylation is shown in Figure 20. The reaction is carried out in a 1/1 *t*-butanol/water mixture to solubilize the potassium ferricyanide/potassium carbonate used as the oxidant. The solvent mixture, normally miscible, separates into two liquid phases upon addition of the inorganic reagents.

As shown, osmium tetroxide bearing the chiral ligand interacts with the olefin to give an Os(VI) ester, which upon hydrolysis releases the chiral diol. The actual oxidant is the metal itself that reduces from Os(VIII) to Os(VI). This reaction was known since the 1930's and in this respect it resembles the Wacker system where ethylene is oxidized to acetaldehyde with reduction of Pd(II) to

Figure 20   *Fundamental steps in the commercial catalytic asymmetric dihydroxylation of olefins catalyzed by osmium tetroxide.*

Pd(0). The real challenge was the reoxidation of osmium to make the process catalytic. When the osmate ester is hydrolyzed at the organic aqueous interface, an osmate(VI) dianion (as a potassium salt) is released in the aqueous phase. Oxidation of osmate(VI) by potassium ferricyanide regenerates osmium tetroxide via an intermediate perosmate(VIII) anion. Loss of two hydroxide groups from the perosmate(VIII) ion gives osmium tetroxide, which then migrates back to the organic phase to restart the cycle.

The new development is that electrochemical oxidation of ferrocyanide to ferricyanide can be coupled with asymmetric dihydroxylation to give a very efficient electrocatalytic process. The electrons necessary to the reoxidation of Fe(II) are provided by water that is reduced to OH^- and H_2. Hydrogen gas, released at the cathode, is the only by-product of this process. This electrochemical device uses iron in catalytic amounts (~ 0.15 equivalents per equivalent of olefin).

2.7.4 Oxidation of Sulfides to Sulfoxides: an Anti-ulcer Medication

The asymmetric oxidation of organic sulfides to chiral sulfoxides was discovered independently by Kagan and Modena in 1984, using the Sharpless system for the asymmetric epoxidation of allylic alcohols. The successful extension of the Sharpless system to sulfides was made possible by the addition of 1 equivalent of water and an excess of diethyl tartrate. Both ingredients are necessary to prepare sulfoxides in good yields and with high *ee*'s, but completely block the activity of the catalyst towards allylic alcohols. Over the years catalyst optimization studies were carried out using a variety of chiral ligands (binaphthol, chiral trialkanolamines, chiral Schiff bases such as salen derivatives), on different metals (vanadium, manganese) and also on iron and manganese porphyrin catalysts. Good catalytic activities and high enantioselectivities were obtained especially when there was a substantial difference in size between the two substituents on sulfur.

Astra–Zeneca commercialized esomeprazole (the *S* enantiomer of omeprazole) obtained by asymmetric oxidation of pyrmetazole (Equation 28) in 2000, using a modified version of the Kagan–Modena protocol. Esomeprazole is an anti-ulcer medicine and one of the largest selling drugs worldwide. It is produced under very mild conditions in a multi tonne per year scale.

The transformation of the laboratory protocol into this successful industrial process was facilitated by *i*) the addition of an organic base (diisopropylethylamine), which substantially increased both the *ee* and the productivity; *ii*) the addition of the appropriate amount of water (0.5/1 with respect to Ti) which directly influenced the stereochemical outcome; *iii*) the use of a 1:2 ratio of [Ti(O-*i*-Pr)$_4$] and diethyl tartrate; *iv*) the use of cumene hydroperoxide as the oxidant; *v*) the replacement of chlorinated solvents with more environmentally friendly ones (toluene, ethyl acetate); and *vi*) carrying out the oxidation at higher and technically more convenient temperatures than before (~ 0–$30°C$ *vs.* $-20°C$).

Discussion Point DP3: *The Sharpless enantioselective catalyst system has some substantial disadvantages with respect to other enantioselective systems: name at least two. Enantioselective oxidation is to some extent underdeveloped with respect to e.g. hydrogenation; can you establish logical connections with: i) the stability of catalysts in the reaction medium; ii) the nature of the oxidants used and their environmental impact; iii) the number of reactions that can successfully be accomplished.*

pyrmetazole   esomeprazole   (28)

2.8 Acrolein and Acrylic Acid from Propylene: for Super-Absorbent Polymers, Paints, and Fibres

Commercial acrolein is an intermediate in the manufacture of several products, in particular D,L-methionine, used as an additive in animal feeds. For the most part however it is directly oxidized to acrylic acid, without being separated and recovered as a pure material. The acid is mainly esterified to methyl and other acrylates, with the remainder being directly used for the manufacture of polymers. Acrylate esters are currently the final destination of most acrolein produced in the world. They readily form homopolymers and copolymerize with methacrylates, styrene, vinyl acetate and acrylonitrile to yield a range of prized products, characterized by excellent clarity, stability to UV light and aging, and good pigmentability.

Polymers based on acrylic acid are highly hydrophilic and are utilized in different applications that include superadsorbent materials, flocculants and dispersants. Polyacrylates and their copolymers range from soft and flexible materials to hard plastics, applied in the production of coatings, paints, binders and adhesives. Their applications include the manufacture of cars (*e.g.*, coatings, upholsteries and adhesives) and the textile (*e.g.*, binders for fiberfill and nonwoven fabrics), paper and leather industries. Methyl acrylate is mainly utilized for copolymerization with acrylonitrile to improve the dyeability of fibres.

In the past, acrolein was produced by the gas phase condensation of acetaldehyde with formaldehyde on sodium silicate, until it was supplanted by the catalytic oxidation of propylene. Early catalysts based on cuprous oxide were only sufficiently selective at low conversions of propylene. The real breakthrough came with the discovery made by Sohio of bismuth molybdate catalysts, developed into formulations specifically optimized for the manufacture of acrylonitrile, acrolein, and methacrolein.

$$CH_2=CH-CH_3 + O_2 \rightarrow CH_2=CH-CHO + H_2O$$
$$(\Delta H = -340.8\,kJ/mol) \tag{29}$$

In current acrolein production (Equation 29), the catalysts are Bi, Fe and Mo mixed oxides, to which other metal (Co, Ni, W, K) and non-metal (P, Sb, B, Si) components are added for performance improvement. Mechanistic studies suggest that the olefin is oxidized by lattice oxygen, in a Mars-van Krevelen type mechanism. Oxygen vacancies are replenished by molecular oxygen in a second step (Figure 21). The feed, composed of a mixture of propylene (5–10%), steam (10–30%) and air, is passed over a packed bed of catalyst, in a multi tubular reactor, at 300–400°C with contact times of a few seconds. Steam can be replaced by an off-gas recycle stream or by other inert diluents. A molten salts fluid circulates on the shell side of the reactor to control the strong exothermicity of the oxidation and to exchange the heat for steam production. High selectivities to acrolein (80–90%) are found at high propylene conversions (> 90%).

Acrolein and condensable by-products, mainly acrylic acid plus some acetic acid and acetaldehyde, are separated from nitrogen and carbon oxides in a water absorber. However in most industrial plants the product is not isolated for sale, but instead the acrolein-rich effluent is transferred to a second-stage reactor for oxidation to acrylic acid. In fact the volume of acrylic acid production (*ca.* 4.2 Mt/a worldwide) is an order of magnitude larger than that of commercial acrolein. The propylene oxidation has supplanted earlier acrylic acid processes based on other feedstocks, such as the Reppe synthesis from acetylene, the ketene process from acetic acid and formaldehyde, or the hydrolysis of acrylonitrile or of ethylene cyanohydrin (from ethylene oxide). In addition to the (preferred) stepwise process, *via* acrolein (Equation 30), a

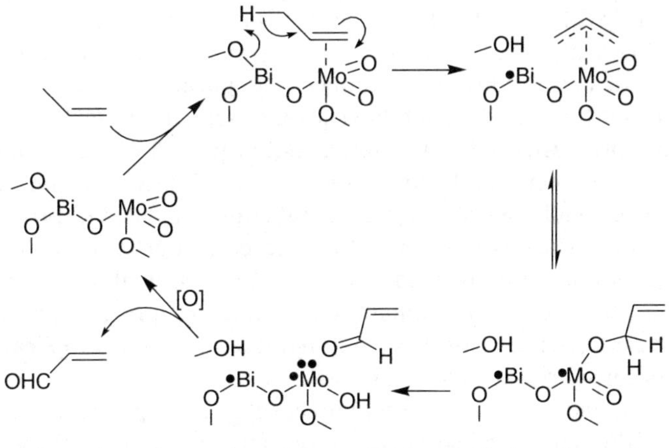

Figure 21 *Schematic representation of the mechanism oxidation of propylene to acrolein on Bi-Mo catalyst.*
(Reprinted from J.D. Burrington, C.T. Kartisek, R.K. Grasselli, *J. Catal.*, 1984, **87**, 363, with permission from Elsevier.)

single-pass oxidation of propylene to acrylic acid (Equation 31) has also been developed.

$$CH_2{=}CH{-}CHO + 1/2O_2 \rightarrow CH_2{=}CH{-}CO_2H$$
$$(\Delta H = -254.1\,kJ/mol) \tag{30}$$

$$CH_2{=}CH{-}CH_3 + 3/2O_2 \rightarrow CH_2{=}CH{-}CO_2H + H_2O$$
$$(\Delta H = -594.9\,kJ/mol) \tag{31}$$

In the two-step process, the second-stage reactor is similar to the first-stage reactor but is packed with an optimized catalyst for aldehyde oxidation, based on Mo:V oxides, and is run under different operating conditions. Care must be exercised during the separation and purification phases to avoid conditions favouring acrylic acid polymerization, *e.g.*, by addition of a radical polymerization inhibitor such as the hydroquinone monomethyl ether. Selectivities to acrylic acid are higher than 90% at total conversion of the aldehyde. Overall yields referred to propylene are in the range 75–85%. Most acrylic acid produced is esterified for the production of acrylate esters.

Even though the two-stage process recommends itself for the flexibility it offers for the optimization of each oxidation step, the single-pass option is attractive as it requires lower investment for fewer reactors and other equipment. The catalysts claimed in the patents are generally multicomponent metal oxides based on Mo with Te as the promoter. However the yields are lower (50-60%) and deactivation faster owing to the volatility of Te oxide.

2.9   Methacrolein and Methacrylic Acid from Isobutene

Methacrylic acid, with a global installed capacity 2.8 Mt/a, is marketed as the methyl ester (MMA), utilized for the production of a variety of polymers (PMMA). These are relatively expensive materials whose utilization is justified by the superior performance in a number of applications. PMMAs are generally harder and tougher than acrylic acid polymers, appreciated for the excellent optical properties (clarity, gloss), stability to UV light and weather conditions, light weight and impact resistance, colourability and good processability. Their end uses are in the manufacture of flat and profiled sheets, impact resistant glass components (for the car, construction and building industries), roofing, aircraft windows, safety glasses, lighting fixtures, optical components (spectacles, soft contact lenses, mirrors, prisms), precision parts (dials and rulers), dental materials, coatings and many other applications, such as additives for lubricants and hydraulic fluids.

Methacrylic acid is produced by a number of different processes, one of which is based on the oxidation of isobutene (or of *t*-butyl alcohol) via the intermediate formation of methacrolein (Equation 32). The general features and the catalyst for the first-stage process are not dissimilar to those for acrolein production, whereas the oxidation of methacrolein to MMA differs in that it is catalyzed by heteropoly compounds of Mo and V. A greater dilution with steam or inert gases than for propylene oxidation is necessary to achieve good selectivities. Overall yields, however, are somewhat lower, probably because of the presence of a second oxidizable methyl group vicinal to the double bond.

$$\text{≡⟨} \xrightarrow{[O]} \text{≡⟨}_{CHO} \xrightarrow{[O]} \text{≡⟨}_{CO_2H} \qquad (32)$$

The competitiveness of the oxidation of isobutene compared to the conventional acetone cyanohydrin route (Equation 33) is not only related to its performance and better environmental standards but has to contend with the demands of other users for isobutene, particularly for MTBE and ETBE production. In fact the predominant methacrylic acid process is still the hydrolysis of acetone cyanohydrin; however, the change of mood on the use of MTBE in gasoline blends in the USA, could signal a future shift of isobutene availability making it a more attractive feedstock for methacrylic acid production.

$$\begin{array}{c}
\underset{H_3C}{\overset{H_3C}{>}}=O \xrightarrow{HCN} \underset{H_3C}{\overset{H_3C}{>}}\underset{CN}{\overset{OH}{<}} \xrightarrow[H^+]{H_2O} \underset{H_3C}{\overset{H_3C}{>}}\underset{CONH_2}{\overset{OH}{<}} \longrightarrow \\
\longrightarrow \text{≡⟨}\overset{CH_3}{\underset{CO_2H}{}}
\end{array} \qquad (33)$$

Other routes to MMA start from ethylene, propylene or propyne and involve metal catalysis at some stage of multi-step transformations; for example by the hydroformylation of ethylene to intermediate propionaldehyde, oxidation to propionic acid, followed by condensation with formaldehyde. The Pd-catalyzed carbonylation of propyne to MMA is a further method. However only the ethylene route has found some industrial application (see Chapter 4, Section 4.3.1).

2.10 Ammoxidation Reactions. Propylene to Acrylonitrile: for Engineering Plastics, Polymers

The initial drive for acrylonitrile (AN) production (6.2 Mt/a in 2004 worldwide) was the discovery, in the late 1930s, of the synthetic rubber Buna N. Today nitrile rubbers represent only a minor outlet for AN which is utilized primarily for polymerization to give textile fibres (50%) and ABS resins (24%), and for dimerization to adiponitrile (10%). Early industrial processes depended on the addition of hydrogen cyanide to acetylene or to ethylene oxide, followed by the dehydration of intermediate ethylene cyanohydrin. Both processes are obsolete and are now supplanted by the ammoxidation of propylene (Equation 34) introduced in 1960 by Standard Oil of Indiana (Sohio). The reason for the success stems from the effectiveness of the catalyst and because propylene, readily available in petrochemical plants, is a cheaper feedstock than ethylene or acetylene (*qv* Chapter 1).

$$2CH_3CH{=}CH_2 + 2NH_3 + 3O_2 \rightarrow 2CH_2{=}CH{-}CN + 6H_2O \qquad (34)$$

The term *ammoxidation* refers to the oxidation of ammonia and propylene (but also of propane and methyl aromatics) with molecular oxygen to yield the nitrile. Current production mainly uses the fluidized bed process originally developed by Sohio, even though fixed bed processes are now available from other producers. A near to stoichiometric mixture of olefin, ammonia and air is contacted with the catalyst, at 400–500°C, at near to atmospheric pressure, and with short residence times. It is a single-pass and highly selective process, with olefin conversion close to 98%. By-products are hydrogen cyanide and smaller amounts of acetonitrile. As in the procedures in other oxidation processes illustrated in this chapter, the heat of reaction is utilized for the generation of high pressure steam.

The early catalyst for AN production was a multicomponent metal oxide, mainly consisting of bismuth and molybdenum oxides. Its composition has evolved over the past 40 years, constantly improved by continuous development work for increasingly better performances. Other catalytic materials that have been used in commercial processes include in their compositions, iron-antimony oxides, uranium-antimony oxides and tellurium-molybdenum oxides.

The ammoxidation mechanism and the role of Bi and Mo sites in the complex multistep reaction taking place on the catalyst surface have been the subject of numerous studies. The currently most accepted mechanism, proposed by Grasselli et al., is illustrated in Figure 22.

The ammoxidation of propylene is the result of the coordinated action of an ensemble of different sites. The rate determining step has been identified as the initial H-abstraction from the methyl group of propylene by lattice oxygen leading to the formation of a π-allyl metal complex. Vicinal Bi and Mo sites cooperate in the reaction. Subsequent steps are the reversible insertion of an imido species into the Mo–C bond producing a σ-bonded allyl imido species, and a second H-abstraction reaction yielding an adsorbed precursor of acrylonitrile. It is significant that these reactions on the surface of a mixed oxide are very reminiscent of reactions occurring in organometallic complexes in solution. Lattice oxygen vacancies are replenished by the dissociative adsorption of molecular oxygen which however involves surface sites different to those involved in the cycle of oxidation reactions. Molybdenum-imido species Mo=NH, indispensable to the ammoxidation path, are produced by the reaction of ammonia with surface Mo=O species. If NH_3 is absent, the attack of Mo=O species at the π-allyl complex bends the oxidation path towards the formation of acrolein.

In other catalytic formulations, the role of Bi^{3+} in the H-abstraction step may be played by Sb^{3+} and Te^{4+}, while that of Mo^{6+} in olefin chemisorption and NH insertion can be taken by Sb^{5+}. The mobility of lattice oxygen, from

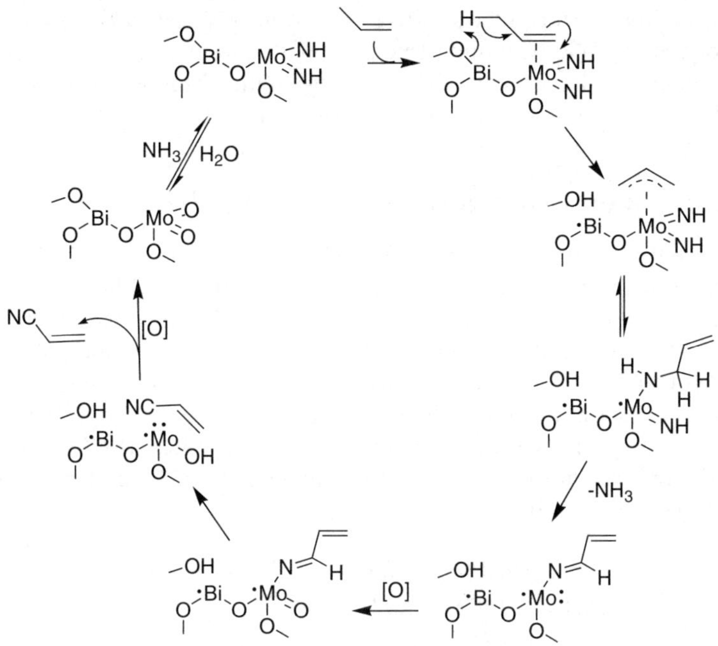

Figure 22 *Schematic representation of the ammoxidation mechanism on Bi:Mo catalysts.* (Adapted from. J.D. Burrington, C.T. Kartisek, R.K. Grasselli, *J. Catal.*, 1984, **87**, 363, with permission from Elsevier.)

oxide surface to lattice vacancies, is assisted by Fe^{2+}/Fe^{3+} or other redox couples added to the multicomponent oxide.

An attractive alternative route to acrylonitrile uses as feedstock propane rather than propylene: it is briefly addressed in Annex 1.

2.10.1 Isophthalonitrile from *m*-Xylene

Ammoxidation is also used for the manufacture of isophthalonitrile from *m*-xylene. The catalysts claimed in the patents are mixed oxides, *e.g.*, of Mo, V, and Sb. The dinitrile is hydrogenated to *m*-xylylene diamine and *m*-diaminodimethylcyclohexane, used for the production of the corresponding diisocyanates that are less toxic than toluene diisocyanate TDI. A different use is for the production of tetrachloroisophthalonitrile, a potent fungicide.

2.11 Maleic Anhydride and Phthalic Anhydride: for THF, Spandex, Swim-suits and Ladies' Tights

2.11.1 Maleic Anhydride

Maleic anhydride (1.17 Mt/a world installed capacity) finds its major use in the synthesis of unsaturated polyester resins (*ca.* 41%), with the remainder going to produce butanediol (14%), maleic copolymers (8%), tetrahydrofuran (7%), fumaric and malic acids, lube additives, and other materials. For many years it was produced by the oxidation of benzene with air (Box 4), until the commercialization in 1974 of the oxidation of *n*-butane. Minor amounts are also formed as a by-product in phthalic anhydride manufacture.

Box 4 Older Process to Maleic Anhydride

In the first commercial process, introduced in 1933, maleic anhydride was produced by the catalytic oxidation of benzene with air. Although its appeal declined after the 1970s the benzene process is still operated, particularly where *n*-butane is not available. The catalyst is a mixed oxide (70% V_2O_5 30% MoO_3) deposited on a low surface area carrier to limit side reactions. Atom efficiency is inherently low, as implied by the stoichiometry of the oxidation in which two carbon atoms out of six are lost as CO_2 (Equation B4). Molar yields however can be relatively high (*ca.* 73%) and are generally higher than those in the *n*-butane processes.

$$C_6H_6 + 9/2\,O_2 \rightarrow C_4H_2O_3 + 2\,CO_2 + 2\,H_2O$$
$$(\Delta H = -1875\,\text{kJ/mol})$$

(B4)

Both the partial oxidation to the product and the combustion to carbon oxides are strongly exothermic. The technical solution devised to prevent hot spots and dissipate the heat consists of a multitubular fixed bed reactor cooled by a circulating molten salts fluid. The production of steam is well above the needs of the plant and the site location is often dictated by the possibilities for heat utilization.

n-Butane oxidation grew rapidly as the preferred process and is now dominant for maleic anhydride production for three reasons: *i*) benzene is a valued petrochemical feedstocks whereas the cost of *n*-butane is effectively that of a fuel; *ii*) the recognition of benzene as a carcinogen now requires the adoption of measures against its release in the workplace and in the environment; and *iii*) two of its carbon atoms are lost as carbon dioxide.

A complex network of reactions is hidden behind the simple stoichiometry of *n*-butane oxidation (Equation 35). Butene, butadiene and furan have been suggested to be intermediates in a cascade of reactions eventually producing the anhydride. Carboxylic acids and carbon oxides are formed in parallel and consecutive oxidations.

$$\begin{array}{c}\text{H}_2\text{C}^{\diagup\text{CH}_3} \\ \text{H}_2\text{C}_{\diagdown\text{CH}_3}\end{array} + 7/2\,O_2 \longrightarrow \text{(maleic anhydride)} + 4\,H_2O \qquad (\Delta H = -1260\,\text{kJ/mol})$$

(35)

Despite the higher atom efficiency, weight yields (*ca.* 95%) are not dissimilar in practice from those of benzene oxidation. However the lower cost of the feedstock tolerates lower conversions (80–85%) and lower selectivities (*ca.* 70%). Unconverted butane is generally burnt, together with by-products, to produce additional recoverable heat. Supported $(VO)_2P_2O_7$, to which promoters are added for improved stability and selectivity, is the catalyst used in commercial processes.

Most industrial installations have adopted the fixed bed technology. The concentration of the hydrocarbon is kept below the lower flammability limits (1.8%), for both inlet and outlet streams. A similar reactor design to that used in benzene oxidation is used to cope with the highly exothermic reaction and heat utilization is a major issue for the economic viability of the plant and its location.

Fluidized and transport bed processes have also been developed for better management of the heat released. The former prevents the occurrence of hot spots in the catalyst bed through a more uniform temperature profile. The concentration of the *n*-butane can also be higher, even within the explosion limits, thanks to the barrier to flame propagation constituted by the fluidized bed of particles. Selectivity is not however dissimilar to that in fixed bed operation due to considerable back-mixing of the products and longer residence times.

In the transport bed process the oxidation of *n*-butane by $(VO)_2P_2O_7$ and the reoxidation of the latter by air are carried out separately in two distinct reactors. Both processes are stoichiometric reactions and the vanadyl pyrophosphate functions as an oxygen carrier. Nonetheless the advantage of a higher selectivity, due to the separation of the product from the oxidant atmosphere, is balanced by the difficulties involved in moving huge volumes of solid between two different reactors.

2.11.2 Phthalic Anhydride

The manufacture of phthalic anhydride (world installed capacity *ca.* 4.4 Mt/a) has several points of similarity to that of maleic anhydride in that there are two alternative feedstocks and a large amount of heat is released. The first process, introduced by BASF at the end of 19th century, was based on the liquid phase oxidation of naphthalene catalyzed by mercury salts. It was later replaced by the cleaner gas phase process, carried out over vanadium and molybdenum oxides. Naphthalene was supplied by coal tar distillation and was used exclusively until the end of 1950s when *o*-xylene, of petrochemical origin, became an abundantly available feedstock (Equation 36). A few production units however can use either feedstock, taking advantage of price fluctuations in coke plants (naphthalene) and in refineries (*o*-xylene).

$$\text{(o-xylene)} + 3\,O_2 \longrightarrow \text{(phthalic anhydride)} + 3\,H_2O \tag{36}$$

$(\Delta H = -1108 \text{ kJ/mol})$

More than 90% of the phthalic anhydride is currently produced by fixed-bed processes. The concentration of *o*-xylene in the feed, *ca.* 60–70 g/m^3, is somewhat above the lower explosion limit (*ca.* 40 g/m^3). This can be raised further through the recycle of part of the exhaust to decrease the oxygen content in the feed. The risk of explosions is minimized by an appropriate reactor design, such as pressure release devices. The catalysts are oxides of vanadium and titanium deposited on low surface area carriers. Less used are fluidized bed reactors in which naphthalene is exclusively used and vanadium oxide supported on silica is the catalyst. As for the maleic anhydride processes, the generation and utilization of steam is an important aspect of the economics of the process.

A third process was the liquid phase oxidation of *o*-xylene, catalyzed by Mn: Co: Br in acetic acid solution, via intermediate phthalic acid production. It was operated for a time in the 1970s and then discontinued.

The major use of phthalic anhydride is for ester derivatives used as plasticizers in the manufacture of flexible poly(vinyl chloride), e.g., wallpaper. The largest volume plasticizer is the di(2-ethylhexyl)phthalate. Other uses of phthalic anhydride are for unsaturated polyester and alkyd resins, for dye intermediates, and for isatoic anhydride (for the production of saccharin).

2.12   Silicalite Process to ε-Caprolactam

Box 5 Conventional Routes to ε-Caprolactam

Current production of ε-caprolactam is mainly based on the reaction of cyclohexanone with hydroxylamine and subsequent rearrangement of the intermediate oxime. Undesirable ammonium sulfate is co-produced well in excess of ε-caprolactam, *i.e.*, 0.9–2.9 kg in the oximation step and 1.5–1.8 kg in subsequent Beckmann rearrangement, for 1 kg of product. The salt, of little commercial value, is sold for use in fertilizers or can be incinerated for the recovery of sulfur (Figure 23).

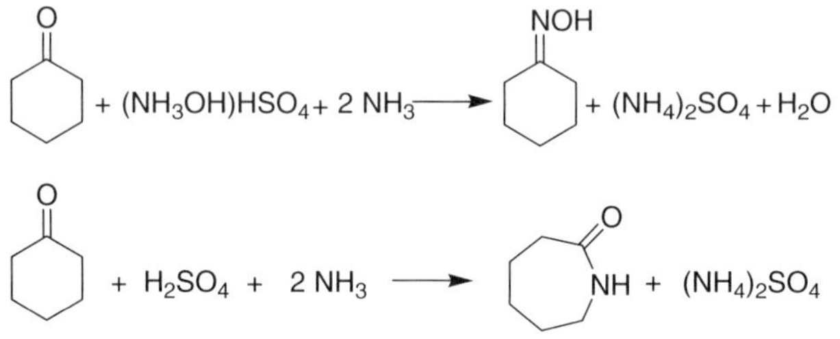

Figure 23   *Synthesis of ε-caprolactam by the Raschig process.*

Alternatives to the use of hydroxylamine include the reaction of nitrosylsulfuric acid with cyclohexanecarboxylic acid, obtained by the hydrogenation of benzoic acid, to produce ε-caprolactam directly, and by the photochemical nitrosation of cyclohexane with NOCl, that yields cyclohexanone oxime hydrochloride. In both cases excess ammonium sulfate is also produced.

ε-Caprolactam (*ca.* 4.0 Mt/a worldwide) is mainly used for the production of nylon-6 fibers and plastics. The ammoximation of cyclohexanone and the catalyzed Beckman rearrangement of the oxime radically changed the traditional mode of production (Box 5), giving access to a new overall process in which ammonium sulfate is no longer co-produced, water is the main by-product and NO_x and SO_x emissions are also greatly reduced or absent (Figure 24). The key was the development in the 1980s of two new zeolites, titanium silicalite-1 (TS-1) and silicalite-1 (S-1), both having the same MFI type of framework, but with different compositions.

2.12.1 Ammoximation of Cyclohexanone on TS-1

The ammoximation of cyclohexanone had been known before the discovery of TS-1, but the performances of conventional catalysts were far below the standards required for development work. In the EniChem process, the reaction is carried out in the liquid phase, at *ca.* 80°C, using a suspension of TS-1 in aqueous t-butanol, with a slight excess of hydrogen peroxide over the ketone. The substrate and the oxidant undergo total conversion with selectivities close to 98% and 94%, respectively. Inorganic by-products comprise minor amounts of ammonium nitrate and nitrite, N_2O, and N_2 produced by the oxidation of ammonia, and O_2 by the decomposition of the oxidant.

In the currently accepted mechanism, ammonia is oxidized to hydroxylamine on Ti sites in TS-1 channels and then reacts rapidly with the carbonyl compounds in solution (even with those that are too bulky to diffuse inside the channels of the catalyst), producing the corresponding oxime in an un-catalyzed reaction (Figure 25).

2.12.2 Gas Phase Rearrangement of Cyclohexanone Oxime to ε-Caprolactam

Ammoximation on its own represented a great simplification in the production of ε-caprolactam, and the catalyzed Beckman rearrangement introduced by

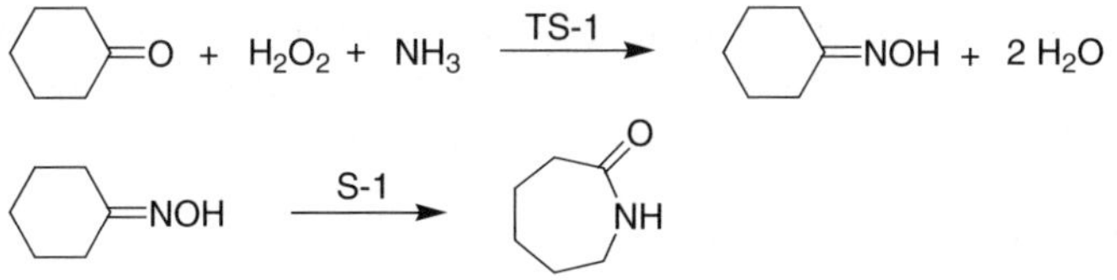

Figure 24 *Synthesis of ε-caprolactam by the EniChem-Sumitomo process.*

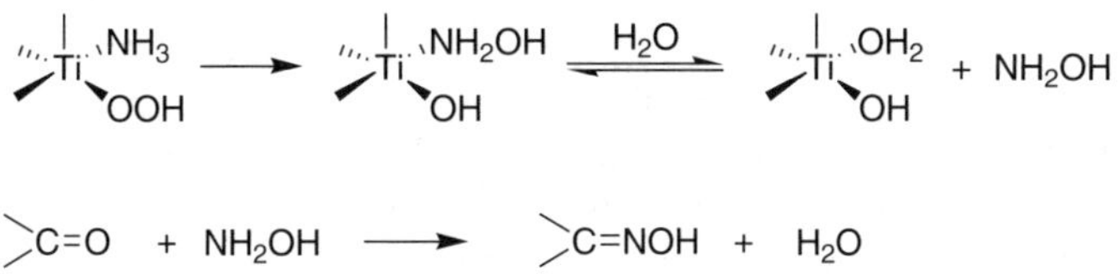

Figure 25 *Schematic representation of the ammoximation mechanism.*

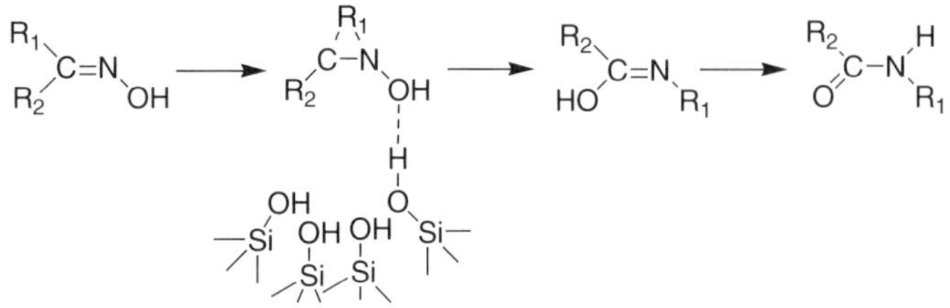

Figure 26   *Proposed mechanism for the S-1 catalyzed oxime rearrangement.*

Sumitomo completed the procedure eliminating the dependence on the use of oleum. The key to success was the all silica zeolite, silicalite-1. Other zeolites have been tested for catalysis, but the presence of strong acid sites reduces the selectivity and increases the deposition of coke. The rearrangement of cyclo-hexanone oxime is typically a high temperature process, carried out at 300–400°C in the gas phase, in a fluidized bed reactor. The latter is required because of the need for frequent regeneration and the process is somewhat reminiscent of FCC, with the catalyst circulating between a reactor, in which the oxime rearranges, and a regenerator in which the coke is burnt off. The selectivity to ε-caprolactam can be above 95%, at total oxime conversion. Although it is lower than if oleum is used (*ca.* 99%), the absence of ammonium sulfate by-product favours the new process.

The reaction mechanism is still a matter of debate. A recent proposal suggested the active species to be nests of silanols, formed by silicon vacancies in the lattice of S-1, in which the rearrangement occurs (Figure 26). This is similar to the reaction with oleum. However, the acidity of S-1 is negligibly low, though the difference in reaction temperature may be enough to close the gap to the activity of oleum.

A process based on the ammoximation of cyclohexanone and on the cataly-zed rearrangement of the oxime went on stream in Japan in 2003; it allows the salt-free production of ε-caprolactam, at lower investment and operating costs than by conventional routes.

Discussion Point DP4*: Hydrogen peroxide is often defined a green oxidant since water is the by-product of its oxidations. Considering its synthesis, how "green" is H_2O_2 in comparison to air or pure oxygen?*

2.13   Oxidation of Phenol to Catechol and Hydroquinone

The hydroxylation of phenol to catechol and hydroquinone with H_2O_2, intro-duced in the 1970s, represented a major advance over earlier methods of production, which utilized the alkaline fusion of *o*-chlorophenol (catechol) and the stoichiometric oxidation of aniline with manganese dioxide (hydroqui-none). The inorganic and organic wastes in both processes were of the order of several kg per kg of product. However, the hydroxylation of phenol too is not free of drawbacks; one is the co-production of two chemicals, destined for two different markets with different growth cycles. Hydroquinone is mainly used in the declining photographic industry, while catechol is mostly required for the production of synthetic vanillin and other fragrances.

$$
\underset{\text{OH}}{\bigodot} + H_2O_2 \longrightarrow \underset{\text{OH}}{\overset{\text{OH}}{\bigodot}} + \underset{}{\overset{\text{OH}}{\bigodot}}\text{OH} + H_2O \longrightarrow \text{tars} \tag{37}
$$

Three processes based on H_2O_2 are commercial; they use different catalysts and show different performances (Table 2). The conversion of phenol is only partial to minimize further oxidations in which the expensive oxidant is also consumed (Equation 37). A compromise, different for each process, is normally made between the need to minimize the energy spent on the separation and recycle of phenol and that of maximizing the selectivities.

In the homolytic process, characterized by relatively low yields, ppm amounts of Fe^{2+} and Co^{2+} are radical initiators. The use of $HClO_4$ and H_3PO_4 as the catalysts allows significantly higher selectivities to be achieved at the expense, however, of conversion. Both processes are operated homogeneously. The third one, catalyzed by titanium silicalite-1 (TS-1), was introduced by EniChem as a replacement for the radical process. It allows analogous or better yields to be achieved than the radical or acid catalyzed processes at a much higher phenol conversion. A relatively expensive oxidant like H_2O_2, is utilized more efficiently and the separation and recycle of phenol is minimized.

The hydroxylation of phenol on TS-1 is normally operated in a slurry reactor, at temperatures close to 100°C, with total consumption of the oxidant. The selectivities on phenol and H_2O_2 are generally in the ranges 90–95% and 80–90%, respectively. The hydroquinone to catechol ratio is well in excess of the statistical value of 1:2, owing to lower steric requirements for *p*-hydroxylation and the faster diffusion of the *p*-substituted product (Table 2). Yields and kinetics are strictly related to the content of lattice Ti. It should be emphasized that any extra-framework Ti species, present as impurities on TS-1, are the major source of unproductive side reactions, such as H_2O_2 decomposition and unselective radical chain oxidations.

Table 2 *Hydroxylation of phenol with hydrogen peroxide on different catalysts[a]*

Catalyst	*o-/p-* ratio	Conversion (% phenol)	Yields (% on H_2O_2)	Yields (% on phenol)
Fe^{2+}/Co^{2+}	2–2.3	9	66	79
H^+	1.2–1.5	5	85–90	90
TS-1	0.5–1.3	30	82	92

[a] Reprinted from U. Romano, A. Esposito, F. Maspero, C. Neri, and M.G. Clerici, Selective Oxidations with Ti-Silicalite, *Chim. Ind. (Milan)*, 1990, **72**, 610, with permission from La Chimica e l'Industria.

It is generally believed that the regularity of shape and the size of the pores in which the Ti sites are located are primary factors for the control of overoxidation to tarry products, permitting high conversions to be achieved while preserving good selectivity. In fact the growth of carbonaceous materials, generally containing fused ring structures, requires more space than the silicalite-1 framework allows.

2.14 Benzene Oxidation to Phenol: Making Phenolic Resins for Building

The world production of phenol, of *ca.* 8.4 Mt/a, is mostly dependent on the cumene process. The yields and selectivities of the process are almost quantitative. However, the per-pass yield is relatively low ($< 8.5\%$) and *ca.* 0.6 tonne of acetone is co-produced per 1 tonne of phenol. The hydrogenation-dehydration of acetone and its recycle has been considered but is not practised commercially.

The direct hydroxylation of benzene to phenol has been intensively investigated, using O_2 and other oxidants. A weakness of the reaction is the increasing activation of the aromatic ring as the degree of hydroxylation increases, favouring further oxidation and the formation of tars and carbon dioxide. Generally a strict limit to the conversion of benzene is required for a high selectivity. Promising in this regard appears the use of H_2O_2 and N_2O on zeolite catalysts (Equations 38 and 39).

$$\text{(38)}$$

$$\text{(39)}$$

TS-1 is a good catalyst for the hydroxylation of benzene when it is used in a highly polar medium, such as sulfolane. This promotes the fast desorption of the product thus hindering its further oxidation (Annex 2). The selectivity of 94% to phenol (plus 6% hydroquinone and catechol), obtained at 9% benzene conversion, is comparable to the per-pass yield of the cumene process.

Closer to industrial application however, is the gas phase hydroxylation with nitrous oxide as the oxidant (Equation 39). The reaction is carried out at 350°C with a selectivity to phenol of 98%, at 27% benzene conversion. The catalyst is Fe-ZSM-5 a zeolite containing Al and Fe in the silicalite-1 framework. Active sites are thought to be binuclear clusters of iron oxide, formed in the channels by the migration of Fe, during thermal treatments of the zeolite. Selectivity is of primary importance for the process, since 14 molecules of nitrous oxide are consumed for the total oxidation of 1 molecule of phenol (Equation 40).

$$\underset{\text{OH}}{\bigcirc} + 14 \ N_2O \longrightarrow 6CO_2 + 3\,H_2O + 14\ N_2 \tag{40}$$

An interesting aspect is the utilization in the process of the N_2O co-produced in the oxidation of KA oil to adipic acid (Section 2.2). This allows a waste product of the integrated production cycle to be reused in the first stage of the cycle also saving on disposal costs. Solutia has recently announced, but has not yet implemented, the commercialization of an integrated adipic acid process based on this concept.

$$\bigcirc \xrightarrow{N_2O} \underset{\text{OH}}{\bigcirc} \xrightarrow{H_2} \underset{O}{\bigcirc} \xrightarrow{HNO_3} \underset{CO_2H}{\overset{CO_2H}{\diagup}} + N_2O \tag{41}$$

Discussion Point DP5: *The conversion of benzene into phenol should simply involve the insertion of one oxygen into a C–H bond, but this seems difficult to achieve, particularly on an industrial scale. Suggest reasons for this and propose a possible new strategy to oxidize benzene directly to phenol.*

2.15 Oxidation Processes in which the Metal Directly Functionalizes the Olefinic Substrate

2.15.1 Ethylene to Acetaldehyde: the Wacker Synthesis

The synthesis of acetaldehyde by oxidation of ethylene, generally known as the Wacker process, was a major landmark in the application of homogeneous catalysis to industrial organic chemistry. It was also a major step in the displacement of acetylene (made from calcium carbide) as the feedstock for the manufacture of organic chemicals. Acetylene-based acetaldehyde was a major intermediate for production of acetic acid and butyraldehyde. However the cost was high because a large energy input is required to produce acetylene. The acetylene process still survives in a few East European countries and in Switzerland, where low cost acetylene is available.

The Wacker process reached a maximum production capacity of 2.6 Mt/a worldwide in the mid 1970's. The cause of the decline in the following years (1.8 Mt/a in 2003) was the increase in the manufacture of acetic acid (the most important product made from acetaldehyde) by the carbonylation of methanol. In future new processes for chemicals, such as acetic anhydride and alkylamines (which were also made from acetaldehyde) will probably further decrease its importance. With the growing use of syngas as feedstock, the one-step conversions of CO/H_2 mixtures to acetaldehyde and other C_2 products may also limit any remaining need (*qv.* Chapter 1).

2.15.2   Chemical Basis of the Wacker Process

The invention of the Wacker process was a triumph of common sense. It had been known since 1894 that ethylene is oxidized to acetaldehyde by palladium chloride in a stoichiometric reaction (Figure 27). However, it was not until 1956 that this reaction was combined with the known reoxidation reactions of palladium by copper and, in turn of copper by oxygen. The total process developed by Wacker and Hoechst between 1957 and 1959 can be depicted as an exothermic catalytic direct oxidation to yield acetaldehyde.

The catalyst is a two-component system consisting of $PdCl_2$ and $CuCl_2$. $PdCl_2$ is the actual oxidant which is reduced from Pd(II) to Pd(0), while $CuCl_2$ reoxidizes the zerovalent palladium to the divalent state. Although numerous other oxidizing agents can also convert Pd(0) into Pd(II), the copper redox system has the advantage that Cu(I) can be easily reoxidized to Cu(II) with O_2.

Olefin oxidation proceeds rapidly, while regeneration reactions are rate determining and can be accelerated by higher HCl concentration. Thus the quantity of palladium salt required can be limited to catalytic amounts by using a large excess of $CuCl_2$.

The Wacker process was a major landmark and a great push towards the development of homogeneous catalysis. The mechanism of acetaldehyde formation differs fundamentally from the other oxidation processes as O_2 itself is not directly involved. As is clear from Figure 28 the actual oxidant is Pd(II) which is reduced to Pd(0). The intimate pathway of the reaction involves nucleophilic attack and was the subject of much debate.

The initial step is replacement of a chloride ion in $[PdCl_4]^{2-}$ (Pd(II) in Figure 28) by ethylene. Other coordinated chloride ions are replaced by water and hydroxy ligands. As a consequence the rate of reaction decreases sharply at high chloride concentrations. The incorporation of oxygen into ethylene occurs by attack of a nucleophile (water) on ethylene made electron-poor by coordination to a Pd(II) centre; this generates a hydroxyethyl-palladium intermediate with displacement of H^+ as HCl. The next step is a β-hydride elimination to generate coordinated vinyl alcohol, followed by hydride migration, proton abstraction from –OH, yielding acetaldehyde and leaving an unstable Pd(II)

$$C_2H_4 + PdCl_2 + H_2O \longrightarrow CH_3CHO + Pd^0 + 2\,HCl$$

$$Pd^0 + 2\,CuCl_2 \longrightarrow PdCl_2 + Cu_2Cl_2$$

$$Cu_2Cl_2 + 2\,HCl + 1/2\,O_2 \longrightarrow 2\,CuCl_2 + H_2O$$

$$C_2H_4 + 1/2\,O_2 \longrightarrow CH_3CHO$$

Figure 27   *The sequence of reactions that, combined together, constitutes the Wacker process.*

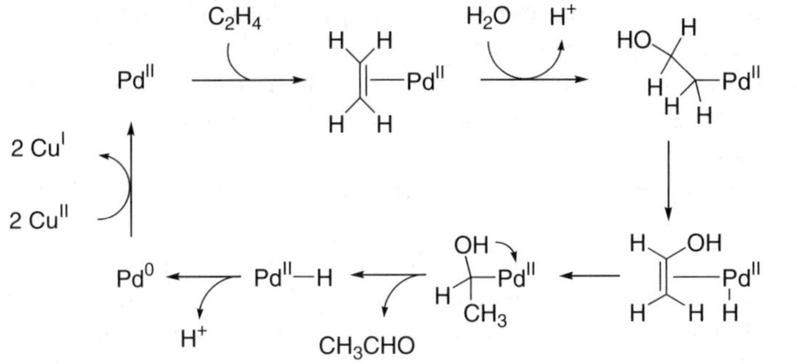

Figure 28 *Representation of the mechanistic cycle involved in the Wacker reaction: the conversion of ethylene into acetaldehyde.*

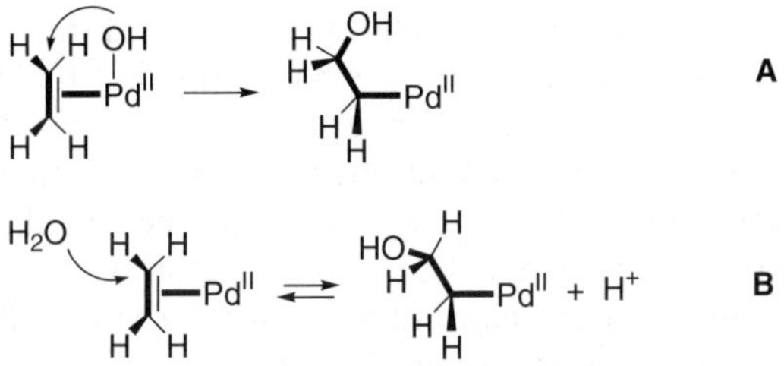

Figure 29 *Illustration of the two modes of formation of the hydroxyethyl palladium intermediate, **A** by ligand migration of a Pd-bonded OH; **B** by nucleophilic attack by external water.*

hydride that decomposes by reductive elimination to Pd(0) and forming a second molecule of HCl.

The major point of controversy was the mechanism of the addition of OH$^-$ to the coordinated ethylene. The hydroxyethyl palladium intermediate has never been observed but is commonly assumed to occur during the reaction. Two mechanisms are possible (Figure 29), both consistent with the kinetic expression: **A**, migration of a coordinated OH$^-$ ligand from palladium to carbon, leading to a *cis* insertion product; and, **B**, external nucleophilic attack by water, leading to a *trans* insertion product.

Thus determination of the geometry of the insertion product differentiates between paths **A** and **B**; the use of *cis*- and *trans*-CHD=CHD allowed workers to establish that the configuration of the inserted product was essentially *trans*, thereby indicating **B** as the prevailing pathway.

2.15.3 Wacker Process Operation

The large scale manufacture of acetaldehyde with the Wacker-Hoechst process takes place in a two-phase gas/liquid system. Ethylene and air (or O$_2$) react with the acidic (pH 0.8–3) aqueous catalyst solution in a corrosion-resistant titanium or lined reactor.

Two versions of the process were developed: the single-step process, in which the reaction and regeneration are conducted simultaneously in a single reactor, and O_2 is used as the oxidizing agent; and the two-step process, in which the reaction and regeneration take place separately in two reactors. Air can be used for the oxidation.

In the single-step process ethylene and O_2 are fed into the catalyst solution at 3 bar and 120–130°C, where 35–45% of ethylene is converted. The heat of reaction is used to distil off acetaldehyde and water from the catalyst solution. Incompletely converted ethylene must be recycled. This requires the use of pure O_2 and ethylene (99.9%) that must be free of inert gases. Inert gas accumulation upon recycling would require venting with consequent ethylene losses.

In the two-step process, the ethylene is almost completely converted by the catalyst solution at 105–110°C and 10 bar. After reducing the pressure and distilling off an acetaldehyde/H_2O mixture, the catalyst solution is regenerated with air at 100°C and 10 bar in the oxidation reactor and then returned back to the synthesis unit. Total ethylene conversion means that the presence of inert gases is no problem and the use of air saves an oxygen plant. The two-stage process also avoids the explosion hazards involved in mixing oxygen and ethylene. However against these advantages must be put the greater capital investment required arising from the double reactor system and the catalyst circulation that is more energy consuming than the gas recycling method.

In both processes the aqueous crude aldehyde is concentrated and by-products are removed in a two-step distillation. Both processes give 94% yields of aldehyde, along with small amounts of 2-chloroethanol, ethyl chloride, acetic acid, chloroacetaldehydes and acetaldehyde condensation products. The Wacker-Hoechst process currently accounts for 85% of the worldwide production capacity for acetaldehyde.

2.15.4 Alternative Catalyst Formulations for Ethylene to Acetaldehyde

The formation of chlorinated by-products and the processing of aqueous chloride solutions are putting heavy ecological constraints on this technology nowadays. Additionally corrosion problems related to the use of highly acidic solutions have always been a major drawback for the Wacker-Hoechst process.

Halide-free catalyst systems have recently been developed by Catalytica Inc. in the USA. Such systems typically comprise a palladium salt along with a heteropolyacid salt such as a phosphomolybdovanadate to reoxidize palladium. In the heteropolyacid the redox couple V(IV)/V(V) replaces Cu(I)/Cu(II) in Figure 27. The reoxidation of V(IV) with oxygen may become so fast that the diffusion of oxygen through the reaction medium becomes rate limiting. Moreover the phosphomolybdate salt constitutes a matrix for vanadium that greatly increases its solubility in the acidic aqueous medium. This catalyst formulation has several advantages, including a higher selectivity and the absence of chlorinated co-products. Another obvious advantage is a reduction in the halide-promoted corrosion of process equipment. This new system has been demonstrated commercially.

Attempts have also been made to heterogenize the homogeneous Pd/Cu catalyst by replacing the anionic chloride medium with a solid cation exchanger, *e.g.* a zeolite. Reaction must be carried out in the vapour phase since metal ion leaching would occur in the liquid phase. Under these conditions water vapour has to be added continuously to the stream of reactants to keep the activity steady. The gas phase Wacker oxidation capability of Pd/Cu zeolites is closely related to the zeolite structure. In fact only the traditional zeolite Y exchanged with appropriate amounts of Pd and Cu shows appreciable activity. While copper can be used as the simple Cu^{2+} ion, palladium must be introduced as the $Pd(NH_3)_4^{2+}$ cation. The Wacker behaviour of a Y zeolite catalyst with optimized Pd/Cu concentration is comparable to that of aqueous homogeneous $PdCl_2/CuCl_2$ solutions and the mechanism is thought to be very similar. Despite the advantages of being halide free with no separation problems this heterogenized catalyst formulation has not yet found applications.

2.15.5 Oxidation of Propylene to Acetone

The Wacker chemistry can also be used to oxidize higher olefins. Terminal olefins are converted to methylketones. In general rates and yields of ketone formation decrease with increasing alkyl chain length. Hence only propylene to acetone has found commercial application.

Three processes are currently used for acetone manufacture:

(i) Wacker-Hoechst direct oxidation of propylene;
(ii) dehydrogenation of isopropanol;
(iii) co-production in the cumene hydroperoxide phenol process.

Most important is the cumene process with an 80–85% share worldwide; cumene (isopropylbenzene obtained from alkylation of benzene with propylene) is oxidized to the corresponding hydroperoxide which is decomposed to a mixture of phenol and acetone. In Japan the second most important process for acetone production is the direct oxidation of propylene with a 12% share.

The Wacker-Hoechst process has been practised commercially since 1964. In this liquid phase process propylene is oxidized to acetone with air at 110–120°C and 10–14 bar in the presence of a catalyst system containing $PdCl_2$. As in the oxidation of ethylene, Pd(II) oxidizes propylene to acetone and is reduced to Pd(0) in a stoichiometric reaction, and is then reoxidized with the $CuCl_2/CuCl$ redox system. The selectivity to acetone is 92%; propionaldehyde is also formed with a selectivity of 2–4%. The conversion of propylene is more than 99%.

As in the acetaldehyde process, this can be carried out commercially in either a single-step or a two-step process. The latter is economically more favourable because a propylene/propane mixture (made by petroleum cracking) can be directly used as the feedstock. Propane behaves like an inert gas and does not participate in the reaction. Acetone is separated from lower and higher boiling compounds in a two-step distillation.

The only plants using this technology were built in Japan. Expansion of the use of this interesting process is hindered by the numerous other processes that give acetone as by- or co-product. Moreover the old process starting from isopropanol still competes favourably. Isopropanol can be converted into acetone either by oxidative dehydrogenation using Ag or Cu heterogeneous catalysts or better by simple dehydrogenation over a ZnO catalyst. In the Standard Oil process, based on ZnO, acetone is obtained in 90% selectivity at an isopropanol conversion of about 98%.

2.15.6 Vinyl Acetate Based on Ethylene (Solution Based Processes)

Figure 28 shows that the chemistry involved in the Wacker process could in principle be extended to other nucleophiles. The modern catalytic manufacturing process making vinyl acetate from ethylene and acetic acid is based on the observation that palladium catalyzed oxidation of ethylene to acetaldehyde can be converted into an acetoxylation reaction if carried out in a solution of acetic acid and in the presence of sodium acetate (Equation 42).

$$\text{CH}_2{=}\text{CH}_2 + \text{PdCl}_2 + 2\,\text{AcONa} \longrightarrow \text{CH}_2{=}\text{CHOAc} + \text{Pd}^0 + \text{AcOH} + 2\,\text{NaCl} \qquad (42)$$

The zerovalent Pd formed can be promptly reoxidized by Cu(II) and the catalytic cycle closed in the same way as for acetaldehyde synthesis (Figure 27).

Based on this catalytic principle, ICI and Celanese developed industrial liquid phase processes which led to the construction of large scale plants. Hoechst independently developed a semi-commercial liquid-phase process.

The liquid phase processes resembled Wacker-Hoechst's acetaldehyde process, *i.e.*, acetic acid solutions of $PdCl_2$ and $CuCl_2$ are used as catalysts. The water produced from the oxidation of Cu(I) to Cu(II) (Figure 27) forms acetaldehyde in a secondary reaction with ethylene. The ratio of acetaldehyde to vinyl acetate can be regulated by changing the operating conditions. The reaction takes place at 110–130°C and 30–40 bar. The vinyl acetate selectivity reaches 93% (based on acetic acid). The net selectivity to acetaldehyde and vinyl acetate is about 83% (based on ethylene), the by-products being CO_2, formic acid, oxalic acid, butene and chlorinated compounds. The reaction solution is very corrosive, so that titanium must be used for many plant components. After a few years of operation, in 1969–1970 both ICI and Celanese shut down their plants due to corrosion and economic problems.

2.15.7 The Gas-phase Ethylene to Vinyl Acetate Process

Ethylene acetoxylation was also developed as a gas phase process following the liquid phase process and has been in commercial use since 1968. There is a notable difference between the two processes: in the liquid phase the presence of palladium salts and redox systems results in the formation of both vinyl acetate and acetaldehyde, whereas in the gas phase process, using palladium metal, vinyl acetate is formed almost exclusively. Additionally there are no corrosion problems leading to cost savings in the use of stainless steel as the construction material.

One version of the gas phase process was developed by National Distillers Products (now Quantum Chemical) in the USA and another independently in Germany by Bayer together with Hoechst. In both versions, ethylene is reacted with acetic acid and oxygen on a palladium-containing fixed-bed catalyst at 5–10 bar and 175–200°C to form vinyl acetate and water. The explosion limit restricts the O_2 content in the feed mixture so that the ethylene conversion is relatively small ($\sim 10\%$). The acetic acid conversion is 20–35% with selectivities to vinyl acetate of up to 94% (based on C_2H_4) and about 98–99% (based on AcOH). The most important side reaction of this process is the total oxidation of ethylene to carbon dioxide and water. Other by-products are acetaldehyde, ethyl acetate and heavy ends. After a multistep distillation the vinyl acetate purity is 99.9% with traces of methyl acetate and ethyl acetate that do not affect the subsequent use in polymerization.

In addition to palladium, the catalysts used commercially always contain alkali salts, preferably potassium acetate. Additional activators include gold, cadmium, platinum, rhodium, barium, while supports such as silica, alumina, aluminosilicates or carbon are used. The catalysts remain in operation for several years but undergo deactivation. The drop in activity is due to a gradual sintering of the palladium particles which causes the catalytically active area to decrease progressively. Under reaction conditions potassium acetate is slowly lost from the catalyst and must continuously be replaced.

Box 6 How Catalyst Manufacture Can Cope with Reaction Conditions

Industrial vinyl acetate synthesis operates at high gas flow-rates to limit contact time and CO_2 formation. A typical Bayer catalyst is prepared by impregnating a medium surface area silica (~ 150 m^2/g) containing small amounts of alumina impurities. Alumina gives a slightly acidic character and greater strength and attrition resistance to the support. The support is impregnated with an aqueous solution of palladium and gold chlorides, dried and then treated with sodium hydroxide to precipitate the corresponding hydroxides from chlorides on site, in a so-called fixing step. Excess chlorides are washed out, the water is removed by drying, and the palladium and gold are reduced to their metallic state with hydrogen. The last step is impregnation with potassium acetate. The resulting catalyst has the precious metals in a shell outside the spherical catalyst pellet. This configuration allows short contact times without wasting precious metals deposited deep inside the catalyst pellet. Potassium acetate is impregnated throughout the catalyst pellet. A typical composition for one of the Bayer-type catalysts is 0.5 wt % palladium, 0.25 wt % gold and about 2 wt % potassium.

The role of gold in the Pd/Au/K acetate catalysts is to stabilize the size of Pd crystallites and avoid sintering. The role of potassium acetate is to maintain the catalyst activity and decrease CO_2 selectivity. Potassium acetate favours a strong adsorption of acetic acid on palladium, lowering the barrier to vinyl acetate formation. Gold by itself is inactive in the catalysis of vinyl acetate. Pd only catalysts produce vinyl acetate at much lower rates than the Pd/Au/K catalyst system and their activity decays rapidly.

The liquid and gas phase catalyst systems for vinyl acetate are based on the same components; no coincidence as the latter was developed after the discovery of the former. They differ mainly in the reoxidation of Pd(0), which is carried out by Cu(II) in the liquid phase process and is not necessary in the gas phase process. It therefore seems tempting to suggest that the chemistry is similar in both cases, at least as far as the vinyl acetate formation is concerned.

The kinetics and the mechanism of the gas phase acetoxylation of ethylene on palladium catalyst has been the subject of many studies. These studies are based on a Langmuir-Hinshelwood type mechanism in which all reacting species are chemisorbed on the Pd surface and reaction occurs between chemisorbed species. Although a complete description is still pending a commonly accepted proposal is shown in Figure 30.

The gaseous reactants interact with the Pd surface giving different chemisorbed species: *i*) ethylene, which chemisorbs dissociatively producing a hydride and a vinyl group; *ii*) acetic acid which initially adsorbs non dissociatively; and *iii*) O_2 which dissociates yielding highly reactive surface atomic oxygen.

In the following step oxygen abstracts a proton from acetic acid, yielding surface hydroxo- and acetato- groups. Hydrogen abstraction from acetic acid and weakening of the Pd–O bond of adsorbed acetate are promoted by the added alkali metal ions (potassium). The rate determining step is the coupling reaction between the surface acetate and vinyl groups with the formation of adsorbed vinyl acetate. The latter is finally desorbed along with water formed by coupling of the surface hydride and hydroxide.

The main source of carbon dioxide, the main by-product, is thought to be the total oxidation of ethylene by the highly reactive surface oxygen.

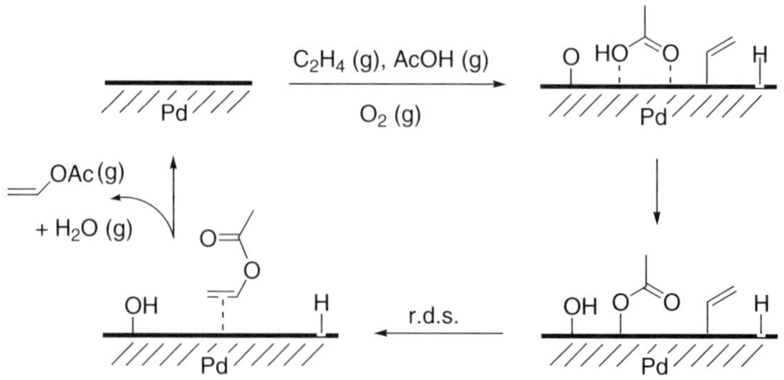

Figure 30 *Mechanism of the acetoxylation of ethylene in the gas phase process.*

The similarities with Figure 28 are, *i*) the formation of a Pd-olefin complex; *ii*) the coupling between the surface acetate and the vinyl group (equivalent to the nucleophilic attack depicted in Figure 29), and *iii*) the formation of a coordinated/adsorbed Pd-vinyl acetate species.

It can be concluded that the understanding of Wacker chemistry with soluble species was a major driving force in developing the heterogeneous catalyst for vinyl acetate synthesis. Unfortunately, a similar heterogeneous catalyst formulation cannot be used for the original Wacker process for the synthesis of acetaldehyde, as that would significantly simplify the engineering of the process, since in the absence of a strong competing nucleophile (such as acetate), the presence of surface oxygen would lead to carbon dioxide formation.

2.15.8 Uses of Vinyl Acetate

Most vinyl acetate is converted into polyvinyl acetate (PVA) which is used in the manufacture of dispersions for paints and binders and as a raw material for paints. It is also copolymerized with vinyl chloride and ethylene and to a lesser extent with acrylic esters. A substantial proportion of vinyl acetate is converted into polyvinyl alcohol by saponification or transesterification of polyvinyl acetate. The main applications for polyvinyl alcohol are either as raw material for adhesives or for fibres. It is also employed in textile finishing and paper glueing, and as a dispersion agent (protective colloid). The world production capacity of PVA was 4.35 Mt/a in 2005, of which 2.1 Mt were converted into polyvinyl alcohol.

2.16 Enzymatic and Microbiological Oxidations. Microbial Hydroxylation of Progesterone

Oxidation reactions are involved in many fundamental biological processes including energy transformation and the biosynthesis of a number of biomolecules (amino acids, vitamins, hormones, *etc.*). This is why, in order to survive, most living organisms need a constant supply of oxygen. Biochemical oxidations are usually catalyzed by enzymes, many of which are actually metalloenzymes so that the fundamental chemical steps they catalyze are essentially the same as those that occur on more traditional metal complexes and centres. Enzymes have been considered as catalysts for a number of reactions of significant industrial interest. However not many large scale applications of enzymes have been developed in the field of oxidation reactions. As discussed below it is usually easier to catalyze oxidations not with isolated enzymes but rather using entire organisms (actually, microorganisms), provided that suitable nutrients are supplied. Transformations in which the whole microorganism cells are used are generally referred to as fermentations and most commercial processes involving biochemical oxidations are actually fermentations. The oldest example of oxidative fermentation is probably the *acetous* fermentation in which any dilute aqueous solution of ethanol (typically, wine) is oxidized with air to acetic acid and water, so producing vinegar:

$$C_2H_5OH + O_2 \rightarrow CH_3COOH + H_2O \tag{43}$$

The "catalysts" are bacteria, typically *Acetobacter*. Already cited in the Bible (see, *e.g.*, *Numbers* 6:3), vinegar is today produced at a rate around 1900×10^6 l/a with an average acetic acid content of 10% (see also Section 4.2.1 *et seq.*).

A specialized field in which several biological oxidations are commonly run on an industrial scale is the chemistry of steroid hormones. The advantages of these technologies (mild reaction conditions and stereospecific reactions at positions difficult to access by chemical reagents) largely compensate for the low substrate concentration (*i.e.*, large reaction volumes) and for the higher costs arising from the need for more sophisticated equipment, skilled personnel, and sterile working conditions. A classical application is found in several glucocorticoids syntheses. Glucocorticoids constitute an important class of steroid drugs with anti-inflammatory, cytotoxic, and immuno-suppressive properties which are exploited in the treatment of a number of syndromes including polyarthritis, rheumatic diseases, allergic reactions such as asthma, anaphylactic and traumatic shock, and cancer. Certain structural features are needed for glucocorticoids to exhibit high pharmacological activities: among them, an 11 β-hydroxyl or keto group. Cortisone was one of the first drugs of this class to be developed and is still a key intermediate in the synthesis of several chemically modified corticoids. It is mainly produced from progesterone or from 16-dehydropregnenolone, both easily and cheaply obtained from diosgenin, a natural product extracted from *Dioscorea* rhizomes. Both syntheses use an elegant microbiological step to insert a hydroxyl at C-11. For instance, the first step of the synthesis starting from progesterone (developed in 1952) is its oxidation by moulds *Rhizopus nigricans* or *R. arrhizus* (Figure 31), with a yield of 90%:

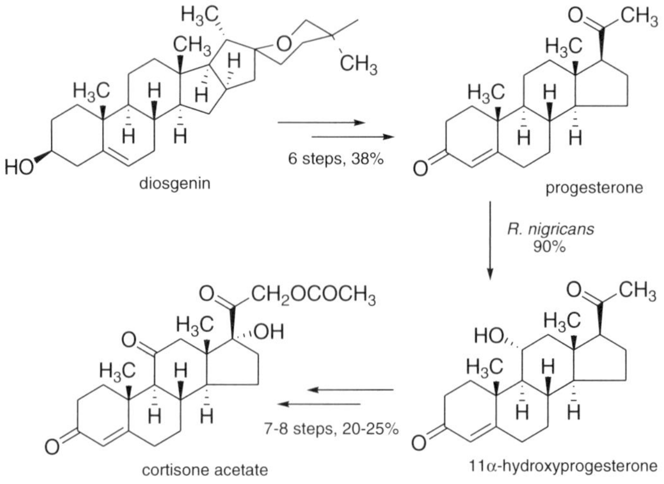

Figure 31 *Synthesis of cortisone acetate from diosgenin.*

The enzyme involved in the reaction is a heme-containing cytochrome P-450. Several relatives of it are involved in a large number of steroids hydroxylations and, overall, play an important role in steroidogenesis.

The discovery of this stereoselective hydroxylation at C-11 and the introduction of diosgenin as a cheap raw material had dramatic effects in reducing the cost of steroid hormones. Thus in the early 1950's, the price of progesterone dropped within just three years from 80 to \$3/g. In its turn cortisone cost about \$200/g in 1949 whereas today its cost is less than \$1/g. Several other steroids are also prepared by processes which include a microbiological oxidation step: among them, cortexone and fluocortolone.

2.16.1 Perspectives of Enzymatic and Microbiological Oxidations

In principle enzymes have several advantages when compared with traditional catalysts:

- most of them are extremelly specific, so that very good selectivities can easily be achieved;
- biotransformations usually occur at room temperature and pressure, so requiring less energy and simpler equipment than most conventional syntheses;
- renewable raw materials can be (and, usually, are) employed;
- there is no need to use corrosive (*e.g.*, acetic acid) or potentially harmful organic solvents;
- wastes do not contain toxic inorganic salts or compounds;
- enzymes are able to catalyze a large variety of oxidations.

Nevertheless, several problems exist: larger equipment is usually required, since relatively low productivities are generally achieved, although there are significant exceptions. But the main problem is that most biochemical oxidations require stoichiometric amounts of co-factors which are quite complex and expensive chemicals, such as flavins, nucleotides, *etc*, that Nature has developed. This is why it is usually easier to carry out oxidations not with isolated enzymes but rather with entire microorganisms able to produce by themselves the cofactors they need.

An interesting attempt to overcome this problem is the design of simplified systems which try to reproduce the activity of natural enzymes (biomimetic catalysts). This approach has produced, *e.g.*, impressive advances in the chemistry of synthetic porphyrins and in understanding the activity of some enzymes (*e.g.* cytochrome P-450) which catalyzes oxidation reactions by an iron-porphyrin centre. Furthermore, interesting similarities have been noticed between enzymes and completely different catalysts. For instance, selective adsorption in the channels of some zeolites provide a confined, relatively hydrophobic medium even in aqueous solvent (Annex 2). This strongly resembles the active sites of several enzymes (including cytochrome P-450) that are deeply buried in hydrophobic pockets where lipophilic substrates are readily oxidized. The more hydrophilic reaction products are promptly released into the surrounding aqueous environment thus reducing further reactions leading to overoxidation. This feature achieves rather unusual selectivities, such as the oxidation (catalyzed by titanium silicalite) of poorly reactive alkanes in the presence of large excess of methanol, used as solvent.

Therefore even if the drawbacks of enzymatic oxidation are not overcome the interplay between chemistry and biochemistry is already producing fruitful cross-fertilization.

Annex 1 Alkane Feedstocks. Alternative Routes to Acetic Acid and Acrylonitrile

The direct synthesis of chemicals from alkanes is an attractive alternative to that *via* olefins. Alkanes are abundant and cheaper, while the elimination of the dehydrogenation unit allows simplified process designs and energy savings. Oxidations of several alkanes are of industrial interest and are being investigated (Table 3); at the most advanced stage are those of ethane to acetic acid and of propane to acrylonitrile.

The appeal of an acetic acid process, based on ethane oxidation, lies mostly in the absence of the need for the energy demanding step for syngas production. On the other hand, it has to compete not only with the well established methanol carbonylation (Section 4.2), but also with the current utilization of ethane in steam crackers for ethylene manufacture. In fact, ethane feedstock becomes attractive for acetic acid production if it is locally abundant and can be supplied at minimal cost, *e.g.*, in a petrochemical complex close to a large gas field. The construction of a semi-commercial plant of 30 kt/a in the Persian Gulf region has been announced.

The oxidation of ethane to acetic acid is believed to proceed via the intermediate formation of ethylene (Equation A1). The catalysts are multicomponent mixed oxides, having optimized compositions for ethane oxydehydrogenation (Mo-V oxides) and ethylene oxidation (Pd, Nb oxide). Reported

Table 3   *Oxidations of alkanes with an industrial interest*

Alkane	Product	Status
Methane	Methanol	Research
Ethane	Acetic acid	Pilot plant
Propane	Acrylonitrile	Pilot plant
	Acrolein, Acrylic acid	
n-Butane	Maleic anhydride	Industrial
Isobutane	Methacrolein	Research
	Methacrylic acid	
n-Pentane	Maleic anhydride	Research
	Phthalic anhydride	

selectivities are close to 50% at 50% ethane conversion, with the balance generally consisting of carbon oxides and minor amounts of ethylene.

$$H_3C-CH_3 \xrightarrow[-H_2]{O_2} H_2C=CH_2 \xrightarrow{O_2} CH_3CO_2H \tag{A1}$$

The direct oxidation of propane has fewer restrictions on plant location since the alkane is easier to ship over long distances as the compressed liquid. Its oxidation to acrolein, acrylic acid and acrylonitrile is the subject of numerous studies. The synthesis of acrylonitrile has already been developed to the stage of a demonstration plant. Catalysts are based on V-Sb mixed oxides, with additional metal promoters. Propylene is generally recognized as the intermediate through which acrylonitrile is obtained. Selectivities are close to 50–60% at *ca.* 20% propane conversion.

Annex 2 Adsorption Effects on the Catalytic Performances of TS-1. Zeolites as Solid Solvents

The molecular dimensions of pores in TS-1 and, more generally, in zeolites, affect the adsorption and, consequently, the reactivity of organic compounds in two opposite ways. Short range repulsions produce the steric constraints discussed in Appendix 2. The longer range interactions, e.g., of van der Waals nature, that are important for physisorption, are magnified by the specificity of the host-guest system. In a micropore, in fact, the entire molecule is interacting with the surface, whereas in mesopores, and even more in macropores, the contact with the sorbate occurs with just a small portion of it. The adsorbed molecule can be seen as enveloped by the walls of the channels, justifying for zeolites the term, "solid solvents", by analogy with the solute-solvent interaction (E.G. Derouane, *J. Mol. Catal.*, 1998, **134**, 29). The effect on physisorption energies can be a major one with respect to an almost flat surface, the extent of it depending on the relative affinity of the surface for the sorbate and the tightness of the fitting. In many cases, the physisorption energies are comparable to the activation energies of reactions that the same molecules may undergo, thus rendering feasible reaction paths otherwise impractical on conventional catalysts. A good example is provided by the smooth and selective hydroxylation of *n*-hexane by a dilute solution of H_2O_2 in methanol, catalyzed by TS-1 at near room temperature (M.G. Clerici, *Appl. Catal.*, 1991, **68**, 249). The physisorption energy of the apolar substrate (critical diameter *ca.* 4.3 Å) in the hydrophobic catalyst (pore diameter *ca.* 5.5 Å) is estimated to be near to 69 kJ/mol, and a similar amount contributes to the apparent activation energy of the reaction. Other linear alkanes are similarly hydroxylated while the polar methanol (see below) is not.

However, not only the nature of the pores but also the nature of the medium, in which the catalyst particle is immersed, determines the extent of adsorption

Table 4   *Partition coefficients of 1-hexene and 1-nonene between TS-1 and three solvents of different polarity*[a]

	CH_3OH	C_2H_5OH	$1\text{-}C_3H_7OH$
1-Hexene	11	3.1	0.7
1-Nonene	–	13	1.8
Water	0.7	0.9	1.6

[a] Adapted from Langhendries et al., Quantitative Sorption Experiments on Ti-zeolites and Relation with α-Olefin Oxidation with H_2O_2, *J. Catal.*, Copyright, 1999, **187**, 453, with permission from Elsevier

(and of reaction). The partition of the solutes, including the solvent itself, between the pore interiors and the external solution is governed by their respective adsorption and solvation energies (Table 4).

It is immediately apparent from Table 3 why hydrophobic TS-1 is an excellent catalyst for the epoxidation of alkenes and why a polar solvent, like methanol or water, is preferable. On a qualitative basis, it can be understood why the apolar hydrocarbon has a greater affinity for the hydrophobic surface of the pores, whereas water (and methanol) has an affinity for the external polar solution. The concentration of the former in the proximity of active sites is high, while that of the latter is small. It is also predicted that a longer chain olefin (or paraffin) is oxidized faster than a shorter one, as is found experimentally.

Discussion Point DP6: *Ti/SiO_2 and soluble Group 4–6 catalysts require completely different conditions for olefin epoxidation, e.g., apolar solvents and organic hydroperoxides. Considering the porosity and surface composition of the former and the coordination properties of the latter, explain why.*

References

General: G. Centi, F. Cavani, F. Trifiró, *Selective Oxidation by Heterogeneous Catalysis*, Kluwer Academic/Plenum Publishers, New York, 2001; G. Ertl, H. Knötzinger, J. Weitkamp, (Eds.), *Handbook of Heterogeneous Catalysis*, Volume 5, Wiley-VCH, Weinheim 1997; R.A. Sheldon, J.K. Kochi, *Metal Catalyzed Oxidations of Organic Compounds*, Academic Press, New York, 1981; G. Strukul (Ed.), *Catalytic Oxidations with Hydrogen Peroxide as Oxidant*, Kluwer, Dordrecht, 1992; K. Weissermel, H.-J. Arpe, *Industrial Organic Chemistry*, 4th Edition, Wiley-VCH, Weinheim 2003.

Epoxidations: J.C. Zomerdijk, M.W. Hall, *Catal. Rev. Sci. Eng.*, 1981, **23**, 163; G. Bellussi, M.G. Clerici, and U. Romano, *J. Catal.*, 1991, **129**, 159.

Asymmetric oxidations: H. U. Blaser, E. Schmidt Eds., *Asymmetric Catalysis on Industrial Scale*, Wiley-VCH, Weinheim, 2004; I. Ojima (Ed.), *Catalytic Asymmetric Synthesis*, 2nd edition, Wiley-VCH, New York, 2000.

Caprolactam, salt-free production: G. Zecchina, G. Spoto, S. Bordiga, F. Geobaldo, G. Petrini, G. Leofanti, M. Padovan, M.A. Mantegazza, P. Roffia, in *New Frontiers in Catalysis*, Stud. Surf. Sci. Catal., L. Guczi, F. Solymosi, P. Tétényi (Eds), Elsevier, Amsterdam, 1993, **75**, 719; H. Ichihashi, M. Kitamura, *Catal. Today*, 2002, **73**, 23.

CHAPTER 3
Hydrogenation Reactions

LUIS A. ORO, DANIEL CARMONA AND
JOSÉ M. FRAILE

Instituto Universitario de Catálisis Homogénea, Instituto de Ciencia de Materiales de Aragón, Facultad de Ciencias, Universidad de Zaragoza-C.S.I.C., Zaragoza E-50009, Spain

3.1 Introduction and Basic Chemistry: Activation of Hydrogen and Transfer to Substrate

Hydrogen is an important raw material for the chemical industry and about 4% of the total consumption is used for the catalytic hydrogenation of organic molecules leading to the formation of new C-H bonds. Hydrogenation can be carried out using heterogeneous catalysts or complexes as homogeneous catalysts in solution. The latter systems have been extensively investigated and much of our understanding of catalysis is based on that work (References and Appendix 1).

3.1.1 Hydrides and Dihydrogen Activation

Various types of complexes containing M-H bonds that are involved in hydrogen-transfer reactions are known; they can be neutral, cationic or anionic. Many are formed by direct reaction with molecular dihydrogen (H_2), although there are numerous other preparative methods. The term *hydride* is generally applied to all such compounds, including those exhibiting either hydridic or protonic behaviour. In each case the H behaves as a uni-negative monodentate ligand. Good σ-donor ligands or an overall negative charge on the complex may favour the hydride form $MH^{\delta-}$, whilst good electron withdrawing ligands or an overall positive charge may favour the form $MH^{\delta+}$.

The formation of hydrides by activation of molecular hydrogen implies the transformation of H_2 into $MH^{\delta+}$, $MH^{\delta-}$ or MH. The hydrogen activation can be either homolytic or heterolytic, and can occur at single or multiple metal centres. Examples of *homolytic* activation are given in Equations (1) and (2).

The formation of a hydride in this way is an example of oxidative addition (formally Rh(I) → Rh(III), or Co(II) → Co(III)).

$$RhCl(PPh_3)_3 + H_2 \rightarrow RhH_2Cl(PPh_3)_2 + PPh_3 \tag{1}$$

$$2[Co(CN)_5]^{3-} + H_2 \rightarrow 2[CoH(CN)_5]^{3-} \tag{2}$$

Equation (3) shows an example of *heterolytic* activation. In this case the intermediacy of a non-classical molecular dihydrogen (η^2-H_2) species has been proposed. This reaction is also an example of a σ-bond metathesis.

$$RuCl_2(PPh_3)_3 \xrightarrow{+ H_2} \overset{\delta^+}{H}\overset{H}{\underset{\delta^-}{\diagdown}}Ru(PPh_3)_3 \xrightarrow{-HCl} RuHCl(PPh_3)_3 \tag{3}$$

Both modes of hydrogen activation, heterolytic and homolytic, may be achieved either directly or through a two-step mechanism involving the intermediate formation of η^2-H_2 complexes.

The heterolytic activation of hydrogen involves no change overall in either the formal oxidation state or the coordination number of the metal. The hydride (H^-) has replaced the coordinated anion (X^-) and release of the proton (H^+) can be promoted by the presence of a base such as triethylamine.

3.1.2 The Reversible Addition of M–H to C=X Bonds on Model Complexes: Olefin Isomerization Reactions

The activity and hence the importance of metal hydrides stems from the facile addition of M–H bonds to organic compounds containing unsaturated groups, such as $H_2C=X$ (X = CR_2, O, NR), leading to the formation of a variety of C–H bonds. The M-H addition to the C=C double bond of an alkene may proceed either in a Markovnikov or anti-Markovnikov fashion, the two being related by a series of equilibria (eq 4). The direction of the addition depends on steric factors (bulkier ligands on M favouring terminal addition) and on the polarity of the M-H bond (negative charge on H favouring anti-Markovnikov addition).

$$MH + RCH_2CH=CH_2 \qquad\qquad MH+RCH=CHCH_3$$
$$\updownarrow \qquad\qquad\qquad\qquad \updownarrow$$
$$RCH_2CH_2\underset{M}{\overset{|}{C}H_2} \quad\rightleftharpoons\quad RCH_2CH=CH_2 \quad\rightleftharpoons\quad RCH_2\underset{M}{\overset{|}{C}HCH_3} \quad\rightleftharpoons\quad RCH=CHCH_3 \tag{4}$$

The reverse reactions are known as β-eliminations where the alkenes are formed by a concerted loss of the metal and the β-hydrogen from the alkyl. This reaction can either reform the original 1-alkene complex or make a new 2-alkene complex. The reversibility shown accounts for the ability of metal complexes to isomerize alkenes. Reactions such as these also account for the high activity of metal hydrides in catalytic hydrogenations.

There are two possible ways to approach the $M(H)_n$(alkene) intermediate in a hydrogenation reaction: the hydride route *a*), implies an initial reaction with hydrogen followed by coordination of the substrate, while the unsaturated (or alkene) route *b*) involves the initial coordination of the substrate followed by reaction with hydrogen. The first has been implicated in the hydrogenation catalyzed by cationic $[Rh(H)_2(PR_3)_2]^+$ while the second is the route followed in the industrial enantioselective hydrogenations shown in Section 3.5 and 3.6. Normally the unsaturated and the hydrides are bound to the metal centre; however Box 1 illustrates a situation where this was not found to be necessary.

Box 1 Catalytic Ionic Hydrogenation

In the classical hydrogenation catalysis the substrate must be coordinated to the metal prior to its insertion into a M–H bond. However, very recently the possibility of ionic hydrogenation of unsaturated substrates (such as ketones or imines) in which the coordination of the substrate to the metal is not required, has been recognized. Equation Box1 shows the proposed cycle for catalytic ionic hydrogenation of ketones.

(Box 1)

The reaction occurs by proton transfer from a cationic metal dihydride (or a metal dihydrogen species), followed by hydride transfer from a neutral metal hydride [R. M. Bullock, *Chem. Eur. J.*, 2004, **10**, 2366].

The stepwise nature of the hydrogenation process, has been elegantly illustrated by the sequence observed in a model for the hydrogenation of coordinated ethylene, where the diffferent steps can be clearly distinguished spectroscopically as the temperature is raised. Thus on raising the temperature from 233K to 273K insertion occurs in the ethylene-dihydride complex $[IrH_2(\eta^2\text{-}C_2H_4)(P^iPr_3)(NCMe)_2]BF_4$ giving the ethyl-hydride, $[IrH(C_2H_5)(P^iPr_3)(NCMe)_2]BF_4$ (Equation 5); above 273K this reductively eliminates ethane giving $[Ir(P^iPr_3)(NCMe)_3]BF_4$ (Equation 6) which, after reaction with hydrogen and coordination of ethylene can start the cycle again [E. Sola, J. Navarro, J.A. López, F.J. Lahoz, L.A. Oro and H. Werner, *Organometallics*, 1999, **18**, 3534].

$$[IrH_2(\eta^2\text{-}C_2H_4)(P^iPr_3)(NCMe)_2]BF_4 \rightarrow$$
$$\rightarrow [IrH(\eta^1\text{-}C_2H_5)(P^iPr_3)(NCMe)_2]BF_4 \tag{5}$$

$$[IrH(\eta^1\text{-}C_2H_5)(P^iPr_3)(NCMe)_2]BF_4 + MeCN \rightarrow$$
$$\rightarrow C_2H_6 + [Ir(P^iPr_3)(NCMe)_3]BF_4 \tag{6}$$

Hydrogenation of ketones ($>C=O$), and aldehydes ($-CHO$), to alcohols and of imines ($>C=NR$) to amines can also be achieved both by heterogeneously and homogeneously catalyzed reactions, The proposed mechanisms are similar to those for olefin hydrogenation except that the initially formed substrate-metal complex has a different structure, normally involving η^1 binding through the hetero-atom, as illustrated in Equation (7) for ketone hydrogenation.

$$H_2M(\eta^1\text{-O}{=}CR_2) \rightarrow HM\text{-OCHR}_2 \rightarrow M + HOCHR_2 \qquad (7)$$

Acetylenes can also be hydrogenated, either heterogeneously or homogeneously, but there both partial (to the olefin) and complete hydrogenation (to the alkane) are possible.

3.1.3 A Typical Homogeneously Catalyzed Hydrogenation Cycle: the Wilkinson Catalyst

Even though heterogeneously catalyzed reactions are the most used in industry because of the ease of product isolation and catalyst recovery, homogeneous catalysis is usually the method of choice if selectivity is called for since high chemo-, regio-, and enantio-selectivities can be achieved through tuning by appropriately designed ligands (for example see Sections 3.5–3.7).

One of the most carefully studied hydrogenations is the one catalyzed by the Rh(I) complex $RhCl(PPh_3)_3$, usually known as the Wilkinson catalyst. It was discovered in 1965 and is easily prepared by the reduction of rhodium trichloride hydrate in the presence of triphenylphosphine.

The generally accepted mechanism for alkene hydrogenation (Figure 1) is mainly due to Halpern and is supported by careful kinetic and spectroscopic studies of cyclohexene hydrogenation. The dominant path of the cycle is inside the dotted line.

The predominant hydride route implies the oxidative addition of a hydrogen molecule prior to alkene coordination. Both pathways, the associative via the 16-electron complex, $RhCl(PPh_3)_3$, and the dissociative via the 14-electron complex, $RhCl(PPh_3)_2$, may function for hydrogenation, depending on the concentration of free PPh_3. However the 14-electron $RhCl(PPh_3)_2$ reacts with H_2 at least ten thousand times faster than the 16-electron $RhCl(PPh_3)_3$, and is therefore the active intermediate. The rapid oxidative addition of hydrogen to $RhCl(PPh_3)_2$, to yield $RhH_2Cl(PPh_3)_2$, is followed by subsequent alkene coordination to form the 18-electron dihydride alkene complex $RhH_2Cl(alkene)(PPh_3)_2$. The unsaturated $RhH_2Cl(PPh_3)_2$ intermediate is also capable of ligand association with PPh_3, and is therefore in equilibrium with $RhH_2Cl(PPh_3)_3$. The rate-determining step for the whole cycle is the intramolecular alkene insertion into the rhodium-hydride bond of $RhH_2Cl(alkene)(PPh_3)_2$, to give the alkyl hydride intermediate, $RhH(alkyl)Cl(PPh_3)_2$. This step is followed by the fast reductive elimination of alkane from the alkyl hydride intermediate to regenerate $RhCl(PPh_3)_2$.

Many cycles similar to Figure 1 involving different catalysts, and different substrates have been elucidated; see References and examples in Sections 3.2.4, 3.4.3, 3.5.2, 3.6.2, etc.

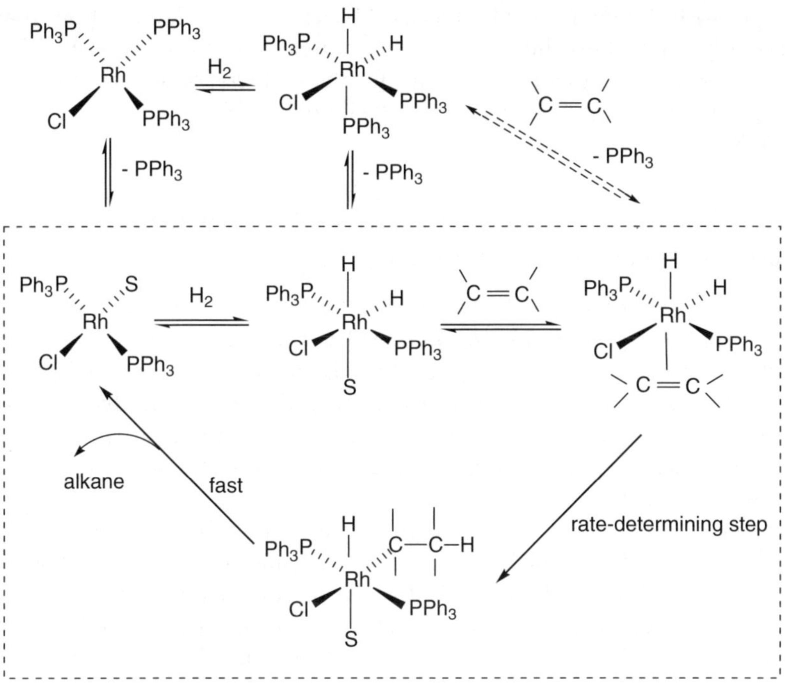

Figure 1 *Mechanism of action of the Wilkinson hydrogenation catalyst; the dashed line encloses the predominant part of the cycle.*

3.1.4 Isomerization of Alkenes

The double bond positional isomerization of alkenes is catalyzed by many transition metals and often occurs as an undesired side reaction. Equations 4 show a simplified scheme of the alkyl-hydride isomerization mechanism in which all the catalytic intermediates have the metal atom in the same oxidation state.

Another isomerization mechanism, observed in some palladium, nickel and iron catalysts, is the allyl-hydride mechanism. This involves coordination of the olefin to a metal centre, which then abstracts an allylic H to form an allyl-metal hydride intermediate. The H can then re-add either to the same carbon or to the carbon on the opposite side (Equation 8).

$$M + RCH_2CH=CH_2 \rightleftharpoons RHC\overset{\displaystyle H}{\underset{\displaystyle MH}{\overset{\displaystyle |}{\overset{\displaystyle C}{\diagup \diagdown}}}}CH_2 \rightleftharpoons MH + RCH=CHCH_3 \qquad (8)$$

In both mechanisms Z- and E- (cis and trans) internal alkenes can be formed, and thus they also account for Z-/E- isomerizations. It is believed that similar mechanisms operate in metal complexes and on metal surfaces.

All isomerization reactions eventually reach their thermodynamic equilibrium, but it is sometimes possible to favour a particular isomer, for example internal alkenes can be made to give the terminal aldehydes in hydroformylation reactions with special liganded catalysts. Alkene isomerization reactions that are industrially important include the enantioselective isomerization of diet-hylgeranylamine and diethylnerylamine for the synthesis of (-)-menthol (Section 3.7), and the isomerization step in the Shell Higher Olefin Process (Chapters 5 and 6, Sections 5.5 and 6.2.2).

3.1.5 Reactions on Metal Surfaces: Heterogeneously Catalyzed Hydrogenation and Isomerization

While the activation of H_2 to give M-H bonds in soluble metal complexes and their use in homogeneous hydrogenations has been the subject of innumerable investigations and is now well understood, thanks to extensive NMR and IR spectroscopic and kinetic measurements, the actual mechanism of the activation of H_2 on a metal surface and its transfer to an organic substrate, although understood in general terms, still lacks some detailed information at the molecular level.

The usual picture (Horiuti-Polanyi mechanism) is based on a Langmuir-Hinshelwood model in which both the hydrogen and the olefin are adsorbed on the surface. As shown in Figure 2 the hydrogen is dissociatively chemisorbed on the metal surface (to form surface hydrides) and can migrate to an adsorbed alkene to give first a surface alkyl and then the alkane. The rate and selectivity of reactions on heterogeneous catalysts depend mainly upon the structure of active sites on the metal. The intrinsic properties of metals, the electronic and steric effects produced by the adjacent surface atoms, including supports, are important. These effects parallel those of ligands and solvent in homogeneously catalyzed hydrogenation.

In each case only the last step, the reductive elimination of the alkane, is irreversible while the reversibility of the other steps leads to isomerization. In

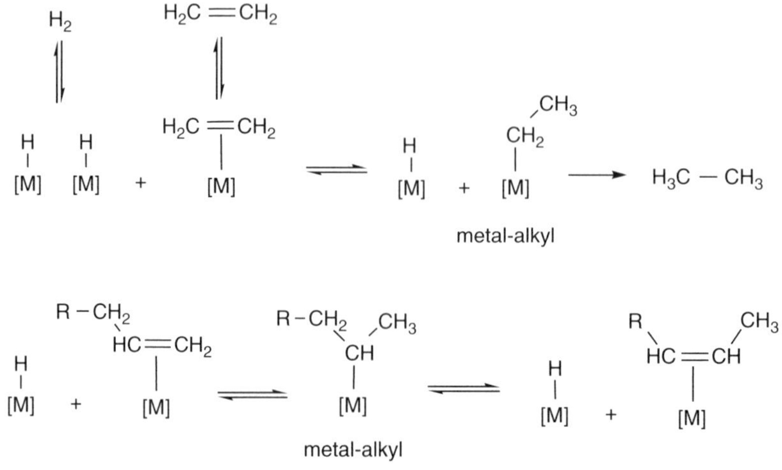

Figure 2 *Mechanism of hydrogenation and isomerization of alkenes by heterogeneous catalysts.*

the case of heterogeneous catalysts, the isomerization is favoured under conditions of low hydrogen availability on the catalyst surface, such as low pressure or slow agitation, when the lifetime of the metal-alkyl becomes significant.

Related reactions can also take place on some metal oxides (eg ZnO in methanol synthesis, Chapter 4, Section 4.7.1) where the H_2 is heterolytically dissociated (eg HZnOH) prior to H-transfer occurring.

Box 2 Cracking and Reforming Processes in Petroleum Chemistry

In a refinery the crude oil is distilled to give fractions of different boiling ranges; the most useful among them are probably light and heavy naphtha (b.p. < 170°C), kerosene (b.p. 170–220°C) and gas oil (b.p. 220–360°C). However, an important fraction of crude oil (40–45% v/v) is composed of heavier hydrocarbons that must be made more useful by converting them into lighter ones. This *cracking* process is an endothermic reaction catalyzed by solid acids. At high temperature, typically 500°C in the presence of an acid catalyst, the hydrocarbons are protonated; the unstable carbonium ions decompose into carbenium ions and lighter alkenes and alkanes. Although several acid catalysts have been used for this purpose, the best results are obtained with particles of microporous zeolite Y (20%) dispersed in a silica-alumina matrix.

Although the hydrocarbons present in naphtha have the right molecular size (both from crude oil distillation and from cracking), they are not suitable as fuels due to self-ignition problems (low octane number). The octane number of gasoline can be increased by converting linear hydrocarbons into branched or aromatic ones, in a process called *reforming*. This is a rather complex process involving several elementary steps, such as dehydrogenation (see Section 3.9), aromatization, hydrogenation, isomerization, and cyclization, and is accompanied by undesirable cracking and coke formation processes. Hence a bifunctional catalyst is needed, with a metallic function (for hydrogenation/dehydrogenation) and an acidic function (for isomerization/cyclization). The most used reforming catalyst is highly dispersed Pt/Al_2O_3.

3.2 Hydrotreating in Petroleum Chemistry

3.2.1 Importance of Hydrotreating in Petroleum Chemistry

Crude oil is probably the most important feedstock in the modern world, not only for fuel production but also as a starting material for making commodity chemicals. Although mainly composed of hydrocarbons, crude oil contains variable amounts of compounds containing heteroatoms (of the order of 1% S,

0.25% N and ppm levels of various metals such as vanadium and nickel). These elements are harmful to the environment, as upon combustion they produce SO_x and NO_x gases (responsible for acid rain), and to the chemical industry, as these molecules can poison the catalysts needed for the subsequent cracking and reforming operations (see Box 2, and Chapter 1, Section 1.2.1).

Hydrotreating, treatment of the distilled fractions of crude oil with hydrogen to eliminate those hetero-atoms, is composed of several individual processes, hydrodesulfurization (HDS), hydrodenitrogenation (HDN) and hydro-demetallation (HDM). These are accompanied by partial hydrogenation of aromatic compounds. This means that in hydrotreating the molecular weight of the compounds present in the oil does not change significantly, by contrast with the effects of the cracking and reforming processes.

This operation has become more and more important with time because of the increasing use of heavy oils which contain more hetero-impurities and the increasingly stringent restrictions on automobile emissions. For example the European limit on the sulfur content of gasoline was reduced from 1000 ppm in 1994 to 500 ppm in 1995, 150 in 2000, and reduces to 50 ppm after 2005. Hydrotreating is probably the most important catalytic process in the petro-leum chemical industry, and, as sulfur is the major impurity, hydrodesulfurizat-ion (HDS) is the most important hydrotreating step. To give some idea of the scale, total world oil production in 2004 was ca. 3.9 billion t/a, almost all of which has undergone some hydrotreatment.

3.2.2 The Catalytic Process

In crude oil sulfur is present as mercaptans (R-SH), sulfides (R-S-R'), disulfides (R-S-S-R') and aromatic compounds, thiophenes and condensed derivatives. The aromatics are much more difficult to desulfurize and require more drastic conditions. The HDS process includes a complex succession of hydrogenolysis and hydrogenation steps, represented for dibenzothiophene in Figure 3.

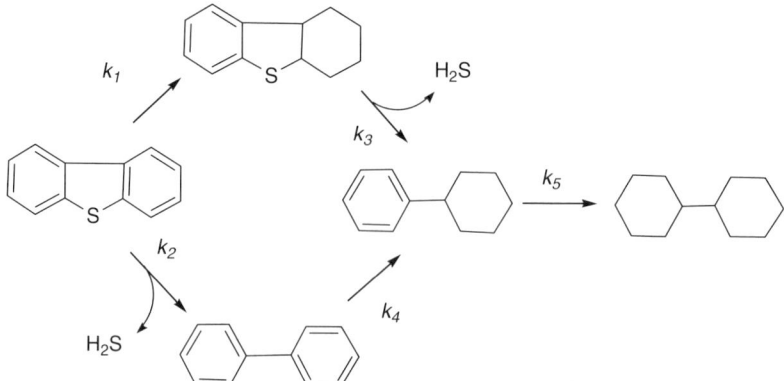

Figure 3 *Hydrogenation and hydrogenolysis steps in HDS of dibenzothiophene.*

The reaction pathway depends on the relative partial pressures of H_2S and H_2. At low H_2S pressure the hydrogenolysis is clearly favoured, given that the pseudo-first-order rate constant of the direct desulfurization step (k_2) is three magnitude orders higher than that of the hydrogenation of the aromatic ring (k_1). However, when the partial pressure of H_2S increases, hydrogenation begins to compete. Typical reaction conditions include high temperature (300–427°C) and pressure (40–100 bar). The type of reactor used depends on the crude oil fraction to be treated. Light fractions are vaporized and passed through a fixed bed reactor, whereas heavier feedstocks that cannot be vaporized require trickle bed reactors.

3.2.3 Composition and Structure of HDS Catalysts

HDS catalysts are solids as are most of the catalysts used in petroleum chemistry, and the active phase of the hydrotreatment process is usually molybdenum disulfide (MoS_2), although other metal sulfides, such as WS_2, are also active. The catalytic process takes place on the surface of the solid: since the surface area increases when the particle size decreases, the surface to volume, or surface to mass, ratio increases. If nanoparticles of the active phase were used, the catalyst would not work in fixed bed or trickle bed reactors, as the particles would be suspended in the gas or liquid phase and would not be retained by normal filters. Thus MoS_2 is dispersed in the form of nanoparticles on an inexpensive support, generally γ-alumina; this gives high catalytic activity from the high surface/volume ratio and also good mechanical properties for industrial applications. The catalyst also includes another metal, usually nickel or cobalt, acting as a promoter.

Commercial catalysts are prepared by impregnation of the alumina support with aqueous solutions of Mo and promoter compounds. Although the amount of Mo added is calculated to be sufficient to form a monolayer on alumina, it has been found that the supported Mo is present as particles containing 7 Mo atoms, together with some Mo–O–Al bonds. That composition reflects that of the major species in solution $[Mo_7O_{24}]^{6-}$. The promoter is found on the edges of the particles and part of the alumina surface remains free. The particles of molybdenum oxide are transformed into the sulfide, usually by treatment with H_2S in H_2, leading to flat particles of MoS_2 with Mo in the centre of trigonal prisms formed by two sulfur layers (Figure 4). The composition of a typical HDS catalyst includes 14% Mo and 3.5% Co on alumina with a surface area of 300 m^2/g.

3.2.4 Mechanistic Studies

The hydrotreating reactions (HDS, HDN) are presumed to take place at vacancies on the edges of the MoS_2 particles (Figure 5); the molecules containing the heteroatom enter the vacant sites, the alkene is formed by β-elimination and the H_2S is lost. This simple picture depicted for an alkylmercaptan becomes more complicated for aromatic sulfur compounds such as

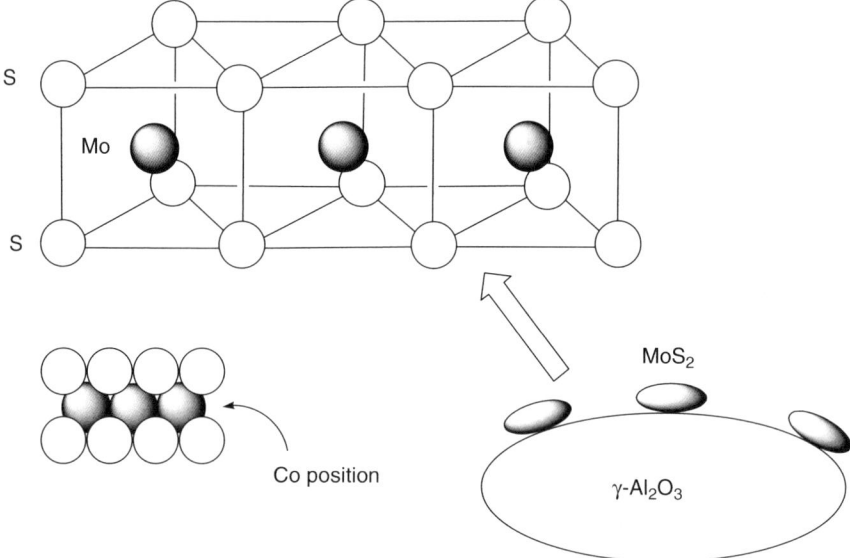

Figure 4   *Representation of the three components of a typical HDS catalyst (alumina, MoS₂, and cobalt promoter) and the spatial relation between them.*

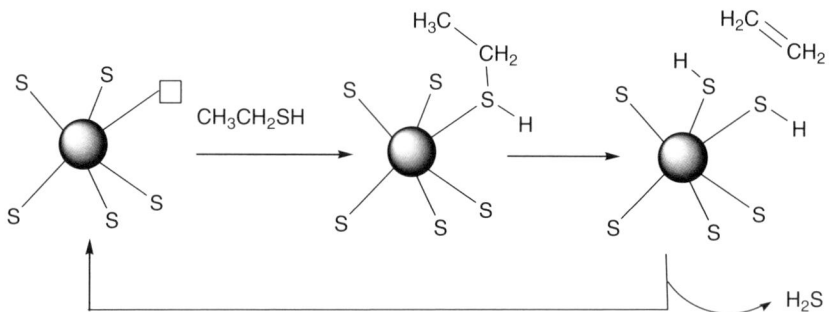

Figure 5   *Schematic representation of the HDS of ethylmercaptan on a MoS₂ catalyst.*

thiophenes, for which a preliminary hydrogenation step to a tetrahydrothiophene is proposed.

The catalytic activity is then not only dependent on the C-heteroatom bond strength but also on the ease of the heteroatom removal from the catalyst, represented by the strength of the metal-sulfur bond. Although unpromoted MoS_2 is not the most active catalyst for HDS, and RuS_2, IrS_2, OsS_2 and Rh_2S_3 are much more active, none of them are used in practice for reasons of cost. The origin of the high activity of the Co- or Ni-promoted MoS_2 is not still clear. Metal-sulfur bond weakening or the presence of unique sulfur-binding sites between two metals are possible explanations for the enhanced activity.

This simple mechanistic picture cannot be applied to thiophene derivatives, whose HDS mechanism is still under study. Research on organometallic models has indicated a rather complex process. First of all, the thiophene derivative can be adsorbed in different ways (η^1-S, η^2, η^5) on the catalyst with different orientations to the surface. The hydrogenation-desulfurization sequence has been commonly accepted for this reaction. However, more recently attention has been directed to the opposite sequence, desulfurization-hydrogenation, which requires a metal insertion in a C-S bond; this has been demonstrated in some model compounds.

3.2.5 Deep Desulfurization

Due to the increasing restrictions on allowable automobile emissions, deep desulfurization will be needed in future. If we consider a normal crude oil feed, with 1% wt sulfur, the reduction down to 50 ppm requires an increase in conversion of 99.5%; that will be increased to 99.9% if the reduction to 10 ppm for 2010 is finally approved. The increase in conversion could be achieved by three strategies:

a) Increase in the hydrogen pressure

This strategy is the simplest in the short term, but it is also very expensive, as it would require new production units able to work at pressures double those currently used. This method would also considerably increase consumption of hydrogen, the availability of which is limited.

b) Enhanced activity of the present catalysts

One of the main problems is the desulfurization of resistant molecules such as the substituted dibenzothiophenes (DBT), especially the 4,6-disubstituted DBTs (Figure 6). It is believed that the first step of HDS involves the coordination of the sulfur or of the C-S bond of the substituted thiophene to the catalyst, and the steric hindrance around the sulphur in DBTs impedes coordination. One possibility is to develop classical HDS catalysts on supports other than alumina. For example on acidic supports such as zeolites, the first reaction could be an isomerization of the alkyl groups to other positions of the aromatic ring, making the subsequent HDS easier. Some results are promising, but it is difficult to prepare the correct CoMoS phase on the microporous support.

c) New catalysts or catalyst combinations

One possibility to enhance the activity for HDS of resistant molecules is to favour the hydrogenation pathway, by doping the catalyst with small amounts of Pt or Rh. Another possibility is to use molybdenum carbide, which has a higher hydrogenating activity than the sulfide.

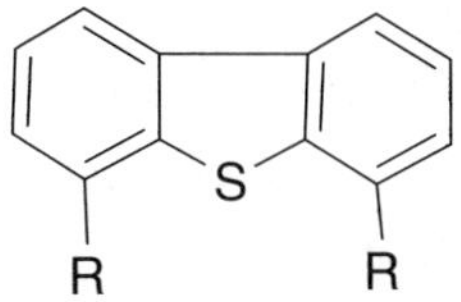

Figure 6   *The structure of a 4,6-dialkyldibenzothiophene (R = alkyl).*

3.3 Mono-unsaturated Fatty Esters by Partial Hydrogenation of Natural Oils

---

Box 3 Margarine

A mechanical mixture of milk with beef tallow under pressure was patented in 1869 with the name "oleo-margarine" as an alternative for butter and pork fats. Its production grew, but it was soon superseded by the introduction of fats made by the catalytic hydrogenation of vegetable oil. Around 120 Mt/a of fats (ie., solids) and oils (ie., liquids) are produced world-wide of which ca. 10 Mt/a is margarine (known as *spread* in the UK), currently produced in more than 60 countries.

3.3.1 Hydrogenation of Fats

Vegetable oils are composed of complex mixtures of triglycerides (triesters of glycerine and fatty acids), whose composition depends on the origin of the oil. This composition is generally described in terms of the proportion of the different component fatty acids and is dominated by acids with 18 carbon atoms (C_{18}) and different degrees of unsaturation (Figure 7). The structures of the fatty acids present in the oil, the number of unsaturations and the stereochemistries, are responsible for the physical properties of the oil.

The first objective of the hydrogenation process is to convert the vegetable oils into materials having the appropriate properties for different applications, such as cooking, food manufacture, animal feed or margarine (spread) production. In addition to good physical and organoleptic properties (solid or liquid state, viscosity, spreadability), the hydrogenation has a second objective, which is to increase the stability of the fat, mainly towards oxidation. Polyunsaturated

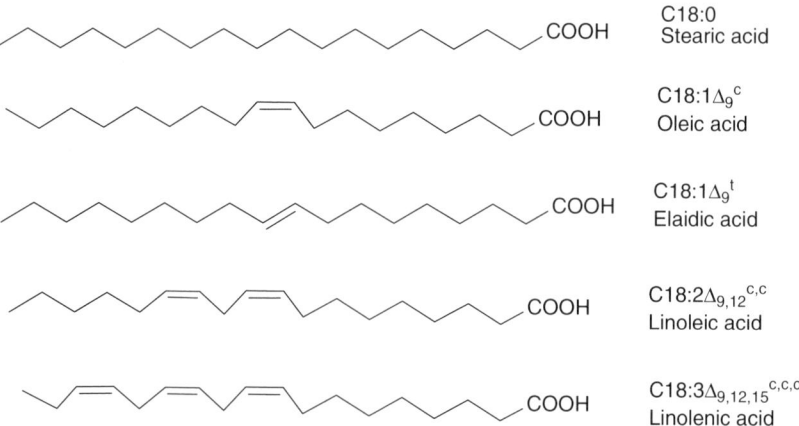

Figure 7 *The structures of some of the C_{18} fatty acids commonly found in vegetable oils.*

fatty acids are very prone to air-oxidation, as demonstrated by the relative rates of oxidation: linolenic (150) > linoleic (100) > oleic (10) > stearic (1).

Wilhelm Norman patented the first liquid phase hydrogenation of fatty acids in 1902, and this led to the first production plants (England 1906, Procter & Gamble in the USA a few years later). The hydrogenation of fats is thus one of the earliest commercial organic processes using heterogeneous catalysis, and it now represents a 15 Mt/a business.

3.3.2   The Selectivity Problem

Although alkene hydrogenation is one of the easiest organic reactions, the partial hydrogenation of vegetable oils presents some selectivity problems to get the desired properties. For example, while it is necessary to reduce the number of unsaturations in order to harden the oil and reduce the sensitivity to oxidation, unsaturated acids are fundamental in our diet and thus it is important to keep at least part of them.

In addition to hydrogenation, two isomerization reactions can take place on the catalyst: i) double bond migration, leading to positional isomers, and ii) *cis-trans* isomerization, leading to stereoisomers. The proportions of *trans* isomers have to be controlled as, although they increase the melting point of the product (which is desirable), dietary studies have emphasized the adverse effects of *trans* isomers on the human LDL/HDL cholesterol ratio. Thus the *trans/cis* selectivity is a critical point to be addressed when considering any new hydrogenation process, see Box 4.

Box 4 Interesterification

The requirement for very low content of *trans*-isomers is likely to be difficult to meet due to the high isomerization activity shown by all the hydrogenation catalysts. In view of that the trend is to go to complete hydrogenation and then to obtain the product with the desired properties through *interesterification*, a transesterification between the starting fat, containing unsaturated acids, and the saturated fat arising from total hydrogenation.

3.3.3   Nickel Catalysts and Catalytic Processes

Although the hydrogenation reaction can be catalyzed by many metals, nickel is most commonly used for three important practical reasons: low cost, good catalytic activity, and metal inertness. The main problems associated with the use of Ni catalysts are the low selectivity and the high activity for isomerization reactions.

The activity is determined by the Ni surface and its accessibility, which also affects the selectivity of the process. Thus the active phase is usually dispersed on a solid support, which also acts as a filter aid and even as promoter. The general method of preparation involves impregnation of the support with a solution of the nickel salt precursor, drying the solid and reducing the precursor, usually with H_2.

Kieselguhr is the most used support. Improvements in supports have been made: synthetic silica or silicates have shown better catalyst reproducibility from batch to batch, with an increase in Ni surface area and a decrease of mean particle size.

With this type of catalyst, the hydrogenation is usually carried out in the ranges 110–190°C, under 1-5 bar H_2 with 0.01–0.15% Ni catalyst (w/w). The hydrogenation of fats is somewhat special due to the need to work in all three phases (gas, liquid, and solid, with corresponding mass and heat transfer problems), and since the natural feedstocks used show significant variations in composition. For these reasons batch reactors are still preferred because of their simplicity, lower cost, and since they have the flexibility to be adapted to different feedstocks or different end products.

Many impurities in the feedstocks may act as poisons for nickel catalysts. Oxygenated products (including water, free fatty acids, and oxidized fats) and sulfur compounds inhibit the ability of Ni to adsorb H_2; this then favours the isomerization reactions. Phosphorus compounds (e.g. phospholipids), are large enough to block pores, with a consequent reduction in catalytic activity. For this reason the feedstocks are now being increasingly purified to allow a reduction in the amount of catalyst needed. This makes re-use of a nickel catalyst economically unattractive. Although other catalysts, including copper and palladium have been considered, they have problems (such as sensitivity to poisons) which have precluded their commercial use.

3.4 Hydrogenation of Adiponitrile to Hexamethylenediamine

3.4.1 The Uses of Hexamethylenediamine

The industrial production of hexamethylenediamine became important with the discovery of the use of polyamides as synthetic fibres, the most important being Nylon-6,6. This polymer, prepared from adipic acid and hexamethylenediamine, was commercialized by DuPont in 1938 and it was virtually the only consumer of the entire production of this diamine until the development of light-stable polyurethanes. Those polymers, e.g. Desmodur N (Bayer) used in paints, are based on aliphatic diisocyanates prepared from the corresponding aliphatic diamine.

Currently the global production of hexamethylenediamine exceeds 1.2 Mt/a and production (e.g. ICI, BASF and Rhône-Poulenc in Europe) is based on the hydrogenation of adiponitrile, largely obtained by catalytic addition of HCN to butadiene. Celanese produced hexamethylenediamine by reaction of ammonia with hexane-1,6-diol, coming from the hydrogenation of adipic acid. However, production by this method was abandoned in 1984.

3.4.2 The Hydrogenation of Adiponitrile

The hydrogenation of any nitrile takes place in two steps (Figure 8), with a primary imine as the intermediate. The selectivity problems come from the ability of the formed primary amine to add the imine intermediate. The elimination of ammonia leads to a secondary imine and the corresponding amine upon hydrogenation. Secondary amines can also act in the same way, although the elimination of ammonia leads to an enamine, precursor of the corresponding tertiary amine. One method to prevent these side reactions is to use a large excess of ammonia, even as the reaction solvent in some cases. In this way the equilibria leading to ammonia evolution are greatly disfavoured, increasing the selectivity to primary amines. The use of liquid ammonia as solvent also has the advantages of good heat transfer and a high solubility of H_2. However, it requires the recycling of the ammonia for economic as well as environmental reasons.

The hydrogenation of adiponitrile has additional complications; in some kinetic studies 6-aminohexanonitrile was detected in solution and is considered a reaction intermediate (Figure 9). Alternatively, it may be a by-product if the hydrogenation is incomplete. Furthermore, the addition of the amines to the imine intermediates leads to an increased number of possible by-products, including cyclic compounds, due to the bifunctional character of the substrate.

Raney nickel is the most commonly used catalyst for the hydrogenation of adiponitrile and seems to be the only one now employed in industry. It is more active and less expensive than the Raney cobalt previously used.

Box 5 Raney Nickel Catalysts

Raney nickel was discovered by Murray Raney in 1927. The sponge-like structure is obtained by treatment of a Ni-Al alloy with concentrated NaOH. The composition of the Ni-Al alloy is important, as three different phases can be produced: a eutectic (6 wt% Ni), as well as $NiAl_3$ and Ni_2Al_3. There are eight categories of Raney Ni: W1 to W8, depending on the treatment parameters used, such as the NaOH/alloy ratio (1:1 or 4:3), time (50 min to 12 h), temperature (50 to 120°C), and washing mode. The surface areas are in the range of 40–120 m^2/g, with 85–90% covered by metal. W2 is the most common type, whereas W1 and W8 are the least active and W6 the most active Raney nickel, with a hydrogenation activity comparable to Pd or Pt for alkenes, alkynes, aromatics and carbonyl compounds.

Aluminium dissolves with H_2 evolution, and this hydrogen remains chemisorbed on nickel, presumably in a dissociated form. Raney nickel catalysts are often doped with other metals in order to improve the catalytic activity; the selectivity decreases in the order, Mo > Cr > Fe $\geq$ Cu > Co. These metals are fused with the Ni–Al alloy and remain on the final catalyst, probably as oxides. It is believed that the role of the doping metals is to strengthen the selective adsorption of nitrogenous substrates.

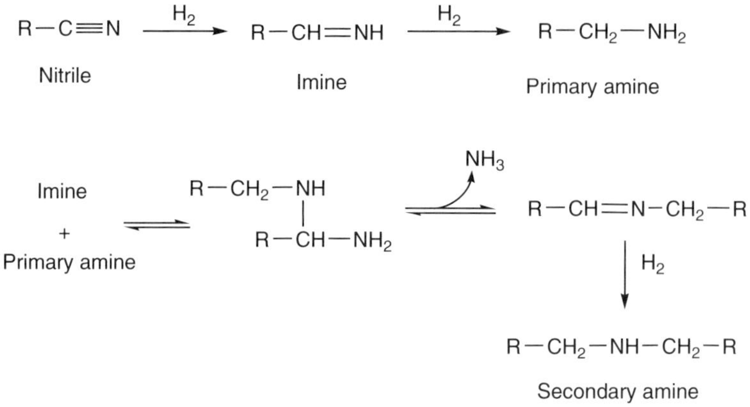

$$R-C{\equiv}N \xrightarrow{\text{H}_2} R-CH{=}NH \xrightarrow{\text{H}_2} R-CH_2-NH_2$$

Nitrile   Imine   Primary amine

Figure 8 *Schematic of the stepwise hydrogenation of a nitrile to a primary amine via an imine and secondary reaction of the product primary amine with the imine intermediate.*

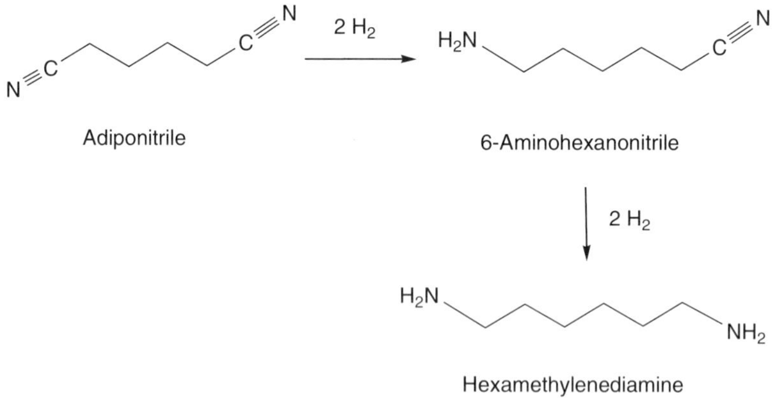

Adiponitrile   6-Aminohexanonitrile

Hexamethylenediamine

Figure 9 *The hydrogenation of adiponitrile, an α,ω-dinitrile, to hexamethylenediamine via 6-aminohexanonitrile.*

Raney nickel is an alternative to dispersing nickel on a support to obtain high surface area particles. It is made by treating of a Ni–Al alloy with a concentrated alkaline solution. Aluminium is selectively dissolved, forming soluble aluminates, and leaving porous nickel metal that retains, at least in part, the structure of the starting alloy with channels easily accessible to the reactants.

3.4.3   Insights into the Reaction Mechanism

The first question in considering the mechanism of nitrile hydrogenation is the mode of adsorption on the catalyst surface. Two adsorption modes are envisaged, corresponding to the end-on (σ, or η^1) and side-on (π- or η^2 Figure 10) modes of coordination of a nitrile to a metal that have been demonstrated in coordination and organometallic compounds. Each adsorption model leads to

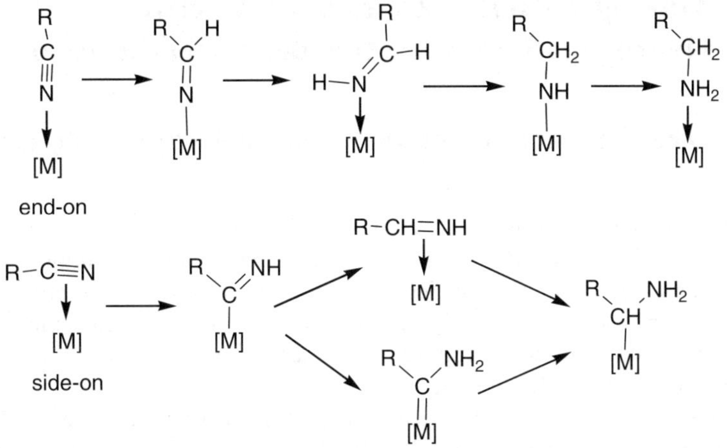

Figure 10 *End-on and side-on binding observed for metal complexes of nitriles, and some probable reaction consequences.*

a different mode of attack of hydrogen on the C–N triple bond. While end-on coordination favours nucleophilic attack at the carbon atom, keeping the M–N bond along the reaction pathway, side-on coordination would favour electrophilic attack at nitrogen, with intermediates that retain the M–C bond. However, these models are over-simplifications. Although no direct experimental proof has yet been obtained, there is probably an equilibrium between the end-on and side-on bonding. Further, the surface of a metallic particle in the heterogeneous catalyst may show both types of adsorption on different centres, as has been observed in some polynuclear organometallic compounds.

These postulated reaction intermediates may also cause some of the by-products detected in this process. For example end-on coordination makes the carbon atom more prone to the nucleophilic attack by the primary amine obtained in the hydrogenation reaction, and favours the formation of secondary amines.

Box 6 Nickel Catalysts: the Future

The main disadvantages of the present industrial process are the use of large amounts of ammonia as solvent and the degradation of the Raney nickel catalyst either by attrition or leaching (solubilization in liquid ammonia). Considerable efforts are currently being made to search for efficient and resistant catalysts for the gas phase hydrogenation of adiponitrile with high hexamethylenediamine selectivity.

Supported nickel catalysts, the analogues Raney nickel prepared from amorphous alloys of Ni-Al with a metalloid such as B or P, and supported Rh catalysts are some of the alternatives currently under study.

3.5 Making *L*-DOPA by Enantioselective Hydrogenation of Acetamidoarylacrylic Acids

3.5.1 The Development of the Enantioselective Hydrogenation Step

In the mid-1960s, it was discovered that *L*-DOPA was useful in treating Parkinson's disease. *L*-DOPA is the common name for the amino acid (*S*)-3,4-dihydroxyphenylalanine, which is the biologically active enantiomer. Its industrial synthesis was formerly achieved by resolution of a racemic intermediate which, in turn, was prepared by heterogeneous hydrogenation of an enamide according to Figure 11a.

At about the same time Wilkinson and his co-workers discovered that the rhodium compound [RhCl(PPh₃)₃] is an efficient homogeneous catalyst for the hydrogenation of unhindered olefins. Inspired by these results Knowles, at the Monsanto Company, conceived the strategy of replacing the PPh₃ of Wilkinson's catalyst with a chiral ligand and using the resulting rhodium compound for the enantioselective hydrogenation of prochiral substrates. First attempts using the chiral methyl(propyl)phenylphosphine as ligand and α-phenylacrylic acid as substrate, gave an enantiomeric excess (*ee*) value of 15%.

Figure 11 a) *Former synthesis of L-DOPA.* b) *(R,R)–(–)–DIOP.*

Shortly after that, Kagan and Dang made an important contribution. They synthesized the chiral diphosphine 2,3-*O*-isopropylidene-2,3-dihydroxy-1,4-bis(diphenylphosphino)butane (DIOP, Figure 11b), derived from tartaric acid, and demonstrated, for the first time, the efficiency of chelating diphosphines with the asymmetry in the side chain, for the catalytic hydrogenation of amino acid precursors.

In the meantime, Knowles and his colleagues had prepared and tested different chiral phosphines with stereogenic phosphorus atoms. It was thought that asymmetry needed to be as near as possible to the metal because the catalytic reaction takes place in the coordination sphere of the metal; the substituents on the phosphorus were varied and the best results were obtained when phosphines such as the *o*-anisyl(methyl)phenylphosphine (PAMP) and *o*-anisyl(cyclohexyl)methylphosphine (CAMP) were used (Figure 12).

The catalyst was made in different ways. A mixture of the rhodium(I)-diene complex [RhCl(1,5-cyclohexadiene)]$_2$ with two chiral ligands per rhodium atom, or RhCl$_3$.3H$_2$O with two phosphines, or isolated [Rh(1,5-cyclooctadiene)L$_2$]A (L = chiral phosphine; A = BPh$_4$, BF$_4$) were quite effective and gave identical results. Hydrogenation with these systems, in an alcohol as solvent, at 25°C, at 1-2 bar generates the active catalyst. In all cases, the chiral phosphine:rhodium ratio is 2:1, suggesting that the catalyst is a cationic complex. Using the CAMP ligand, the L-DOPA precursor Z-4-acetoxy-3-methoxyacetamidocinnamic acid (Figure 13) was hydrogenated with yields in the 85–90% range with *ee's* as high as 88%, after 3 hours of treatment in 2-propanol, at 25°C, under 500 mm of H$_2$, with 0.05% of metal. Both benzoyl or acetyl groups can be used as protecting groups for the amine substituent without losing catalytic efficiency. This homogeneous hydrogenation process was the first to be used in a large-scale production process.

Subsequently the Monsanto group dimerised PAMP to DIPAMP, the rhodium complexes of which gave better results than their CAMP counterparts and therefore the commercial process was adapted to use them. It is shown in Figure 13. The starting vanillin aldehyde is converted to an oxazolidinone by reaction with *N*-acetylglycine. The resulting heterocycle is hydrolyzed to give the acetamidocinnamic derivative which is enantioselectively hydrogenated. This is the key step of the process. Isolated [Rh(COD)(DIPAMP)]BF$_4$ (COD = 1,5-cyclooctadiene) was used as catalyst precursor and methanol as solvent. The catalytic reaction was conducted at 50°C/3 bar

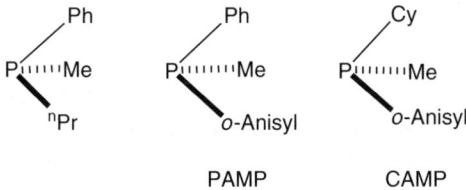

Figure 12 *Stereogenic at phosphorus phosphines.*

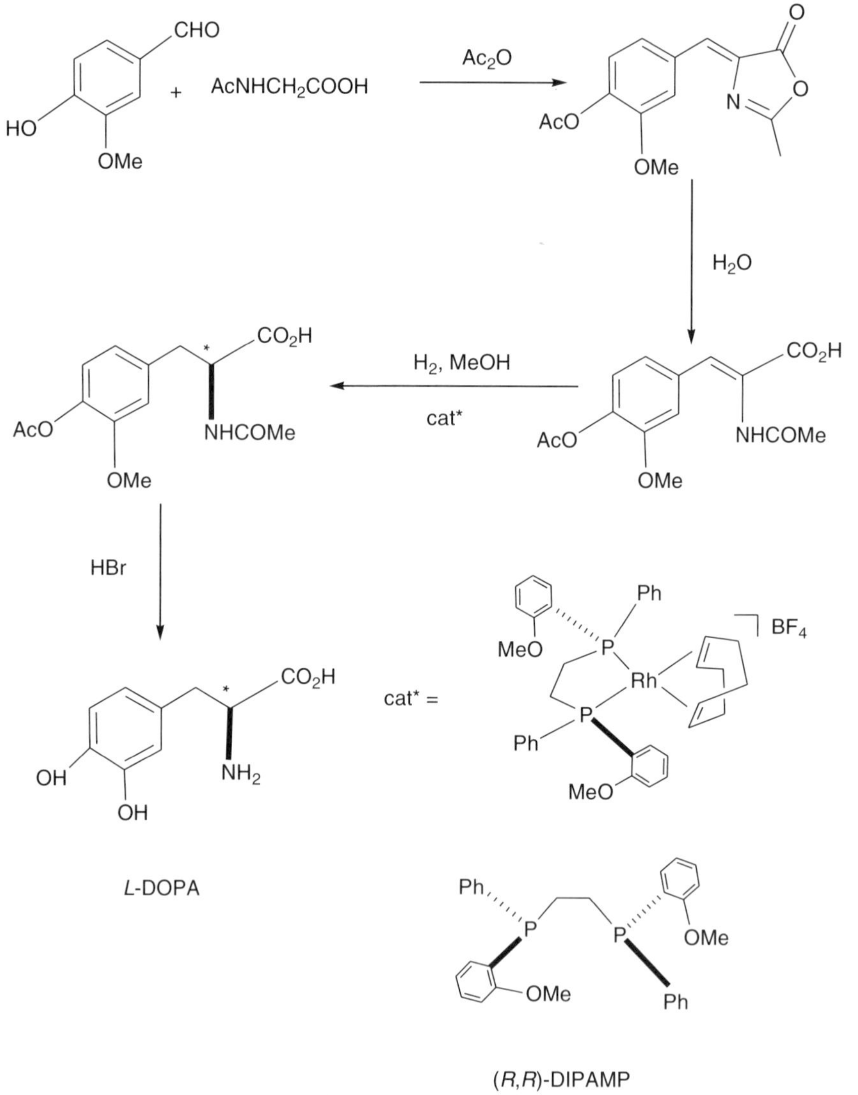

Figure 13   *Enantioselective synthesis of L-DOPA.*

H$_2$, and with a substrate:catalyst ratio > 10.000:1. Typically, an *ee* of 95% is achieved. Subsequent hydrolysis of the hydrogenation product gives *L*-DOPA.

These findings showed for the first time that synthetic catalysts are capable of rivalling natural enzymatic catalysts. The commercial synthesis of *L*-DOPA by this route promoted intensive worldwide research on asymmetric catalysis. The area has undergone spectacular development which continues to the present.

3.5.2 Mechanism of the Asymmetric Catalytic Hydrogenation

The mechanism of homogeneous asymmetric hydrogenation of prochiral olefins has been extensively studied since the 1970s and aspects are still debated. The first significant experimental results in the hydrogenation of amino acid precursors catalyzed by a Rh/chiral diphosphine system, obtained by Halpern and Brown, merit comment. Through careful spectroscopic, kinetic, and, when possible, crystallographic studies, the mechanism depicted in Figure 14 (applicable to the *L*-DOPA case) was established. The substrate methyl (Z)-α-acetamidocinnamate is used in the scheme, as a representative example. The *cis*-chelating diolefin pre-catalyst $[Rh(diolefin)(PP^*)]^+$ reacts with H_2 to form a solvate $[Rh(PP^*)S_2]^+$ complex which does not further react with molecular hydrogen. This species reacts with the substrate to give two diastereomeric enamide complexes in equilibrium. Diastereoisomerism stems from the coordination of the rhodium atom to the *Re* or to the *Si* face of the prochiral olefin. When DIPAMP was the chelating ligand the enamide complexes were in *ca.* 10:1 ratio, the minor isomer having the olefin *Re*-coordinated (right-hand side in Figure 14). The next step is the oxidative addition of dihydrogen, and is the rate-determining step of the cycle. An (undetected) dihydride intermediate is formed which transfers a hydrido ligand to the rhodium-bound face of the coordinated olefin to produce two diastereomeric rhodium alkyl hydride complexes. The subsequent reductive elimination of the product regenerates the catalytically active square-planar species, $[Rh(PP^*)S_2]^+$, which re-starts the cycle. As the alkene association precedes the addition of dihydrogen this mechanism is called the *alkene route*. If the addition of dihydrogen precedes the alkene association, the reaction is said to occur by the *hydride route*.

The intermediate derived from DIPAMP provided the most surprising result of this study; the rate of the reaction of hydrogen with the minor enamide isomer is so much faster than that with the major isomer that it dominates the kinetics of the overall process. Hence the configuration of the minor enamide isomer determines the chirality of the final product. Assuming intramolecular hydrogen transfer to the face of the coordinated olefin, the minor isomer gives rise to the *S* product. It has been experimentally shown that the *S* product predominates by about 95% *ee*. This means that the initially formed major diastereoisomer must gradually convert into the minor one through the equilibria shown in Figure 14.

3.6 Enantioselective Hydrogenation of *N*-Arylimines in the Synthesis of the Chiral Herbicide, (*S*)-Metolachlor

3.6.1 The Synthesis of Metolachlor

Metolachlor is one of the most active herbicides used on maize and other crops. Its chemical formula is depicted in Figure 15. It is a sterically hindered tertiary

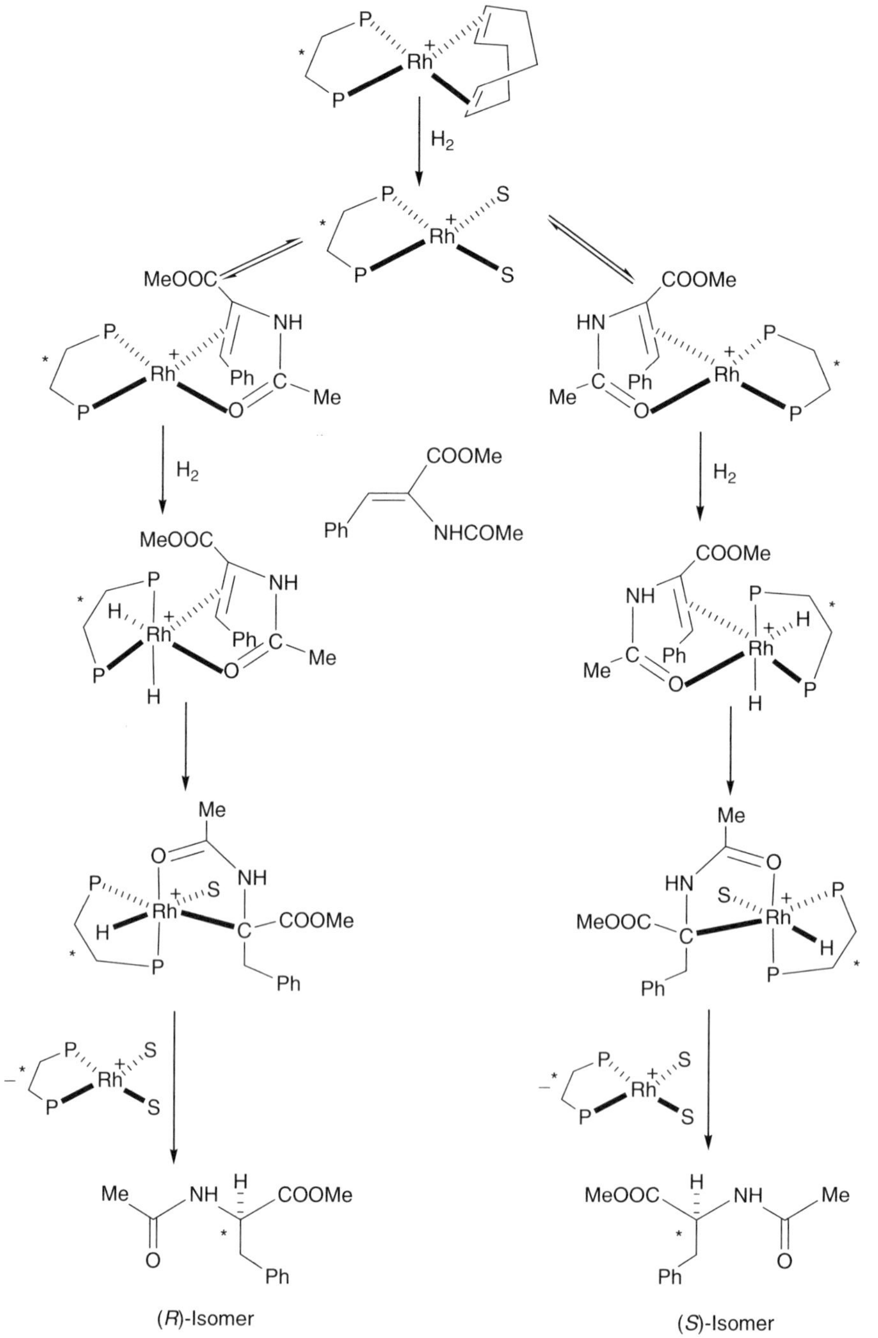

Figure 14 *Mechanism of the enantioselective hydrogenation of dehydroamino acids and their ester by cationic Rh/DIPAMP complexes.*

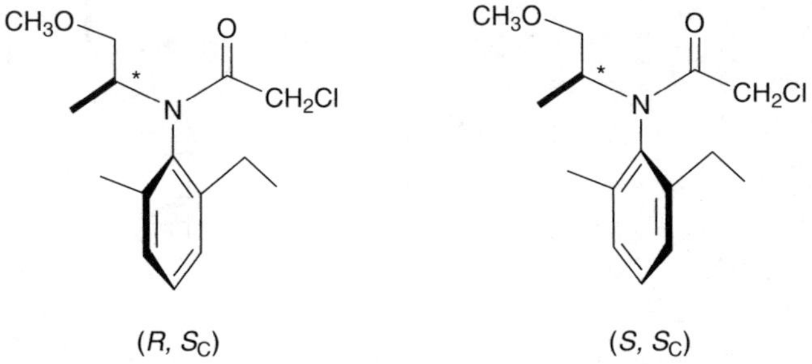

Figure 15 *The two active isomers of metolachlor.*

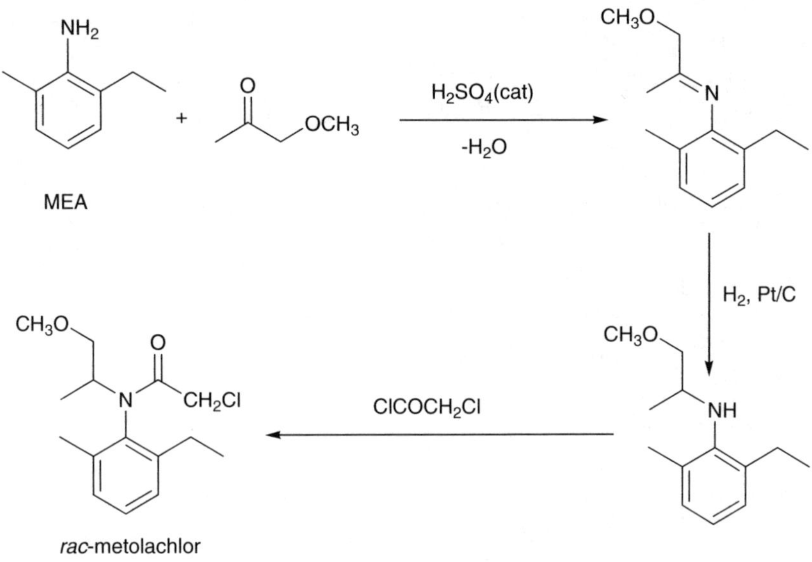

Figure 16 *Racemic synthesis of metolachlor.*

amine with two elements of chirality: a chiral axis (atropoisomerism) and a stereogenic carbon centre. In the 1970s it was brought to market as a mixture of all four possible stereoisomers; however, at the beginning of the 1980s it was realised that about 95% of its herbicide activity resides in the two S_C diastereomers.

The synthesis of racemic metolachlor was accomplished by condensation of 2-methyl-6-ethyl-aniline (MEA) with methoxyacetone, followed by Pt/C reduction and chloroacetylation according to Figure 16.

The chiral switch of the metolachlor was achieved in 1997. It was put on the market with a content of approximately 90% of the S_C active diastereomers. The key step of the large-scale enantioselective synthesis is the catalytic hydrogenation of the MEA imine shown in Figure 17. A mixture of [IrCl(1,5-cyclooctadiene)]$_2$, the chiral diphosphine (*R,S*)-xylyphos, iodide (as tetrabutylammonium or sodium salts) and acetic (30%) or sulfuric (at low

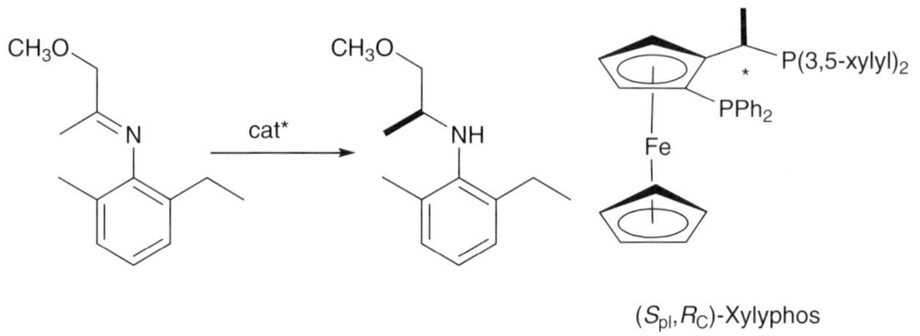

Figure 17 *Enantioselective imine hydrogenation.*

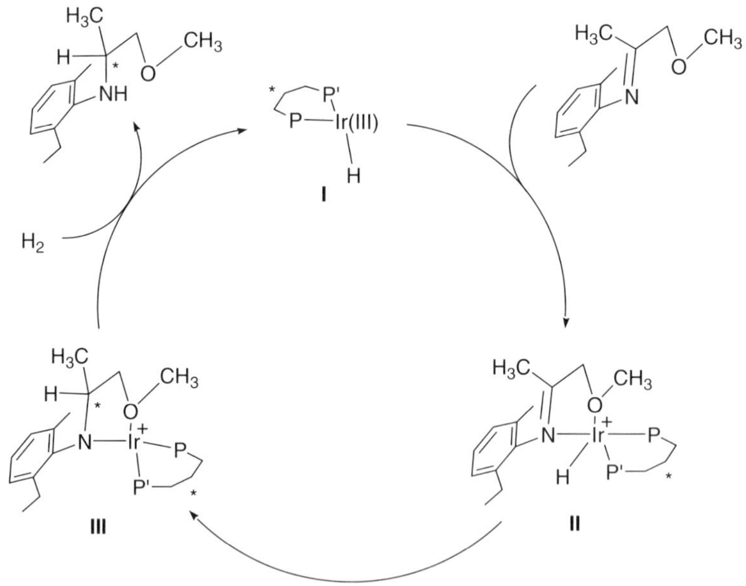

Figure 18 *Schematic catalytic cycle proposed for the enantioselective imine hydrogenation.*

concentration) acid is used as the chiral catalyst. At a hydrogen pressure of 80
bar and at a reaction temperature of 50°C, with a substrate/catalyst ratio of
about 10^6, an enantiomeric excess of 79% is achieved. This is one of the fastest
homogeneous systems known with an initial turnover frequency (TOF) > 1.8
$\times 10^6$ h^{-1}. With a capacity of more than 10 kt/a it is currently the largest-scale
enantioselective catalytic process.

3.6.2 Mechanistic Studies

The reaction mechanism of the homogeneous hydrogenation of imines has
scarcely been investigated. From the available data, the catalytic cycle depicted
in Figure 18, applied to the imine precursor of metolachlor, can be postulated.

An iridium(III) hydride, **I**, coordinates the imine in a bidentate fashion to give complex **II**. The coordinated C=N imine bond inserts into the Ir–H bond to give the iridium complex **III** which after hydrogenolysis, probably *via* heterolytic splitting of the H–H bond, eliminates the amine adduct and regenerates the Ir(III) hydride.

In contrast to the generally accepted mechanism for the homogeneous hydrogenation of C=C double bonds catalysed by rhodium or iridium complexes, which is assumed to occur through M(III)/M(I) species, the postulated cycle for the hydrogenation of imines involves only Ir(III) species. Many aspects remain unclear; for example the simple cycle neither explains the origin of the enantioselectivity nor the effect of acids and iodide used as promoters.

3.7 Isomerization Reactions: Diethylgeranylamine and Diethylnerylamine for the Production of (−)-Menthol

3.7.1 The Synthetic Route to Menthol

(−)-Menthol is widely used as an additive in many consumer products including tobacco flavours, chewing gum, toothpaste, and pharmaceuticals. Its chemical formula is depicted in Figure 19. As the molecule contains three stereogenic carbon centres, there are eight possible stereoisomers. Among them, the only one with the desired biological properties is the isomer with (1*R*,2*S*,5*R*)-configuration. Its world market was about 4.5 kt/a at the beginning of the 1990s and increased to *ca.* 12 kt/a by the end of the decade. At present about 70% of (−)-menthol is obtained from natural products isolated from *Mentha Arvensis*, which is cultivated mainly in India and China. The remaining 30% market is made synthetically.

Evolution of synthetic production of (−)-menthol parallels the development of methods for producing enantiopure compounds. Chiral pool and resolution methods were initially employed and, since 1984, Takasago International Co. has been producing (−)-menthol by the asymmetric synthesis shown in Figure 19. The reaction of myrcene or isoprene (obtained from natural or petroleum feedstocks, respectively) with diethylamine in the presence of lithium diethylamide, gives (*N,N*)-diethylgeranylamine ((*E*)-isomer) or (*N,N*)-diethylnerylamine ((*Z*)-isomer), respectively, in nearly quantitative yield. Both allylic amines are catalytically isomerized to the optically active (*R*)-citronellal enamine using a chiral catalyst. Hydrolysis of the amine produces the aldehyde citronellal with retention of the configuration. Cyclisation of (*R*)-citronellal, promoted by ZnBr$_2$, gives isopulegol diastereoselectively with the three correct stereogenic centres. Hydrogenation of the double bond completes the synthesis of (−)-menthol.

The key step of this process is the asymmetric isomerization of (*N,N*)-diethylgeranylamine or diethylnerylamine to give (*R*)-citronellal (*E*)-diethylenamine.

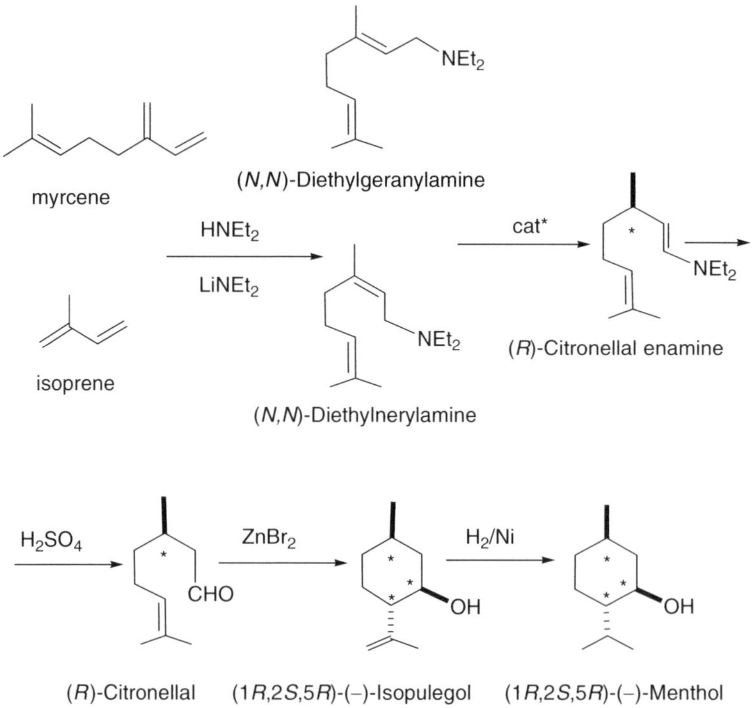

Figure 19 *Takasago process for the synthesis of (−)–menthol.*

The isomerization involves the intramolecular 1,3-hydrogen shift outlined in Figure 20. From deuterium labelling experiments it was established that this migration occurs in a suprafacial manner.

The first catalysts used were Co(I)-DIOP (DIOP, see Figure 11b) based systems. Up to 32% *ee* was achieved with 39% yield in the isomerization of (*N*,*N*)-diethylgeranylamine to (*R*)-citronellal. Subsequently, some rhodium (I)-DIOP or BINAP (BINAP = 2,2′-bis(diphenylphosphino)-1,1′-binaphthyl) systems proved to be very active. In particular, cationic rhodium(I)-BINAP complexes show very high selectivities and catalytic activities for this isomerization. BINAP is an atropoisomeric diphosphine (Figure 20) which was first synthesized by Noyori and Takaya and since then its metal complexes have been extensively used as catalysts in a variety of asymmetric syntheses.

One additional favourable feature of the use of rhodium-BINAP catalysts is that they are stereospecific: the (*E*)-enamine gives the (*R*)-amine and the (*Z*)-enamine gives the (*S*)-amine with catalysts containing (*S*)-BINAP. Obviously, the use of (*R*)-BINAP catalysts affords the opposite enantiomers of the amine (Figure 21). Thus to obtain a high enantiomeric purity, it is essential to start from isomerically pure amine. Almost perfect enantioselectivity (> 96% *ee*) and quantitative yields were obtained by all routes. Initially, [Rh(BINAP)(-COD)]$^+$ (COD = 1,5-cyclooctadiene) was used as catalyst precursor, and TON up to 8000 were achieved. Further improvements in TON were achieved with

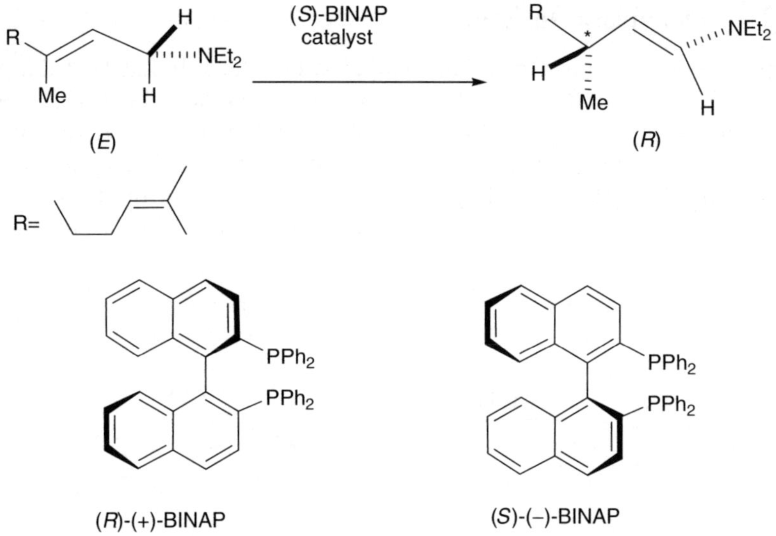

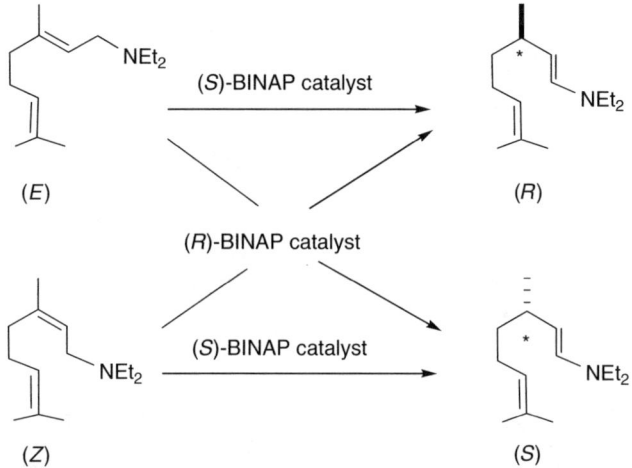

Figure 20 *Isomerization process.*

Figure 21 *Stereochemical correlation of the asymmetric isomerization.*

the rhodium-bis-BINAP complex $[Rh(BINAP)_2]^+$. It enabled the repeated use of the catalyst and TON as high as 400,000 can be achieved. The stereochemical correlation is maintained implying that an identical catalyst species determines the enantioselectivity. Thus, dissociation of one of the BINAP ligands of $[Rh(BINAP)_2]^+$ had to occur during the catalytic process. Industrially, the new catalyst $[Rh(p\text{-Tol-BINAP})_2]^+$, in which the phenyl groups of the BINAP are substituted by p-tolyl groups, is used due to its better solubility in organic solvents and to its better crystallization properties both in the resolution and in the preparation of the rhodium complexes.

3.7.2 Mechanistic Insights

The proposed mechanism, based on detailed studies [S.-I. Inoue, H. Takaya, K. Tani, S. Otsuka, T. Sato and R. Noyori, *J. Am. Chem. Soc.*, 1990, **112**, 4897] is schematically represented in Figure 22. Dissociation of one of the BINAP ligands gives the solvated species **I** which coordinates an allylamine molecule affording **II**. Complex **II** undergoes β-elimination to form the transient iminium-rhodium hydrido π complex **III**. This complex isomerizes to the aza-allyl

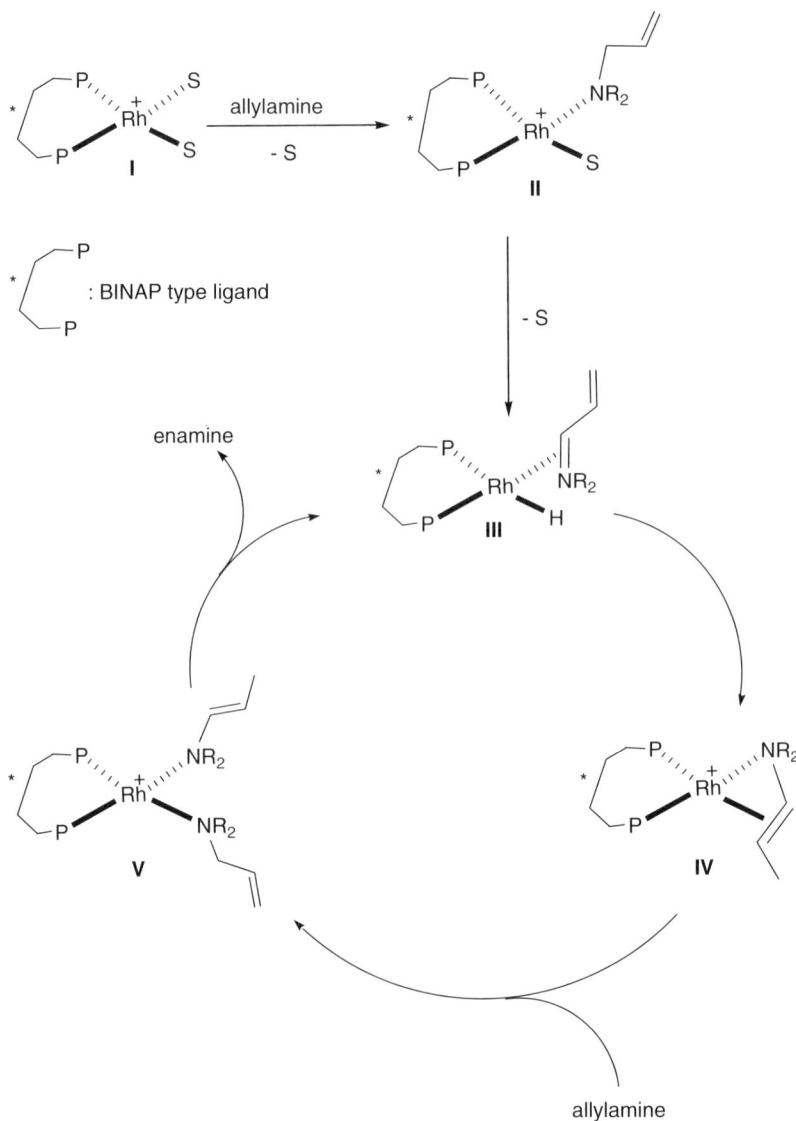

Figure 22   *Proposed mechanism for the enantioselective isomerization process.*

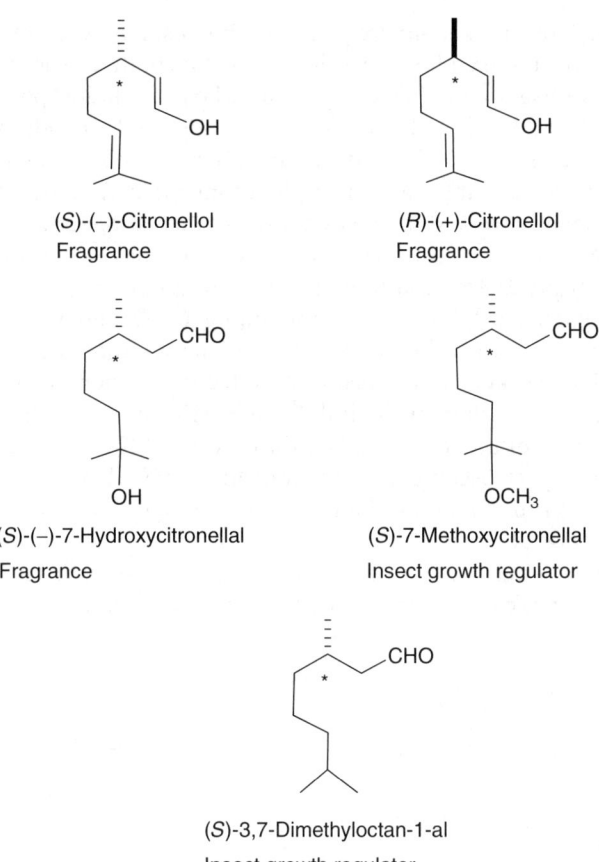

Figure 23 *Optically active terpenoids manufactured by Takasago.*

derivative **IV** which coordinates a new allylamine molecule to give the mixed allylamine-enamine complex **V**. Elimination of the enamine product from it is followed by the regeneration of the hydride **III** which restarts the catalytic cycle.

Based on the enantioselective isomerization of allylamines, Takasago has been manufacturing, besides (−)-menthol and the intermediates citronellal and isopulegol, a series of optically active terpenoids which are summarized in Figure 23.

3.8 Enantioselective Hydrogen Transfer

Hydrogen transfer from donor molecules in the presence of a metal complex catalyst is an alternative methodology to the usual hydrogenation procedures using H_2 as hydrogen source. Equation 9 illustrates a general equilibrium between the hydrogen donor (DH_2) and the unsaturated substrate which acts as hydrogen acceptor (A).

$$DH_2 + A \rightleftharpoons AH_2 + D \tag{9}$$

The catalysts for transfer hydrogenations are usually late transition metal complexes with tertiary phosphine ligands or bidentate nitrogen ligands, and the donors are usually organic compounds whose oxidation potential is sufficiently low to tolerate hydrogen transfer under mild conditions. Suitable donors are secondary alcohols such isopropanol. This alcohol is the most convenient since it is stable, non-toxic, environmentally friendly, easy to handle (bp 82°C), inexpensive and dissolves many organic compounds.

Recent developments in transfer hydrogenation have concentrated on the asymmetric transfer to unsaturated substrates, not only containing C=C bonds, but particularly substrates containing C=O and C=N bonds. The breakthrough in transfer hydrogenation came in the mid nineties when Noyori and coworkers developed very effective ruthenium catalysts with diphosphine and/or diamine chiral ligands (Figure 24). They established that the presence of an NH or NH_2 group is crucial for the catalytic activity. From kinetic, spectroscopic, and theoretical studies they proposed the new metal-ligand bifunctional mechanism for the hydrogen transfer reaction shown in Figure 24.

The most striking feature of this mechanism is that, in contrast to the currently accepted catalytic mechanisms, it does not require the formation of

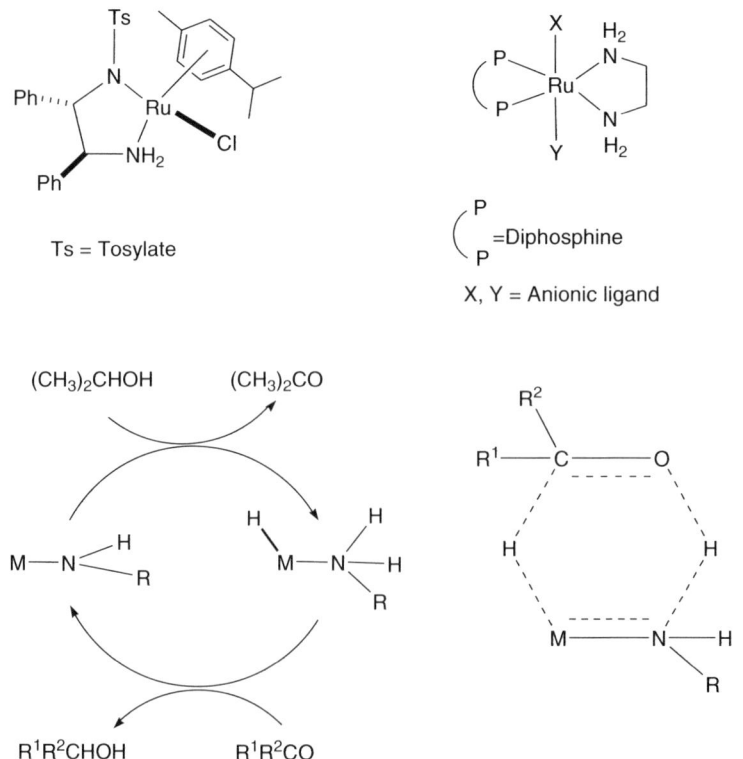

Figure 24   *Metal-ligand bifunctional mechanism.*

direct metal-substrate linkages. The key intermediate is the six-membered species depicted in Figure 24. The NH and MH functionalities interact with the oxygen and carbon atom of the ketone, respectively. Therefore the ketone reduction does not require any coordinative unsaturation at the metal. From this intermediate the protons are transferred to the ketone and the formed alcohol is subsequently eliminated. Theoretical and experimental studies on this mechanism have been reported [R. Noyori, M. Yamakawa, and S. Hashiguchi, *J. Org. Chem.*, 2001, 66, 7931].

3.9 Ethylbenzene Dehydrogenation to Styrene

3.9.1 The Styrene Market

Styrene is at the centre of an important industry, with a value of some 66 billion euros. The styrene production capacity is ca. 20 Mt/a worldwide. Most is obtained by ethylbenzene dehydrogenation and all the production is used for the synthesis of polymers (polystyrene, styrene-acrylonitrile, styrene-butadiene) used as plastics and rubbers in the manufacture of household products: packaging, tubes, tires, and endless other applications (see also Chapter 7).

3.9.2 Ethylbenzene Non-Oxidative Dehydrogenation

The ethylbenzene dehydrogenation (Figure 25) is a reversible endothermic reaction ($\Delta H_{(600°C)} = 124.9$ kJ/mol) that takes place in the gas phase. As one mole of ethylbenzene yields 2 moles of products (styrene and hydrogen), to drive the equilibrium forwards requires the use of as *low* a pressure as possible.

In view of these constraints, the reaction is carried out at 600–650°C in a current of superheated steam. Steam is used both to provide the high temperature and to dilute ethylbenzene to a low partial pressure. The use of a catalyst

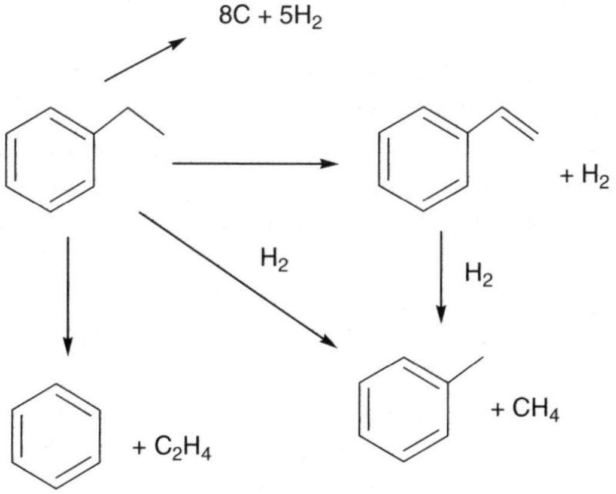

Figure 25 *Non-oxidative ethylbenzene dehydrogenation and side reactions.*

is essential to reach high conversion and under such conditions the catalyst must be heterogeneous, working in a fixed-bed reactor.

Other thermal reactions compete with dehydrogenation, such as the total dehydrogenation to carbon or some cracking processes. The importance of these reactions increases with temperature and time on stream and hence the equilibrium position (representing about 80% conversion), is seldom reached and most plants work at conversion levels in the range of 50–70%.

Although all the side reactions affect the selectivity, from the point of view of the catalyst, the production of carbon (coke) is the biggest problem as it remains *in situ* and acts as a poison. The use of superheated steam plays a vital role here, as it helps to clean the catalyst by gasification of carbon at the working temperature (Equation 10).

$$C + 2H_2O \rightarrow CO_2 + 2H_2 \qquad (10)$$

3.9.3 Catalysts for Ethylbenzene Dehydrogenation: Mechanism and Deactivation

The "classical" catalyst for this reaction is iron as the oxide, with the addition of an alkali metal as promoter, generally potassium (as KOH or K_2CO_3). This type of catalyst was originally patented in 1947 by Standard Oil [K. K. Kearby, US Patent 2,426,829 (1947)]. In order to obtain higher activity, this catalyst is improved by addition of other promoters, Cr_2O_3 in the first generation catalysts. The market was dominated for a long time by the Shell 105 catalyst, composed of 83.4% Fe_2O_3, 13.3% K_2CO_3, and 2.4% Cr_2O_3.

In the second generation catalysts, deactivation problems have led to chromium being substituted by other metals that play various roles. The newer patents report rather complicated catalyst compositions, e.g. in 1989 United Catalysts described an optimum catalyst with composition, 78.6% Fe_2O_3, 9.5% K_2O, 5.0% CeO_2, 2.5% MoO_3, 2.2% CaO, 2.2% MgO and 1000-5000 ppm Cr. To prepare a catalyst of this type, a slurry of the components is extruded as pellets which are then dried, and subsequently calcined at high temperature, usually 800–1000°C.

Studies point to Fe^{3+}-containing species (e.g., Fe_2O_3, $K_2Fe_{22}O_{34}$) as the active sites for dehydrogenation; here ethylbenzene is bound by the π-aromatic system, and adjacent sites are then responsible for the abstraction of 2H leading to styrene and the creation of two Fe^{2+} sites (Figure 26). Although K does not play any role in this process, it is important in extending the catalyst lifetime. As replacement of the catalyst, typically every 1-2 years, is a costly operation given the large capacities of styrene production units (>400 kt/a), many studies on the deactivation of dehydrogenation catalysts have been carried out both in industry and academia.

Four deactivation processes can be considered for dehydrogenation catalysts. First is coke deposition, which can be minimized by gasification with steam at high temperature. In this regard potassium is very important as K_2CO_3 is believed to be the gasification catalyst.

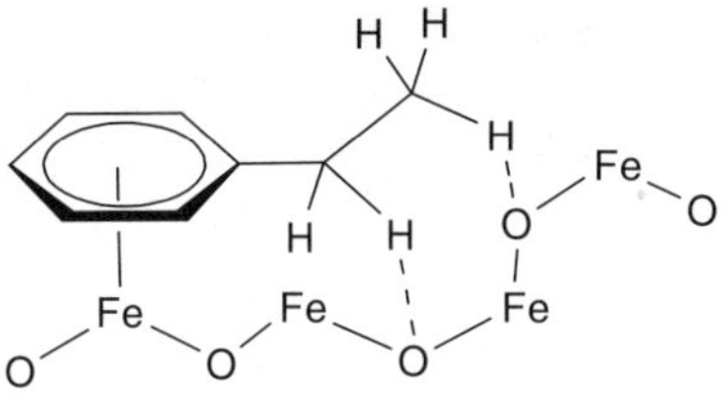

Figure 26 *Proposal illustrating the π-complexed ethylbenzene and the mode by which the dehydrogenation involving the migration of 2 H occurs.*

A second deactivation mechanism involves modification of the oxidation state of the catalyst. The initial Fe_2O_3 phase (hematite) is easily reduced to the less active (or even inactive) Fe_3O_4 (magnetite). Transformations between the ternary phases, $K_2Fe_{22}O_{34}$ and $KFeO_2$, are envisaged as responsible for the regeneration of the active surface Fe^{3+} species.

The third deactivation process involves loss or migration of the promoters, due to the slight mobility of KOH under the reaction conditions. The migration of potassium to the core of the particles produces a potassium-free surface, which easily deactivates.

Finally, a deactivation process shared by all the heterogeneous catalysts is the physical degradation or pulverization of the particles by attrition. Hence the important role played by additives such as MgO, or CaO which can act as structure stabilizers.

3.9.4   Alternatives for Non-Oxidative Dehydrogenation

The main drawbacks of the non-oxidative dehydrogenation reaction can be summarized as, the thermodynamic limitation, the low conversion rate, the need for recovery of unreacted ethylbenzene, the high energy consumption, and deactivation of the catalyst. Thus in recent years several alternatives to overcome those problems have been investigated.

a) Coupling with oxidation of H_2.

The strategy of not releasing H_2 but of adding an oxidizer to remove it, allows the equilibrium to be shifted by continuous elimination of one of the products. In addition the oxidation produces water and part of the heat necessary for the endothermic dehydrogenation. A set of two different catalysts or a bifunctional catalyst is required for this approach.

b) Oxidative dehydrogenation.

This is a direct dehydrogenation with oxygen (Figure 27), with the advantage of being an exothermic process. Several companies (BASF, Eni, FINA) are currently developing multi-component oxide catalysts for this reaction, based on V-Mg-O systems with addition of other metals such as Bi, Na, Ca, Fe, Si or K. The main disadvantage of this strategy is a danger of total oxidation of the aromatic molecule, although the selectivity can be improved by passing the two reactants sequentially over the catalyst.

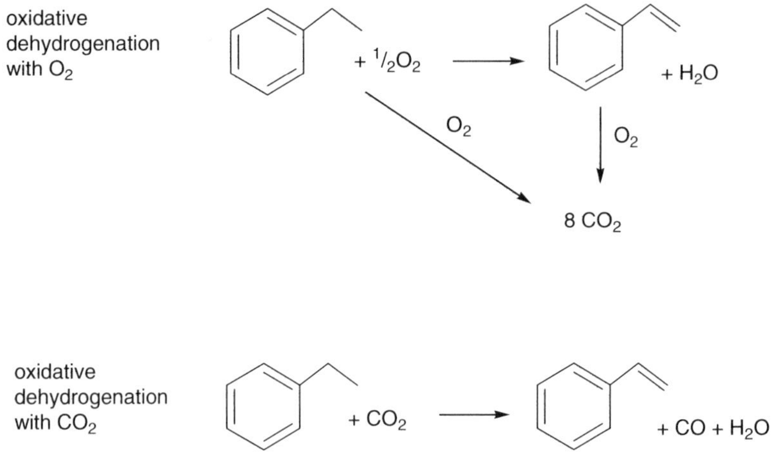

Figure 27 *Oxidative dehydrogenation paths that can be employed to facilitate ethylbenzene to styrene conversion.*

c) Membrane reactors.

Permselective membranes can be used to remove the hydrogen produced in the reaction, thus shifting the equilibrium to higher conversion. However, cost and engineering issues at present preclude the industrial application of this approach.

d) Dehydrogenation with CO_2.

In this case the oxidant for hydrogen is CO_2, leading to CO and water (Figure 27). This strategy shows the same advantages as the oxidative dehydrogenation.

Discussion Points

Discussion Point DP1: While there are many examples of industrial processes that depend on heterogeneously catalyzed hydrogenation reactions, there are very few (apart from the enantioselective processes) which are homogeneously catalyzed. Suggest some reasons for this.

Discussion Point DP2: Heterogeneous catalysis is mainly used in industry for bulk chemicals (e.g. petrochemicals) whereas homogeneous catalysis is used in fine chemicals manufacture as the selectivity is higher and the complexes are tuneable. Would it be possible to combine the advantages of both types of catalysis? How could we obtain a catalyst with high selectivity (including enantioselectivity) but easily recoverable and reusable? Discuss the possible disadvantages of that type of catalyst and the industrial viability.

Discussion Point DP3: Suggest new ligands for enantioselective hydrogenation based on what we understand about presently known systems.

Discussion Point DP4: Some of the most successful processes involve H-transfer (for example from a secondary alcohol to an unsaturated substrate) rather than hydrogenation with H_2 gas. Why may this be so?

References

General: R. H. Crabtree, *The Organometallic Chemistry of the Transition Metals*, 4th Edition, Wiley, New York, 2005.

P. A. Chaloner, M. A. Esteruelas, F. Joó and L. A. Oro, *Homogeneous Hydrogenation*, Kluwer Academic, Dordrecht, 1994.

R. L. Augustine, *Heterogeneous Catalysis for the Synthetic Chemist*, Marcel Dekker Inc, New York, 1996, Chapter 15.

Oil refining, petrochemistry and hydrotreatment: I. Chorkendorff and J. W. Niemantsverdriet, *Concepts of Modern Catalysis and Kinetics*, Wiley-VCH, Weinheim, 2003, pp. 349–369; Models for hydrodesulfurization mechanisms; R. J. Angelici, *Acc. Chem. Res.*, 1988, **21**, 387: Hydrogenation of natural oils: J. W. Veldsink, M. J. Bouma, N.-H. Schöön and A. A. C. M. Beenackers, *Catal. Rev.–Sci. Eng.*, 1997, **39**, 253.

Kinetic models of vegetable oil hydrogenation: R. J. Grau, A. E. Cassano and M. A. Baltanás, *Catal. Rev.–Sci. Eng.*, 1988, 30, 1.

Hydrogenation of nitriles: C. De Bellefon and P. Fouilloux, *Catal. Rev.–Sci. Eng.*, 1994, 36, 459.

Asymmetric catalytic hydrogenation: *L*-DOPA, W. S. Knowles, *Angew. Chem. Int. Ed.*, 2002, 41, 1998 (Nobel Lecture). Mechanism: J. Halpern, *Science*, 1982, **217**, 401; J. M. Brown, *Comprehensive Asymmetric Catalysis*, E. N. Jacobsen, A. Pfaltz and H. Yamamoto, Eds.; Springer: Berlin, 1999; Vol. I, Chapter 5.1, pp. 121–182; I. D. Gridnev and T. Imamoto, *Acc. Chem. Res.*, 2004, **37**, 633. Metolachlor: H.-U. Blaser and F. Spindler in *Comprehensive Asymmetric Catalysis*, E. N. Jacobsen, A. Pfaltz and H. Yamamoto, Eds.; Springer, Berlin, 1999; Vol. III, Chapter 41.1, pp. 1427–1437. Industrial enantioselective catalytic synthesis: H.-Blaser, R. Hanreich, H.-D. Schneider, F. Spindler and B. Steinacher in *Asymmetric Catalysis on Industrial Scale*, H.-U. Blaser and E. Schmidt, Eds.; Wiley-VCH: Weinheim, 2004; Chapter 3, pp. 55–70.

Hydrogenation of imino groups: H.-U. Blaser and F. Spindler in *Comprehensive Asymmetric Catalysis*, E. N. Jacobsen, A. Pfaltz and H. Yamamoto, Eds.; Springer, Berlin: 1999; Vol. I, Chapter 6.2, pp. 247–265.

Asymmetric isomerization of allylamines: S. Akutagawa and K. Tani in *Catalytic Asymmetric Synthesis*, 2nd ed., I Ojima, Ed., Wiley-VCH: New York, 2000, Chapter 3, pp. 145–163.

Hydrogen transfer reactions: S. E. Clapham, A. Hadzovic and R. H. Morris, *Coord. Chem. Rev.* 2004, **248**, 2201. Asymmetric transfer hydrogenations, G. Zassinovich, G. Mestroni and S. Gladiali, *Chem. Rev.* 1992, **92**, 1051.

Dehydrogenation: E. H. Lee, *Catal. Rev.*, 1973, **8**, 285.

Deactivation of dehydrogenation catalysts, G. R. Meima and P. G. Menon, *Appl. Catal. A*, 2001, **212**, 235.

CHAPTER 4

Syntheses Based on Carbon Monoxide

PETER MAITLIS AND ANTHONY HAYNES

Department of Chemistry, University of Sheffield, Sheffield S3 7HF, UK

4.1 Introduction

In this chapter we discuss *carbonylation* reactions in which carbon monoxide (CO, Box 1) is incorporated into organic molecules. CO is one of the major building blocks of the organic commodity chemicals industry today, and is of increasing importance in fine chemicals too (Chapter 1). It can readily be obtained from a variety of sources, it is relatively cheap, and there are many efficient catalyzed carbonylation processes that use it to build a range of important chemicals. For example, the reaction of carbon monoxide and hydrogen over heterogeneous metal oxide catalysts gives methanol (and is the basis of a major industry producing ca. 30 Mt/a; Section 4.7.1). If the CO hydrogenation is carried out over a supported Group 8 or 9 metal (*e.g.* Fe or Co) the products are linear hydrocarbons (1-alkenes, and alkanes; Section 4.7.2) and form the basis of an industry producing ca. 8 Mt/a as automobile fuel (especially diesel) or as a feedstock for higher added value chemicals. Other reactions involve the carbonylation of simple organic molecules (alcohols, olefins, *etc.*) making, for example, acetic acid (9 Mt/a), propionic acid, acetic anhydride; *n*-butyraldehyde and *n*-butanol (2 Mt/a). Production figures must be regarded as approximate, especially as old plants are phased out, and new ones come on stream over the next few years in countries not traditionally associated with the chemicals industry, such as China, India, Taiwan, Malaysia, Saudi Arabia and the Gulf States.

Methanol is currently the largest volume carbonylation product and is made by passing syngas (CO + H$_2$; Section 4.1.2) over a solid Cu–Zn oxide catalyst. Most of the other carbonylation reactions are catalyzed by the later *d*-block transition metals, often under homogeneous conditions in solution. This is despite a public perception that the use of heavy metals (such as the complexes of the *4d* and *5d* transition metals) is generally undesirable. However their extremely effective catalytic properties now make their use mandatory in many organic syntheses; see for example Box 6, which demonstrates an example of how the commercial syntheses of some speciality chemicals for the pharmaceutical, agrochemical, and food industries depend on such processes, such as the manufacture of ibuprofen and related anti-inflammatories. The cobalt catalyzed hydroformylation of acrylonitrile was even at one time used in Japan as a route to N-acetylglutamic acid, an intermediate to MSG, monosodium glutamate.

Box 1 Properties of CO

Carbon monoxide is a colourless, odourless, flammable and highly toxic gas (b.p. $-192°C$). It is formally a compound of divalent carbon and hence may be expected to show very high reactivity. However, although CO is thermodynamically unstable (e.g. $\Delta H = -284$ kJ/mol for the oxidation: $CO + \frac{1}{2}O_2 \rightarrow CO_2$), it is kinetically inert as the activation barrier for reaction is large. Thus CO can conveniently be stored in cylinders under pressure for later use. CO is usually made, together with hydrogen, as syngas (Section 4.1.2). CO acts as both a σ-donor (via the lone pair of electrons on carbon) and a π-acceptor ligand in transition metal complexes, the *metal carbonyls*. The reactivity of CO in a metal carbonyl is considerably modified: it is for example attacked by nucleophiles (at the carbon), and undergoes migration reactions in which metal bonded alkyls transfer to the CO, giving metal-acyls M-CO-R, and similar species. Such reactions lie at the heart of the utility of CO in chemical transformations. The bonding in CO and in metal carbonyls is discussed in Appendix 1, page 257.

We consider the industrially important reactions of carbon monoxide under the following headings: (Section 4.2) carbonylation of alcohols and esters; (Sections 4.3–4.5) hydroxycarbonylation of alkenes and related processes; (Section 4.6) hydroformylation of alkenes; and (Section 4.7) hydrogenation of CO to *a*) methanol and *b*) α-olefins. The CO hydrogenations in Section 4.7 are dominated by heterogeneously catalyzed processes whereas the other Sections generally involve homogeneous catalysis by transition metal complexes in solution. The order has been chosen to develop a rational understanding on the basis of our knowledge of the details of the various reactions involved: thus the homogeneous methanol carbonylation involves only CO and the alcohol, in addition to the catalyst, and is the simplest process to understand. Alkene hydroformylation, and the other reactions discussed in Sections 4.2–4.6 involve an organic substrate, CO and hydrogen and are mostly straightforward solution processes that we can understand on the basis of organometallic and organic chemistry. Indeed the successful investigation of these reactions has contributed greatly to our understanding of catalysis. This knowledge can then allow the extrapolation to some of the other processes where less "hard" information is available, in order to enable the student and the teacher to develop insights to plan new processes.

4.1.1   Carbonylation Reactions: Historical and General Perspectives

The first metal catalyzed hydrocarbonylation was discovered by Sabatier and Senderens in 1902 who found that nickel catalyzed the reaction of CO and hydrogen (syngas) to give methane (Equation 1):

$$CO + 3H_2 \rightarrow CH_4 + H_2O \tag{1}$$

This is essentially the inverse of *methane reforming* used to make syngas from natural gas (Section 4.1.2). In 1923, Fischer and Tropsch found that when syngas was passed over other metals, especially iron or cobalt, substantial amounts of long chain hydrocarbons were formed, in addition to methane. The *Fischer-Tropsch* (F-T) reaction (Section 4.7.2) is again attracting widespread attention for its huge commercial potential, especially as the oil price rises, as well as for the engineering and the chemistry involved. F-T is a major part of the processes known collectively as *gas-to-liquids* (GTL), *methane-to-liquids* (MTL) or *coal-to-liquids* (CTL) technology.

A number of other metal catalyzed carbonylation reactions for organic syntheses were discovered in Germany in the 1920s and 1930s. The first industrial methanol synthesis, known as the *high pressure* process, was developed by BASF in the 1920s (Section 4.7.1). Since the synthesis gas employed at that time was based on low-grade coal (lignite) a relatively poison resistant catalyst (ZnO/Cr_2O_3) was developed. The availability of cleaner syngas later allowed ICI to develop the *low pressure* process using a copper-zinc oxide-alumina catalyst.

The cobalt-catalyzed reaction of olefins with CO and hydrogen to give aldehydes (Equation 2) was discovered by Roelen at Ruhrchemie in 1938, and is now known as *hydroformylation* (or, sometimes, as the "oxo" synthesis, Section 4.6) and is the basis of another major industrial process.

$$R\diagdown\diagdown + CO + H_2 \longrightarrow R\diagup\diagdown\diagup CHO + \underset{R}{\diagup}\overset{CHO}{\diagdown} \tag{2}$$

Since the aldehydes are easily hydrogenated to alcohols, which are the most frequently required end-products, the reactions are often run to make the alcohol. Much later it was discovered that a rhodium triphenylphosphine complex, developed by Wilkinson, was a much better catalyst for many hydroformylation reactions as it required milder conditions (lower temperature and pressure) and gave a higher selectivity.

A cobalt/iodide catalyzed process to make acetic acid from methanol, introduced by BASF around 1960, grew out of the carbonylation studies by

Reppe. This was quickly superseded by the rhodium/iodide catalyzed process, introduced by Monsanto a little later. Although the cobalt catalyst was a lot cheaper than rhodium, that process suffered from the disadvantages that it required more drastic operating conditions and was less selective. More recently BP Chemicals introduced the Cativa^{TM} process, based on an iridium/iodide catalyst, promoted by a ruthenium complex. This used similar conditions, but since the Cativa process operated at low water levels considerable energy savings were possible as less water had to be removed to make the anhydrous acid. This is a remarkable story in that all three Group 9 metals are active and there has been a progressive improvement in the iodide promoted process from Co *via* Rh to Ir (+ Ru) (Section 4.2.1). Many other *d*-block metals are good at promoting carbonylation reactions and Pd especially is widely used in making fine chemicals (see for example Section 4.2.10).

4.1.2   Syngas as a Feedstock

The chemicals industry was originally based on coal, in particular to make chemicals such as aromatics that were accessible by destructive distillation. This was the pattern until well into the 20th century when the feedstocks moved to oil-based materials. Chief amongst these was ethylene which was obtained by cracking higher hydrocarbons; alkanes such as butane or naphtha (also derived from petroleum) were also used as feedstocks. Today one of the chief building blocks for the petrochemical industry is CO. This has the advantage that it can be made from a variety of carbon sources: coal, methane, naphtha (C_5–C_9 hydrocarbons), and even biomass. Methane (from natural gas, often associated with crude oil) is a favoured feedstock; the only impurities are some higher alkanes, hydrogen sulfide, nitrogen and other inert gases. These are relatively easy to remove: thus methane is relatively clean and unwanted reaction by-products can be minimized. However, since it is a very low boiling gas, transportation on a large scale is a problem and its extensive use by the chemicals industry requires it to be available directly to the plant via a pipeline (see Section 1.2.3).

A convenient source of CO is in admixture with hydrogen; when necessary both can then be used in the reaction. Mixtures of CO and H_2 are referred to as *syngas* (synthesis gas) or *water-gas* from the involvement of water in its manufacture.

One route to syngas is by steam-reforming methane, over a nickel catalyst at high temperature (Equation 3):

$$CH_4 + H_2O \rightarrow CO + 3H_2$$
$$(\Delta H_{298K}\,206\,kJ/mol;\ \Delta G_{1073K} - 24\,kJ/mol)$$

(3)

another is by passing steam over coal (Equation 4):

$$C + H_2O \rightarrow CO + H_2$$
$$(\Delta H_{298K}\,131\,kJ/mol;\ \Delta G_{1073K} - 12\,kJ/mol)$$

(4)

As both reactions are quite strongly endothermic, they may be coupled to an exothermic partial oxidation (Equations 5 and 6):

$$CH_4 + {}^1/_2O_2 \rightarrow CO + 2H_2 \quad (\Delta H_{298K} \; -36 \, kJ/mol) \tag{5}$$

$$2C + H_2O + O_2 \rightarrow CO + CO_2 + H_2 \quad (\Delta H_{298K} \; -285 \, kJ/mol) \tag{6}$$

Since world coal reserves are so large (for example in China), reforming coal will be a major source of syngas, as it was previously for Fischer-Tropsch reactions in Germany. Much syngas continues to be made this way in South Africa. Interestingly, therefore, the wheel is turning full circle and the chemical industry may eventually come back to coal as the ultimate feedstock. Other routes to syngas include the partial oxidation of methane or naphtha. The components of syngas can be separated and if only the CO is needed for a particular process then the hydrogen can be applied elsewhere, for example to make ammonia.

4.1.3   The Water-Gas Shift Reaction

The ratio of CO to H_2 in syngas can be controlled by the water-gas shift reaction (WGSR, Equation 7) and it is possible to make either hydrogen or carbon monoxide, or any ratio of the two, by suitable adjustment of conditions.

$$CO + H_2O \rightleftharpoons CO_2 + H_2$$
$$(\Delta H_{298K} \; -41 \, kJ/mol; \; \Delta G_{298K} \; -28 \, kJ/mol) \tag{7}$$

The WGSR is normally practised as a heterogeneously metal-catalyzed reaction; Fe is the most commonly used catalyst. However other metals are also active, for example the homogeneous Rh/I^- catalyst in the Monsanto acetic acid process (Section 4.2.4) concurrently catalyzes the WGSR via a Rh(I)/Rh(III) cycle (Equations 8 and 9),

$$[Rh(CO)_2I_2]^- + 2HI \rightarrow [Rh(CO)_2I_4]^- + H_2 \tag{8}$$

$$[Rh(CO)_2I_4]^- + H_2O + CO \rightarrow [Rh(CO)_2I_2]^- + CO_2 + 2HI \tag{9}$$

In these reactions, CO coordinated to a transition metal (particularly in a high oxidation state) becomes susceptible to nucleophilic attack by water, leading to release of CO_2 [mechanism: E. C. Baker, D. E. Hendricksen, R. Eisenberg, *J. Am. Chem. Soc., 1980, **102**, 1020*].

4.2   Carbonylation Reactions of Alcohols and Esters

The main large-scale metal catalyzed reactions involving addition of CO (rather than $CO + H_2$) to an organic substrate are the manufacture of acetic acid from methanol and the related production of acetic anhydride from methyl acetate. The syntheses of some other carboxylic acids and the conversion of a substituted benzyl alcohol to ibuprofen involve similar reactions.

4.2.1 Manufacture of Acetic Acid from Methanol

4.2.2 Acetic Acid Historical and Background

Acetic acid is one of the major commodity chemicals, with a current world production capacity of ca. 9 Mt/a. It has long been a mainstay of the organic chemicals industry, its manufacture increasing in sophistication and selectivity over the years. From being a simple agrochemical made in small quantitites as a foodstuff additive it has become a major bulk chemical with many important downstream applications (Box 2).

Box 2 Uses of Acetic Acid

Acetic acid is used in the manufacture of a wide variety of products including adhesives, polyester fibres, plastics, paints, resins and solvents. About 40% of the acetic acid made industrially is used in the manufacture of vinyl acetate monomer for the plastics industry; other large uses are to make cellulose acetate, a variety of acetate esters that are used as solvents, as well as monochloracetic acid, a pesticide. Acetic acid is also used as a solvent for the oxidation of *p*-xylene to terephthalic acid, a precursor to the important polyester, polyethylene terephthalate (PET). A minor, but important use is as *non-brewed condiment*, a vinegar substitute widely used in British fish and chip shops; this is made using food-grade industrial acetic acid and is less expensive than fermentation vinegar.

Vinegar, a dilute solution of acetic acid in water, has been a vital ingredient of foodstuffs since time immemorial. Fermentation of sugars, first to dilute ethanol and then to vinegar, was one of the first routes to acetic acid and still remains the usual method for making acetic acid for the food industries. The sour taste of wine left open to the air is caused by the unintentional synthesis of vinegar! However as vinegar typically contains only 5–10% acetic acid in water, the water must be removed to make the anhydrous "glacial" acetic acid that is needed for industrial use. This is an extremely energy demanding and expensive operation and is one reason why this route is not used, despite the attraction of it being based on easily renewable natural products from plants such as sugar cane. The destructive distillation of wood which gives methanol and acetone in addition to acetic acid was another process that was run for a while but is no longer viable. A number of synthetic routes to acetic acid based on oxidation of acetaldehyde were later developed: these included making acetaldehyde by ethanol oxidation, but were later superseded by the palladium(II) catalyzed oxidation of ethylene in water (Wacker reaction Section 2.17.1). The largest production of acetic acid came from the direct liquid phase air-oxidation of naphtha (from petroleum refining, and used by BP Chemicals) or butane (from natural gas, used by Celanese and Hüls). These routes had the advantage that they did not need any catalyst or solvent and thus allowed relatively easy access to the anhydrous acid. These direct alkane oxidation reactions also gave (smaller) amounts of formic acid as well as some propionic acid, and other products of partial oxidation. Thus the acetic acid still needed to be purified by demanding fractional distillation procedures.

Today most of the world's production of acetic acid uses the catalytic carbonylation of methanol (Equation 10),

$$CH_3OH + CO \rightarrow CH_3CO_2H \qquad (10)$$

By comparison to the old oxidation processes, the use of carbon monoxide as a feedstock gives significant cost advantages arising from a cheaper raw material and from the outstanding selectivity of the methanol carbonylation. Since methanol itself is made from syngas, both carbons in acetic acid arise from CO. If natural gas is used as the feedstock for syngas production, the conversion to acetic acid can be represented in three steps based only upon methane and water as reagents (Equations 3, 11, and 10):

$$CH_4 + H_2O \rightarrow CO + 3H_2 \qquad (3)$$

$$CO + 2H_2 \rightarrow CH_3OH \qquad (11)$$

$$CH_3OH + CO \rightarrow CH_3CO_2H \qquad (10)$$

with the overall stoichiometry (Equation 12):

$$2CH_4 + 2H_2O \rightarrow CH_3CO_2H + 4H_2 \qquad (12)$$

The excess H_2 produced can be utilized in other processes; thus an integrated chemical plant might produce acetic acid and use the hydrogen to make ammonia.

Commercial methanol carbonylation processes have employed each of the group 9 metals, cobalt, rhodium and iridium as catalysts. In each case acid and an iodide co-catalyst are required to activate the methanol by converting it into iodomethane ($CH_3OH + HI \rightarrow CH_3I + H_2O$); catalytic carbonylation of iodomethane into acetyl iodide is followed by hydrolysis to acetic acid. A problem common to all these processes arises because the mixture of HI and acetic acid is highly corrosive; this necessitates special techniques for plant construction involving the use of expensive steels. We discuss each catalyst system in turn below.

4.2.3 Cobalt-Catalyzed Carbonylation of Methanol

The first methanol carbonylation process, commercialized in the 1960's by BASF, used an iodide promoted cobalt catalyst but required very high pressures (~ 700 atm) as well as high temperatures ($\sim 250°C$), and gave only ca. 90% selectivity. Few mechanistic studies have been published and little is known for certain about the mechanism.

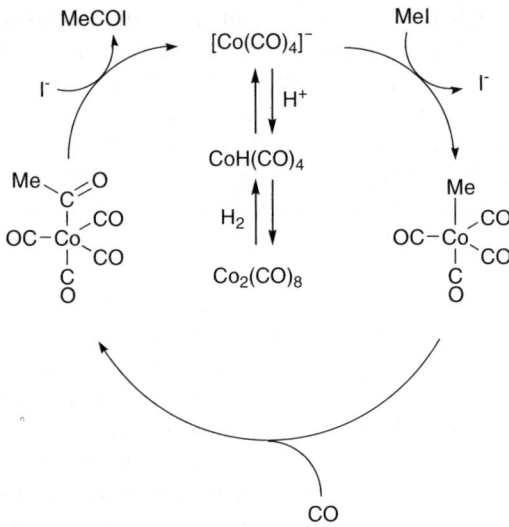

Figure 1 *Schematic representation of a possible mechanism for the Co/I⁻ catalyzed carbonylation of methanol. Organic reactions involving MeI and MeCOI are omitted (see Fig. 3).*

At elevated temperature and under a pressure of CO most cobalt(II) salts (such as the acetates) will give $Co_2(CO)_8$ stoichiometrically. With hydrogen this gives the hydride $CoH(CO)_4$, which is a strong protic acid that readily yields the anion, $[Co(CO)_4]^-$ in ionizing solvents. The Co(-I) anion is a powerful nucleophile and reacts readily with MeI to give $Co(Me)(CO)_4$. This is in contrast to the cycles for the Rh/I⁻ and Ir/I⁻ processes where M(I) and M(III) carbonyl iodide complexes are involved (Sections 4.2.5 and 4.2.7); as Co is too strongly oxidizing to form Co(III) iodides, reaction occurs via Co(-I) and Co(I) intermediates. The methylcobalt(I) complex, $Co(Me)(CO)_4$, is then believed to undergo methyl migration, carbonylation, and reductive elimination (as illustrated in Figure 1) rather analogously to the similar steps identified from the Rh and Ir catalysts. Although the reaction of $[Co(CO)_4]^-$ with MeI is known to be rapid, it is likely that only a small proportion of the cobalt catalyst exists as the anion, in equilibrium with $Co_2(CO)_8$ and $CoH(CO)_4$. This may account for the relatively low activity of cobalt (relative to rhodium and iridium catalysts) and the more severe conditions of temperature and pressure required for the cobalt-based process.

4.2.4   Rhodium-Catalyzed Carbonylation of Methanol

The BASF cobalt/iodide catalyzed process for methanol carbonylation was quite quickly superseded by a rhodium/iodide catalyzed process discovered at Monsanto and first commercialized in 1970 at a plant in Texas City. The Monsanto process was a significant advance and became one of the few large tonnage processes to use a homogeneous transition metal catalyst. It was later taken over by **BP** Chemicals; closely related rhodium catalyzed processes are also operated by other companies (e.g. Hoechst-Celanese, Eastman). Compared to the BASF process, the rhodium-based processes use significantly lower pressure and temperature (30–60 bar pressure and 150–200°C). This allows substantial savings in construction costs and hence in capital expenditure. In addition the excellent selectivity (>99% based on methanol) allows further savings on both running and capital costs and virtually all new acetic acid plants commissioned during the 1970s and 1980s employed the rhodium-based technology.

The major units of a methanol carbonylation plant are illustrated in Figure 2. The MeOH and CO feedstocks are fed to the reactor vessel on a continuous basis. In the initial product separation step the reaction mixture is passed from the reactor into a "flash-tank" where the pressure is reduced to induce vaporization of most of the volatiles. The catalyst remains dissolved in the liquid phase and is recycled back to the reactor vessel. The product stream from the flash-tank is directed into a distillation train which removes iodomethane, water and heavier by-products (e.g. propionic acid) from the acetic acid product.

As well as being the product, acetic acid also acts as the major solvent component for the Monsanto process. This means that the methanol feedstock is largely esterified into methyl acetate (Equation 13) under process conditions:

$$MeOH + MeCO_2H \rightarrow MeCO_2Me + H_2O \qquad (13)$$

As well as the water produced by esterification, quite a high concentration of water (ca. 10 M) is required to maintain high rates and prevent deactivation by precipitation of the rhodium catalyst (see Box 3). Separation of water from the acetic acid product by distillation incurs substantial costs. In addition, high water levels increase the rate of the water gas shift reaction (Section 4.1.3), catalyzed in competition with carbonylation by the rhodium/iodide system

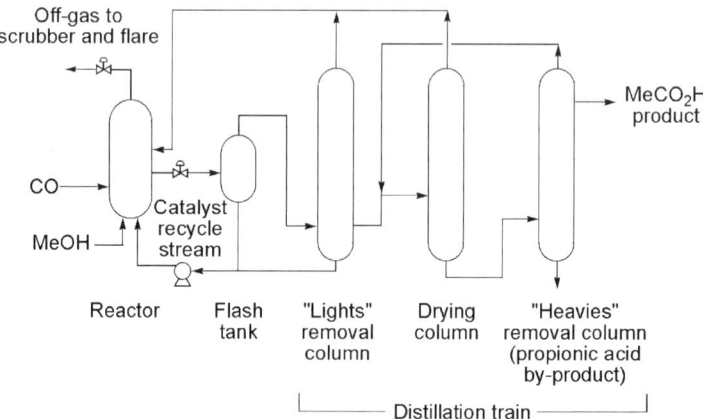

Figure 2 *Schematic representation of a commercial scale plant for the Rh/I⁻ catalyzed carbonylation of methanol.*

leading to loss of significant amounts of CO as CO_2. In view of these difficulties, most attempts to improve upon the original Monsanto technology in recent years have involved strategies to allow operation at lower water concentration. Notably, Hoechst-Celanese have developed a modified catalytic system which uses an iodide salt (e.g. LiI) to stabilize the rhodium catalyst at relatively low water concentration (<4 M). Their Acid Optimization (AO) technology was used to increase production at a plant in Clear Lake, Texas from its original (1978) capacity of 270 kt/a to 1200 kt/a in 2001. A new 500 kt/a plant using the AO low-water process has also been commissioned in Singapore. The LiI promoted low-water process is operated at higher methyl acetate concentration, which helps to stabilize the Rh(I) catalyst by moderating the equilibrium HI concentration, via the reaction (Equation 14):

$$MeOAc + HI \rightarrow MeI + AcOH \tag{14}$$

A lower HI concentration results in a decreased tendency for $[Rh(CO)_2I_2]^-$ to be oxidised to $[Rh(CO)_2I_4]^-$ (see Box 3) hence keeping the catalyst in its active form.

Box 3 Catalyst Deactivation in the Monsanto Process

A problem in all catalyzed processes is that the catalyst slowly loses its efficacy – it *deactivates*. In the rhodium catalyzed methanol carbonylation this occurs at high acid levels by oxidation by HI to a Rh(III) tetraiodide anion, $[Rh(CO)_2I_4]^-$ (Equation 8).

This complex has a tendency to lose CO leading, in several steps, to precipitation of RhI_3 which appears as an insoluble and intractable sludge (schemes in B3).

$$[Rh(CO)_2I_4]^- \rightarrow [Rh(CO)I_4]^- \rightarrow [RhI_4]^- \rightarrow [RhI_3]_n \downarrow \tag{B3}$$

This occurs in parts of the plant where the CO pressure is lower, such as in the flash tank and recycle loop (Figure 2). Water promotes reduction of the Rh(III) complex into the active Rh(I) form via the water-gas shift reaction (Section 4.1.3), and helps to minimize the proportion of the catalyst existing as the problematic $[Rh(CO)_2I_4]^-$, thus reducing precipitation.

Another approach, developed by Chiyoda/UOP, uses a rhodium catalyst heterogenized on a polymeric cation exchange resin. This takes advantage of the fact that the rhodium catalyzed carbonylation involves anionic complexes (see Section 4.2.5 below). The Chiyoda/UOP Acetica process employs a cross-linked polyvinylpyridine which is quaternized by methyl iodide to generate cationic pyridinium sites and which hold the anionic rhodium complexes by electrostatic interactions. The polymer support is tolerant of elevated temperatures and the ionic attachment of the catalyst is quite robust, resulting in only very low Rh losses in the effluent from the reactor. Since the heterogenized catalyst is essentially confined to the reactor, problems with precipitation in the product separation and catalyst recycle stages are minimized. The removal of solubility constraints allows operation at lower water concentration with increased catalyst loading, and rates comparable to the homogeneous reaction can be achieved. The lower water concentration is also claimed to result in reduced by-product formation. Mechanistic studies have indicated that the catalytic cycle for the supported catalyst is essentially identical to the homogeneous process (A. Haynes, P.M. Maitlis, R. Quyoum, C. Pulling, H. Adams, S. E. Spey, and R. W. Strange, *J. Chem. Soc., Dalton Trans.*, **2002**, 2565 and references therein). The Acetica process has been offered for licence by Chiyoda/UOP but so far does not operate on a full scale plant.

4.2.5 Mechanism of Rhodium/Iodide Catalyzed Methanol Carbonylation

Detailed mechanistic studies on the rhodium/iodide catalyst were first carried out by Forster and co-workers at Monsanto; they resulted in the elucidation of a mechanism which is now regarded as a classic example of a homogeneous catalytic cycle. More recently the Sheffield group of Maitlis and Haynes, in collaboration with BP Chemicals, have obtained quantitative kinetic data for some of the individual steps in the cycle and have used spectroscopic methods to detect reactive intermediates. *In situ* IR spectroscopy has shown that the rhodium catalyst exists mainly as $[Rh(CO)_2I_2]^-$, although $[Rh(CO)_2I_4]^-$ also forms under certain conditions (see Box 3). Kinetic studies showed the overall reaction to be first order in [MeI] and [Rh], and zero-order in [CO] and [MeOH], consistent with a rate determining step involving reaction of MeI with the rhodium catalyst. The well-established mechanism, shown in Figure 3, actually comprises two cycles: (i) the "rhodium cycle" which involves reactions of organometallic complexes, and (ii) the "iodide cycle" which involves organic reactions of the iodide co-catalyst.

Initially, methanol (or methyl acetate) reacts with the hydrogen iodide promoter to give iodomethane. The iodomethane undergoes an oxidative addition reaction with the square planar rhodium(I) complex, *cis*-$[Rh(CO)_2I_2]^-$ to give a product where both methyl and iodide ligands bind to the rhodium centre. This is the slowest reaction in the cycle, i.e. the rate determining step. The product is an octahedral rhodium(III) complex in which the methyl ligand is coordinated *cis* to two CO ligands. This complex is highly reactive (and only present in small concentration) due to a rapid migratory insertion reaction which brings together the methyl group and a CO to give an acetyl (C(O)Me) ligand. This is a key step in the cycle because it results in a new C–C bond being formed between ligands derived from the two C_1 feedstocks. The migratory insertion opens a vacant coordination site on rhodium, which can take up another molecule of CO to give an acetyl dicarbonyl complex. The final step in the organometallic cycle is a reductive elimination reaction that releases acetyl

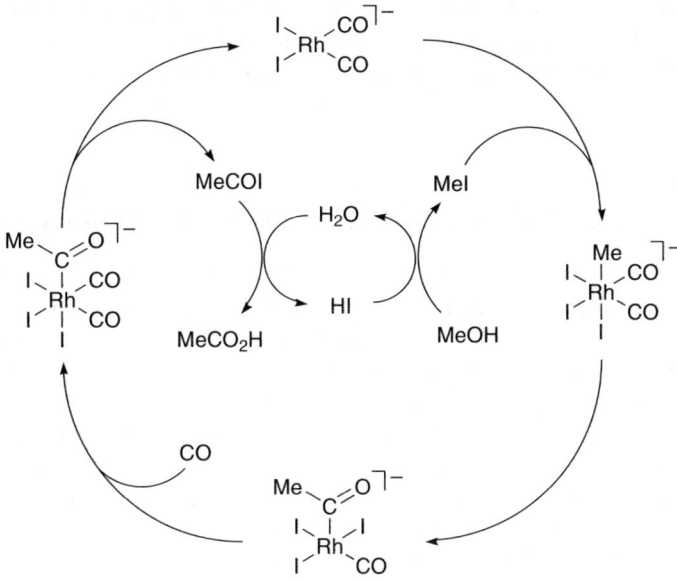

Figure 3 *Schematic representation of the mechanism of the Rh/I⁻ catalyzed carbonylation of methanol.*

iodide and regenerates the active Rh(I) complex. Hydrolysis of acetyl iodide gives the acetic acid product and regenerates the co-catalyst, HI, to complete the iodide cycle.

On the basis that oxidative addition of iodomethane to Rh(I) is rate determining in the catalytic cycle, many attempts have been made to accelerate the overall process by promoting this step. Since the oxidative addition is favoured by an electron rich metal centre, many workers have proposed the use of good σ-donor ligands to increase the nucleophilicity of the metal centre; unfortunately these complexes degrade under the harsh conditions of the carbonylation reaction and so offer no longer term advantage (see Box 4). It has also been speculated that coordination of an additional ionic ligand to give a di-anion, $[Rh(CO)_2I_2X]^{2-}$ (e.g. X = I or OAc) might accelerate the reaction. While iodide and acetate salts do affect the carbonylation rate (especially at low water concentrations), this may be largely due to inhibition of catalyst deactivation, and the proposed di-anions have not been observed experimentally.

Discussion Point DP1: *Oxidative addition of methyl iodide to Rh(I) complexes can be greatly accelerated using electron donating phosphine ligands. Why has this effect not been taken advantage of in a commercial rhodium-catalyzed methanol carbonylation process, whereas phosphine ligands are employed in hydroformylation?*

Discussion Point DP2: *Water is potentially an ideal solvent for metal catalyzed reactions. It is plentiful and cheap; it is polar and dissolves many polar and ionic compounds, and it facilitates polar reactions. By contrast, organic compounds of low polarity have only limited water-solubility. Why then is water generally used rather infrequently and, for example, only as a co-reactant in the acetic acid processes?*

Box 4 Ligand Promotion of Methanol Carbonylation

The oxidative addition of MeI to a Rh(I) centre occurs via nucleophilic attack by the metal, and can often be promoted by coordination of strongly electron donating ligands, such as phosphines. The effects of many phosphine ligands and derivatives have been tested, in a search for improved catalytic performance in "modified" systems, as demonstrated in hydroformylation (Section 4.6.6). A good example is the triethylphosphine modified catalyst, $Rh(PEt_3)_2(CO)I$, which almost doubles the initial carbonylation rate. With time however, this effect dies away as the catalyst reverts to the conventional $[Rh(CO)_2I_2]^-$, the phosphine ligands being lost as Et_3PI, Et_3PO, Et_3MeP^+ and Et_3PH^+. Long term stability is a recurring problem in ligand-modified catalyst systems of this kind. A large number of other ligands have been tried, including bidentate diphosphines and *hemilabile* phosphines (containing one phosphine and one heteroatom donor, e.g. O or S). These generally form more stable complexes due to the chelate effect, but unfortunately still tend to degrade under the harsh conditions of the actual process, thus offering no long term advantage. Lead references: K. G. Moloy, J. L. Petersen, *Organometallics*, 1995, **14**, 2931; J. Rankin, A. C. Benyei, A. D. Poole, D. J. Cole-Hamilton; *Dalton* **1999**, 3771; L. Gonsalvi, H. Adams, G. J. Sunley, E. Ditzel, and A. Haynes, *J. Am. Chem. Soc.*, 2002, **124**, 13597.

4.2.6   Iridium-Catalyzed Carbonylation of Methanol

Monsanto also discovered significant catalytic activity for iridium/iodide catalysts; however, they chose to commercialize the rhodium-based process due to its higher activity under "conventional" high water conditions. Despite this, detailed mechanistic studies by Forster and his colleagues were undertaken at Monsanto and revealed a catalytic mechanism for iridium which is similar to the rhodium system in many respects, but with additional complexity due to participation of both anionic and neutral complexes (see below).

Interest in iridium-catalyzed methanol carbonylation was rekindled in the 1990's when BP Chemicals developed and commercialized the Cativa process, which utilizes an iridium/iodide catalyst and a ruthenium promoter. This process has the important advantage that the highest catalytic rates occur at significantly lower water concentration (ca. 5% wt) than for Monsanto's rhodium process which requires $> 10\%$ wt H_2O to attain a plateau in catalytic activity. As discussed above, operation at low water allows substantial cost savings associated with product purification: specifically, plants using the Cativa™ process employ two rather than three distillation columns, since the "light ends" and drying columns (Figure 2) can be combined.

The Cativa™ process was first commercialized in 1995, with the retro-fitting to an existing rhodium-based plant in Texas City (USA), and several other acetic acid plants now use the Cativa™ technology.

An advantage of iridium, compared to rhodium catalysts is that a broad range of conditions is accessible without precipitation of IrI_3 occurring. This is because CO loss from iridium iodocarbonyls is inhibited by stronger metal-ligand bonding for the $5d$ transition metal. The rate of iridium-catalyzed carbonylation displays a rather complicated dependence on a range of process variables such as pCO, [MeOAc] and [H_2O], but is zero order in [MeI] and independent of CO partial pressure above a limiting threshold.

A range of compounds enhance the activity of an iridium catalyst. The promoters fall into two categories: (i) carbonyl or halocarbonyl complexes of W, Re, Ru, Os and Pt; and (ii) simple iodides of Zn, Cd, Hg, Ga and In. The preferred ruthenium promoter is effective over a range of water concentrations the maximum rate being attained at ca. 5% wt H_2O, as in the absence of promoter. By contrast, ionic iodides such as LiI and Bu_4NI are strong catalyst poisons.

The high activity of ruthenium-promoted iridium catalysts has allowed greatly improved productivity by retro-fitting the Cativa™ process to plants which previously used rhodium catalysts. Another benefit is higher selectivity, with smaller amounts of both gaseous and liquid by-products. The WGSR does occur, but at a somewhat lower rate than for rhodium, resulting in reduced formation of CO_2. Since the process is less sensitive to CO partial pressure, the reactor can operate with a lower vent-rate, despite an increased formation of methane derived from long-lived methyl-iridium species. This, as well as other modifications, results in an increase in CO utilization from ca. 85% (Rh) to $> 94\%$ (Ir). Acetaldehyde (thought to result from hydrogenolysis of a metal acetyl complex) is present in the reactor at lower levels than in the Rh process, since iridium effectively catalyzes hydrogenation of acetaldehyde to ethanol, which is subsequently carbonylated to propionic acid. Significantly, for iridium, the formation of propionic acid is decreased which, together with the lower water concentration, leads to a substantial decrease in the product purification costs. For example, steam and cooling water requirements are reduced by 30% by comparison with the rhodium system. Acetaldehyde can also participate in condensation reactions, eventually leading to longer chain iodoalkanes. This can be problematic for low-water rhodium systems since iodine-containing compounds can poison downstream processes such as vinyl acetate manufacture, and their removal requires further treatment steps. The overall CO_2 emissions (both direct and indirect) for each tonne of product are estimated to be ca. 30% lower than for the rhodium system. All of these factors improve the environmental impact of the Cativa™ process [G. J. Sunley, D. J. Watson, *Catal. Today* 2000, **58**, 293].

4.2.7 Mechanism of the Iridium/Iodide Catalyzed Methanol Carbonylation

The catalytic cycle involves the same fundamental reaction steps as the rhodium system: oxidative addition of MeI to Ir(I), followed by migratory CO insertion to form an Ir(III) acetyl complex, from which acetic acid is derived. However, there are significant differences in reactivity between analogous rhodium and iridium complexes which are important for the overall catalytic activity. *In situ* spectroscopy indicates that the dominant active iridium species present under catalytic conditions is the anionic Ir(III) methyl complex, $[IrMe(CO)_2I_3]^-$, by contrast to the rhodium system where the dominant complex is $[Rh(CO)_2I_2]^-$. $[IrMe(CO)_2I_3]^-$ and an inactive form of the catalyst, $[Ir(CO)_2I_4]^-$ represent the resting states of the iridium catalyst in the "anionic cycles" for carbonylation and the WGSR respectively. At lower concentrations of water and iodide, $[Ir(CO)_3I]$ and $[Ir(CO)_3I_3]$ are present due to the operation of related "neutral cycles".

Oxidative addition of iodomethane to $[Ir(CO)_2I_2]^-$ is considerably (ca. 100x) faster than to $[Rh(CO)_2I_2]^-$, such that this step is not rate determining in the iridium process. By contrast, $[IrMe(CO)_2I_3]^-$ is much less reactive than its rhodium analogue, and carbonylation of this species to give an iridium acetyl complex becomes the rate determining step. The reaction of $[IrMe(CO)_2I_3]^-$ with CO, key to the overall performance of the catalyst, has been studied in considerable detail. Evidence from kinetic and spectroscopic investigations indicates that dissociation of one iodide is required. This allows coordination of an additional CO ligand to give a neutral tricarbonyl complex, $[IrMe(CO)_3I_2]$ which undergoes migratory CO insertion much more readily than the anionic precursor (see Figure 4).

The necessity of interconversion between anionic and neutral species means that the concentration of ionic iodide is crucial in controlling the catalytic rate. If the iodide concentration is too high, then the steady state concentration of the active neutral species, $[IrMe(CO)_3I_2]$, is decreased, which slows the overall carbonylation rate. However, if the iodide concentration is too low, a build-up of $[Ir(CO)_3I]$ can occur since this neutral complex is much less reactive towards iodomethane than $[Ir(CO)_2I_2]^-$. The iodide level is also affected by the water concentration via the equilibrium (Equation 15):

$$MeOAc + H_3O^+ + I^- \rightleftharpoons MeI + H_2O + AcOH \qquad (15)$$

Thus, raising $[H_2O]$ tends also to increase $[I^-]$, with the outcome that the catalytic rate passes through a maximum and then decreases on increasing the water concentration.

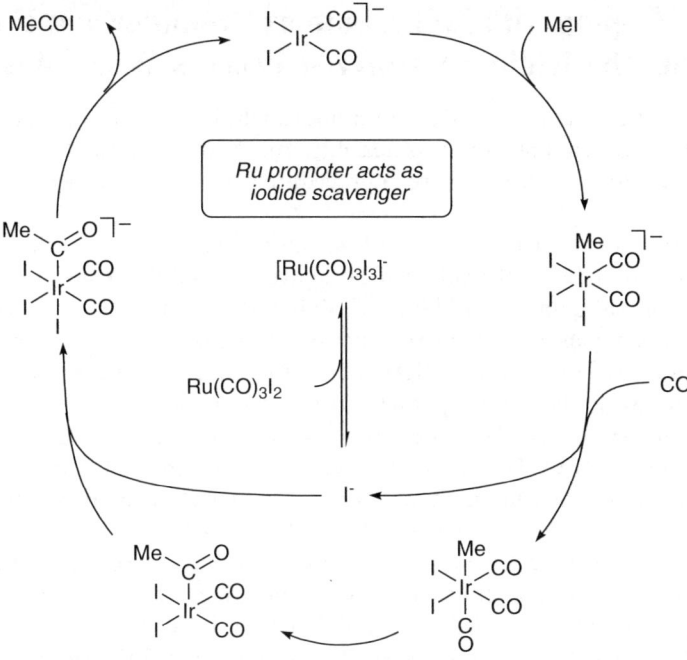

Figure 4 *Simplified mechanism for the Ir/I$^-$ catalyzed, Ru promoted carbonylation of methanol.*

The ruthenium promoter is thought to operate by moderating the iodide concentration. Neutral ruthenium(II) iodocarbonyls (e.g. Ru(CO)$_4$I$_2$ and [Ru(CO)$_3$I$_2$]$_2$) are effective at scavenging iodide to give the anion [Ru(CO)$_3$I$_3$]$^-$, which is observed as the major form of the ruthenium promoter during catalysis. In addition, the significant concentration of [Ru(CO)$_3$I$_3$]$^-$ H$_3$O$^+$ raises the Brønsted acidity and accelerates the reaction of methyl acetate with HI$_{(aq)}$ by acid catalysis. This is important, since each turnover of the catalytic cycle generates one equivalent of HI$_{(aq)}$ which has to be recycled into MeI by reaction with methyl acetate or methanol. Promotion of this step lowers the steady state concentration of iodide which in turn accelerates the carbonylation cycle. Other promoters are thought to operate in the same way; model studies have demonstrated iodide transfer from [IrMe(CO)$_2$I$_3$]$^-$ to neutral promoter species such as [Ru(CO)$_3$I$_2$]$_2$ and InI$_3$, and kinetic measurements showed that the stoichiometric carbonylation of [IrMe(CO)$_2$I$_3$]$^-$ is accelerated by addition of the iodide abstracting promoters [See Box 5 and A. Haynes, P. M. Maitlis, G. E. Morris, G. J. Sunley, H. Adams, P. W. Badger, C. M. Bowers, D. B. Cook, P. I. P. Elliott, T. Ghaffar, H. Green, T. R. Griffin, M. Payne, J. M. Pearson, M. J. Taylor, P. W. Vickers and R. J. Watt, *J. Am. Chem. Soc.*, 2004, **125**, 2847].

Box 5 Coping with Iodide both as Promoter and Poison: Making the Iridium Catalyzed Process Work Well

Although the studies at Monsanto indicated that an iodide promoted iridium catalyzed reaction was feasible, much more basic research was needed before a viable commercial procedure (the Cativa$^{®}$ process) became possible.

   Key was a detailed appreciation of the rates of various steps of the cycle by workers at BP Chemicals and Sheffield University and the realization *a*) that although initial oxidative addition of MeI to the metal centre was slow for Rh(I) it was much faster for Ir(I), *b*) that the migration step was very much slower for Ir(III) than for Rh(III), and *c*) that although the migration step was very slow in the anionic $[IrMe(CO)_2I_3]^-$ it was substantially faster in the neutral $[IrMe(CO)_3I_2]$. The latter complex was readily accessible by carbonylation of $[IrMe(CO)_2I_2]$, and the problem then resolved itself to find a system which permitted the iridium to be in the neutral Ir(III) form at the migration step and yet to be in the anionic Ir(I) form, $[Ir(CO)_2I_2]^-$, to facilitate the initial oxidative addition step. To accomplish this promoters were needed that can scavenge iodide into a form in which it no longer acts as a poison for carbonylation. Model studies showed that a range of neutral metal iodide co-promoters could play this role; especially attractive were InI_3 and $RuI_2(CO)_4$, which also moderate the steady state concentration of $HI_{(aq)}$. In the proposed mechanism (Figure 4) the neutral promoter species $[Ru(CO)_3I_2(sol)]$ scavenges $HI_{(aq)}$ to give $[Ru(CO)_3I_3]^-H_3O^+$. Such species can act as Brønsted acid catalysts for the reaction of $HI_{(aq)}$ with MeOAc, consequently lowering the standing concentration of $HI_{(aq)}$ and enhancing turnover in the Ir cycle. Since the neutral promoter species accelerate the stoichiometric carbonylation of the catalyst resting state $[Ir(CO)_2I_3Me]^-$ by acting as iodide acceptors, they can also be thought of as sources of "iodide holes".

4.2.8   Rhodium-Catalyzed Carbonylation of Methyl Acetate to Acetic Anhydride

Acetic anhydride is another important commodity chemical made in large quantities. The classic organic synthesis of acetic anhydride involves pyrolysis of acetic acid to ketene ($MeCO_2H \rightarrow CH_2 = C=O$ at 700–800°C) which is then trapped in acetic acid ($CH_2=C=O + MeCO_2H \rightarrow (MeCO)_2O$). This is a highly energy demanding route and has now been largely superseded by the carbonylation of methyl acetate (Equation 16):

$$\underset{\substack{\\ Me \quad\quad OMe}}{\overset{\substack{O\\\|}}{\diagup\!\!\diagdown}} \;+\; CO \;\longrightarrow\; \underset{\substack{\\ Me \quad O \quad Me}}{\overset{\substack{O\quad\; O\\\|\quad\;\;\|}}{\diagup\!\!\diagdown\diagdown\!\!\diagup}} \qquad\qquad (16)$$

Thermodynamically, the carbonylation of methyl acetate (ΔG_{298} –10 kJ/mol) is considerably less favourable than that of methanol (ΔG_{298} –74 kJ/mol). This means that the reaction does not reach completion but attains an equilibrium which is dependent on the temperature and the CO pressure. Two variants are currently practised commercially: that developed by Tennessee Eastman, based on a Halcon process, and a BP process in which acetic acid and the anhydride are co-produced in proportions which can be varied according to demand. Syngas for the Eastman process is made from coal which is mined close to the plant in Tennessee and the acetic anhydride produced is used to make cellulose acetate for film production. The BP process uses syngas generated from North Sea gas which is piped directly to the BP plant in Hull. [Acetic anhydride manufacture: M. J. Howard, M. D. Jones, M. S. Roberts, S. A. Taylor, *Catalysis Today*, 1993, **18**, 325].

The basic organometallic reaction cycle for the Rh/I⁻ catalyzed carbonylation of methyl acetate is the same as for methanol carbonylation. However some differences arise due to the absence of water in the anhydrous process. As described in Section 4.2.4, the Monsanto acetic acid process employs quite high water concentrations to maintain catalyst stability and activity, since at low water levels the catalyst tends to convert into an inactive Rh(III) form. An alternative strategy, employed in anhydrous methyl acetate carbonylation, is to use iodide salts as promoters/stabilizers. The Eastman process uses a substantial concentration of lithium iodide, whereas a quaternary ammonium iodide is used by BP in their combined acetic acid/anhydride process. The iodide salt is thought to aid catalysis by acting as an alternative source of iodide (in addition to HI) for activation of the methyl acetate substrate (Equation 17):

$$MeOAc + LiI \rightleftharpoons MeI + LiOAc \qquad (17)$$

The lithium acetate can trap acetyl iodide (formed by reductive elimination from $[Rh(COMe)(CO)_2I_3]^-$) to give acetic anhydride (Equation 18):

$$LiOAc + MeCOI \rightleftharpoons LiI + Ac_2O \qquad (18)$$

Even with added iodide salt formation of the inactive $[Rh(CO)_2I_4]^-$ can be a problem, since under anhydrous conditions this Rh(III) species cannot be reduced to the active $[Rh(CO)_2I_2]^-$ by reaction with water. In the Eastman process, this problem is addressed by addition to the CO gas feed of some H_2 which can reduce $[Rh(CO)_2I_4]^-$ by the reverse of Equation 8. However, the added H_2 does lead to some undesired by-products, particularly ethylidene diacetate (1,1-diacetoxyethane) which probably arises from the reaction of acetic anhydride with acetaldehyde (Equation 19; from hydrogenolysis of a rhodium acetyl):

$$Ac_2O + MeC(H)O \rightarrow MeC(H)(OAc)_2 \qquad (19)$$

Another by-product is acetone (possibly arising from MeI+Rh-COMe or decarboxylation of Ac_2O) which undergoes condensation reactions that lead to heavier products, which can take the form of intractable tars.

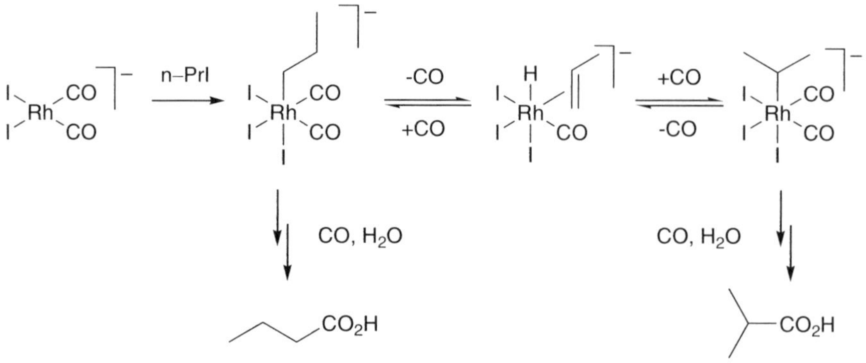

Figure 5   *Possible mechanism of formation of n- and i-butyric acids in the Rh/I^- catalyzed carbonylation of n-propanol.*

4.2.9   Carbonylation of Higher Alcohols: Higher Carboxylic Acids

Higher linear alcohols are also carbonylated to the corresponding acids under similar conditions using rhodium/iodide catalysts, although the rates are substantially slower than for methanol carbonylation; the relative rates are 21:1:0.5 for MeOH, EtOH and n-PrOH respectively. n-PrOH gives both the linear and branched chain acid products (butyric and isobutyric acids), with the selectivity depending upon the reaction conditions. The catalytic cycle is proposed to be similar to that of the methanol carbonylation; the rate determining step remains oxidative addition of the alkyl iodide to $[Rh(CO)_2I_2]^-$ but this is slower for bulkier alkyl groups, explaining the lower overall rates. The formation of both linear and branched butyric acids from n-propanol is explained by isomerization of the alkyl intermediate via a hydrido olefin species as shown in Figure 5. Raising the CO pressure inhibits this isomerization and favours formation of the linear acid. An alternative possible mechanism involves initial dehydration of the alcohol to give an olefin, which reacts with $[Rh(H)(CO)_2I_3]^-$ (resulting from oxidative addition of HI to $[Rh(CO)_2I_2]^-$) to give the same isomeric alkyl complexes, although detailed studies indicate this route is less likely. However, alkenes themselves can be convenient substrates for hydroxycarbonylation to carboxylic acids, see Section 4.3.

   Discussion Point DP3: *You are charged with designing a process to make isobutyric acid from basic C_1 starting materials. Draw a scheme illustrating how you would go about this task, and what catalytic reactions you would choose to employ.*

4.2.10   Carbonylation of Benzyl Alcohol to Phenylacetic Acid; Manufacture of Ibuprofen

The profens are non-steroidal anti-inflammatory agents based on arylpropionic acids; the best known is ibuprofen. The BHC company (Boots-Hoechst-Celanese) developed a commercial route to ibuprofen which involves a

$PdCl_2(PPh_3)_2$ catalyzed homogeneous carbonylation of 1-(4'-isobutylphenyl) ethanol, in which some of the isomeric linear acid, $4\text{-}i\text{-}Bu\text{-}C_6H_4CH_2CH_2CO_2H$, is also formed; typical yields are 92% of ibuprofen and 6% of the linear acid (see also Section 4.3.1). The carbonylation is performed at 50 bar, 130°C, in an acidic aqueous medium containing chloride ions, a ketone (such as methyl ethyl ketone) as solvent, and at least 10% of water [V. Elango *et al.*, US Patent 4 981 995 (1991)]. A convenient ratio of substrate: palladium catalyst is 10^5:1. A 3.5kt/a plant in Bishop, Texas has been operated by BHC since 1992. The palladium catalyst can be precipitated from the organic phase of the carbonylation reaction and re-used without further treatment.

The carbonylation reaction is the final part of a three step route to ibuprofen (shown in Figure 6a) which has superseded a less efficient six-step pathway from isobutylbenzene (Box 6). A related profen, naproxen is made by a hydroxycarbonylation route (Figure 6b).

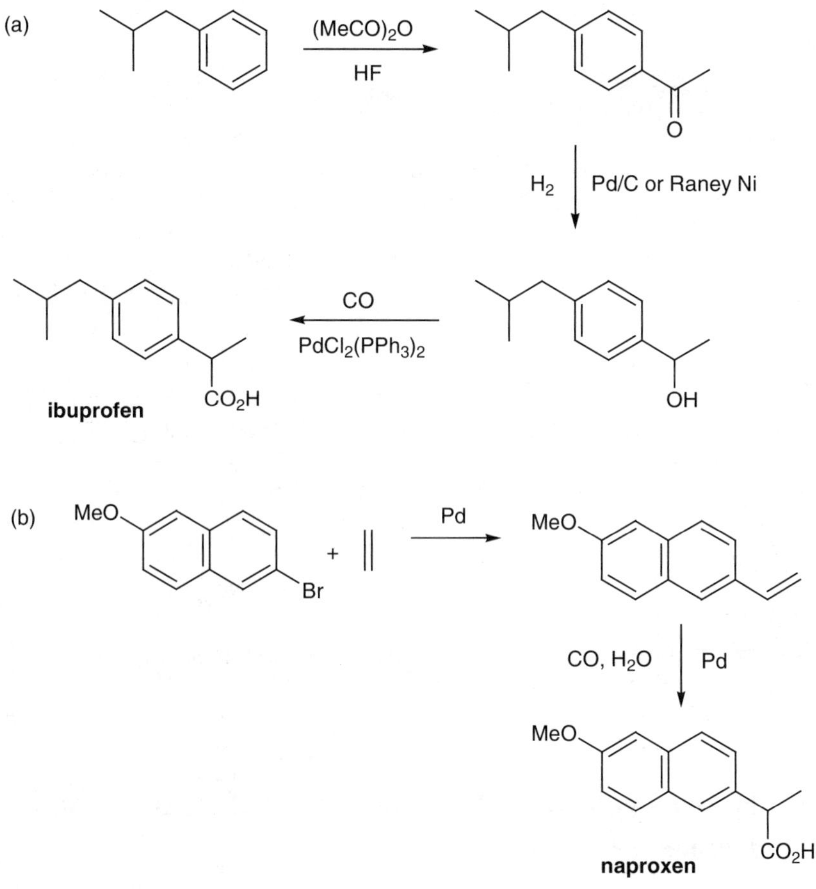

Figure 6 *Synthetic routes to anti-inflammatories (a) ibuprofen and (b) naproxen involving palladium catalyzed carbonylation steps.*

Box 6 Atom Economy and Green Chemistry

The three step BHC route to ibuprofen supersedes a previous synthetic pathway which also started from isobutylbenzene but required six steps and was environmentally less friendly, i.e. less *green* (see Figure 7).

It is instructive to compare the atom economies of the two pathways. Atom economy is a measure of the efficiency of a chemical process, defined in percentage terms as x (formula wt. of atoms utilized)/(formula wt. of all reactants). For the old six-step ibuprofen synthesis the atom economy was only 40% (with $MeCO_2H$, EtOH, NaCl, $EtOCO_2H$, $2H_2O$ and NH_3 as waste). This is dramatically improved to 77% for the new three-step route with only $MeCO_2H$ as a by-product from the first step. Recovery and use of this increases the atom economy to 99%. Additionally, the catalytic amounts of HF and Pd complex used in the BHC process are recovered and reused, whereas stoichiometric quantities of $AlCl_3$ hydrate were produced as waste by the old route.

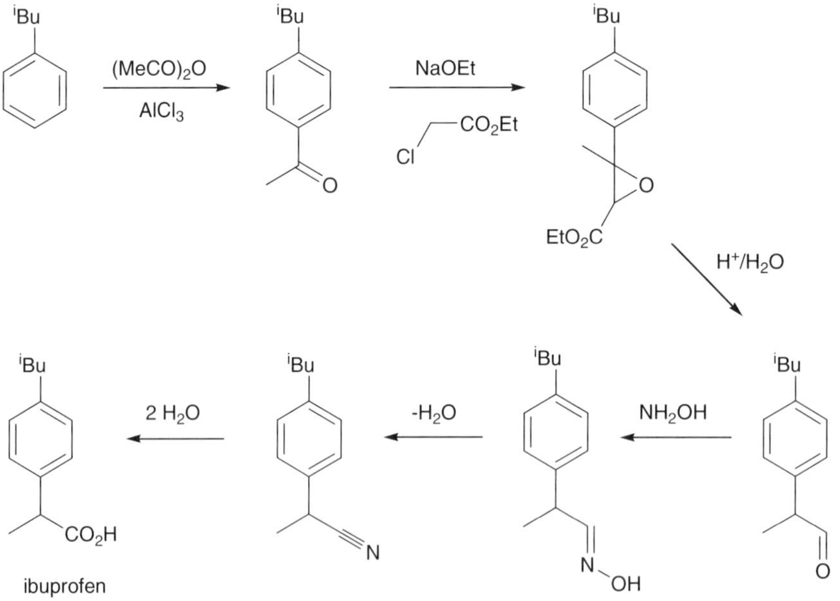

Figure 7 *The older, classical six-step synthesis of ibuprofen not involving metal-catalyzed carbonylation.*

4.3 Hydroxy/Alkoxy-Carbonylation of Alkenes and Dienes

An alternative route to carboxylic acids (and their esters) involves carbonylation of an olefin in the presence of water (to give the acid) or an alcohol (to give the ester). These reactions, in which H and CO_2H (or CO_2R) units are added across the double bond, are often referred to as *hydrocarboxylations* or *alkoxycarbonylations* (Equations 20 and 21):

$$R\diagup\!\!\!\diagdown + CO + H_2O \longrightarrow R\diagdown\!\!\diagup\!\!\diagdown\!\!\diagup^{CO_2H} + \underset{R}{\diagup}\!\!\overset{CO_2H}{\diagdown} \qquad (20)$$

$$R\diagup\!\!\!\diagdown + CO + MeOH \longrightarrow R\diagdown\!\!\diagup\!\!\diagdown\!\!\diagup^{CO_2Me} + \underset{R}{\diagup}\!\!\overset{CO_2Me}{\diagdown} \qquad (21)$$

Although in principle any olefin can undergo such a transformation, problems often arise with higher alkenes, because of the ease of double bond isomerizations that lead to a variety of by-products.

The most common catalysts are cobalt based. However although palladium/phosphine systems are more expensive they generally work under milder conditions and are preferred, especially when catalyst recovery is good. A mild (1 bar, 20°C) enantioselective catalytic hydroxycarbonylation of 4'-isobutylstyrene to (S)-ibuprofen has been developed using a $PdCl_2/CuCl_2/HCl$ catalyst system with (S)-1,1'-binapthyl-2,2'-diyl hydrogen phosphate as chiral ligand (see Section 4.2.10). However, the benefit of an enantioselective synthesis of ibuprofen is doubtful, since an isomerase enzyme converts the (R)-form into the (S)-enantiomer *in vivo*. The commercial product is therefore usually the racemate.

A commercial synthesis (Albemarle) of naproxen (a 2-aryl propionic acid anti-inflammatory related to ibuprofen) involves palladium catalyzed hydroxycarbonylation of an aryl olefin which is itself made in a palladium catalyzed Heck coupling reaction (Figure 6b). Resolution is needed to obtain the (S)-enantiomer of naproxen since its optical isomer is a liver toxin.

4.3.1 Ethylene to Propionic Acid; Methyl Propionate and Methyl Methacrylate (MMA)

Propionic acid is a significant intermediate in the production of a number of plastics, textiles, solvents, fragrances, pharmaceuticals and pesticides. Several transition metals (e.g. Ni, Pd, Rh, Ir, Ru, Mo) can catalyze the conversion of ethylene into propionic acid (Equation 22), and it is made by BASF by a nickel catalyzed hydroxycarbonylation of ethylene (100–300 bar; 250–320°C):

$$= + CO + H_2O \longrightarrow \diagup\!\!\!\diagup^{CO_2H} \qquad (22)$$

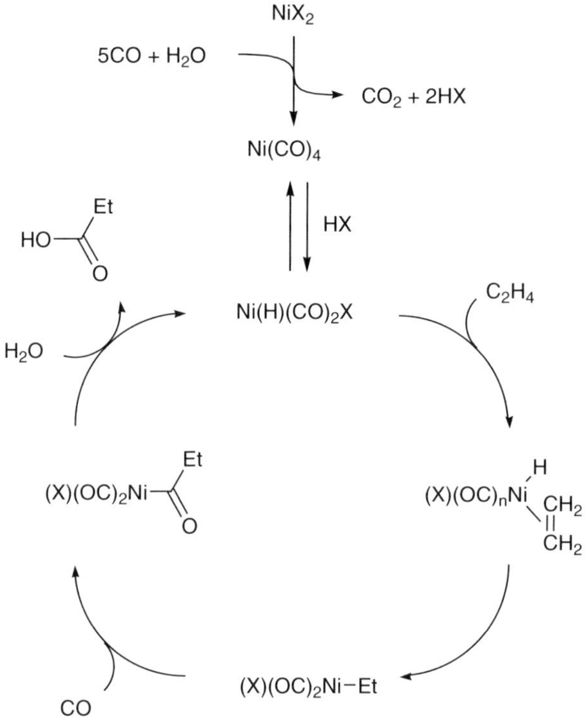

Figure 8 *Representation of a possible mechanism for the Ni(CO)$_4$ catalyzed carbonylation of ethylene to give propionic acid (X = OH or halide).*

The system is halide-free and the actual catalyst is probably a nickel hydride with the extremely toxic and volatile Ni(CO)$_4$ as precursor. No detailed mechanistic studies have been published, but a possible scheme for the reaction is in Figure 8.

Methyl methacrylate (MMA) is an important commodity since it is polymerized to give poly methylmethacrylate (PMMA), a strong, durable and transparent polymer sold under the trade-names Perspex and Plexiglas. Since the conventional routes to MMA involve either the reaction of acetone with HCN to give the cyanohydrin (which has environmental problems), or the oxidation of isobutene, alternative carbonylation routes to MMA are being developed. One of these is the Lucite Alpha process which is claimed to decrease production costs by ca. 40%. This first synthesizes methyl propionate by a methoxycarbonylation of ethylene (Equation 23), using a palladium catalyst with very high (99.8%) selectivity. In the second step, MMA is formed in 95% selectivity by the reaction of methyl propionate with formaldehyde (Equation 24).

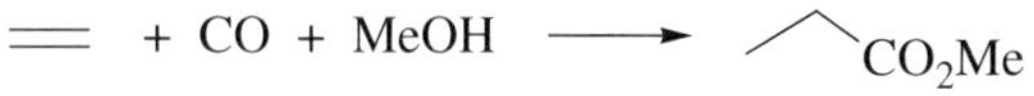

$$\text{==} \quad + \text{ CO } + \text{ MeOH} \quad \longrightarrow \quad \diagup\!\!\diagdown\text{CO}_2\text{Me} \qquad (23)$$

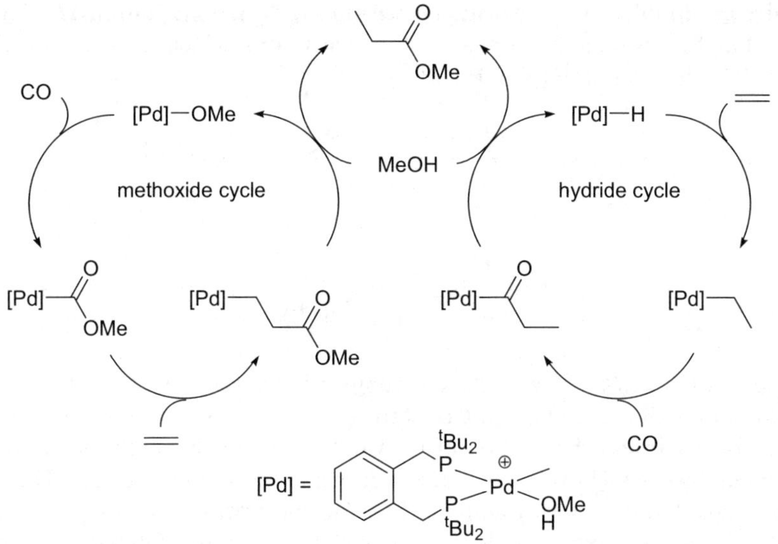

Figure 9 *Catalytic cycles proposed for the palladium/diphosphine catalyzed methoxycarbonylation of ethylene.*

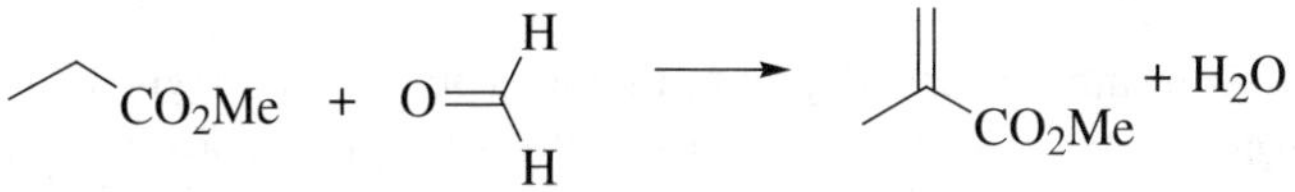

$$\text{CH}_2\text{=CHCO}_2\text{Me} + \text{O=CH}_2 \longrightarrow \text{CH}_2\text{=C(CO}_2\text{Me)CH}_3 + \text{H}_2\text{O} \qquad (24)$$

Production facilities in the USA and in Singapore based on this technology are scheduled to come on stream in 2007.

Palladium catalysts with simple monodentate phosphine ligands (e.g. PPh_3) can catalyze the methoxycarbonylation of ethylene. However, the Lucite process employs a bulky diphosphine, 1,2-($^t\text{Bu}_2\text{PCH}_2)_2\text{C}_6\text{H}_4$, and is highly active and selective under quite mild conditions (10 bar/80°C). Two alternative catalytic cycles are possible, based either upon a palladium hydride or a palladium methoxide complex (Figure 9), and mechanistic and spectroscopic studies indicate that the hydride cycle is dominant. The alkene and CO insertion steps are the same as those in the Pd-catalyzed co-polymerisation of CO and alkenes to polyketones (Section 4.4).

4.3.2 Cobalt-Catalyzed Butadiene Dimethoxycarbonylation to Dimethyl Adipate

BASF makes dimethyl adipate via the dimethoxycarbonylation of butadiene, using $Co_2(CO)_8$ as catalyst in methanol in the presence of an organic base such as quinoline. The reaction is carried out as a two-step process, the first (Equation 25; at 600 bar, 120°C) gives $MeCH=CHCH_2COOMe$ (98% selectivity). The second carbonylation step leading to the final product formally involves a double bond shift to the terminal position (Equation 26; 30 bar, 185°C) and gives dimethyl adipate (85% selectivity), which is then hydrolyzed (acid) to adipic acid ($HO_2C(CH_2)_4CO_2H$).

$$\diagup\!\!\!\!\diagup \xrightarrow{\text{CO/MeOH}} \diagup\!\!\!\!\diagup\text{CO}_2\text{Me} \qquad (25)$$

$$\diagup\!\!\!\!\diagup\text{CO}_2\text{Me} \xrightarrow{\text{CO/MeOH}} \text{MeO}_2\text{C}\diagdown\!\!\!\!\diagup\text{CO}_2\text{Me} \qquad (26)$$

The active species in the methoxycarbonylation is presumably $CoH(CO)_4$ (in equilibrium with $Co_2(CO)_8$ and $Co(CO)_4^-$) which adds Co–H 1,4- to the diene; this is followed by carbonylation of the Co–C bond, methanolysis of the RCO–Co bond by MeOH or OMe$^-$, and regeneration of the hydride. The methoxycarbonylation route to adipic acid is an alternative both to the du Pont (Ni(II)/Lewis acid (BPh_3)) catalyzed double hydrocyanation of butadiene (Section 5.4.4) and to the process based on the oxidation of cyclohexane (Section 2.2).

4.4   Polyketones

Olefin-carbon monoxide co-polymers of the type $(-RCH-CH_2-CO-)_n$, known as polyketones, have a wide variety of interesting properties, such as thermoplasticity, flexibility, durability and high impact strength. The ketone function in the polymer makes them sensitive to UV radiation and as a consequence they are photodegradable and hence environmentally acceptable plastics. However their light sensitivity has limited their applications. They can be made by the co-polymerisation of an alkene with carbon monoxide (Equation 27; see also Chapter 7, Section 7.7):

$$^n R\diagup\!\!\!\!\diagup + nCO \longrightarrow \left(\!\!\begin{array}{c} R \quad O \\ \diagdown\!\!\diagup\!\!\diagdown\!\!\diagup\!\!\diagdown \end{array}\!\!\right)_n \qquad (27)$$

This was first achieved at very high pressures using free radical initiation but the polymers produced were irregular in structure and of rather low molecular weight. Transition metals can catalyze the co-polymerisation reaction; some activity is found for nickel and rhodium complexes but the vast majority of catalysts which have been used are of palladium.

In the 1980s Drent and colleagues at Shell found a remarkable change in catalytic behaviour when monodentate phosphines were replaced with certain bidentate diphosphines. Cationic palladium(II) catalysts generated in methanol from $Pd(OAc)_2$, PPh_3 and the acid of a weakly coordinating anion (e.g. tosylate, triflate) catalyzed the methoxycarbonylation of ethylene to

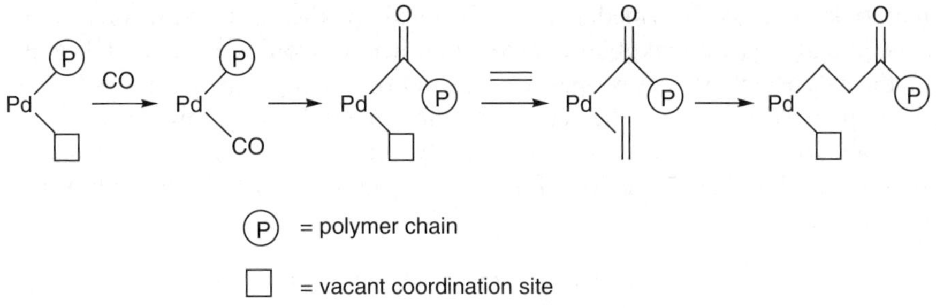

(P) = polymer chain

▢ = vacant coordination site

Figure 10 *Chain growth steps for alternating CO/ethylene co-polymerization in polyke-tone formation.*

methyl propionate. By contrast, switching to a diphosphine (in particular $Ph_2P(CH_2)_3PPh_2$, dppp) but retaining the other components, resulted in formation of a perfectly alternating co-polymer of ethylene and CO with essentially 100% selectivity. Mechanistically, the reactions are closely related to the methoxycarbonylation of ethylene shown in Figure 9, with hydride and methoxide cycles both being feasible. The major difference, of course, is that multiple insertions of CO and ethylene must occur to give the polyketone rather than methyl propionate. Thus insertion of ethylene into a Pd-acyl bond, and of CO into a Pd-alkyl bond must occur in preference to termination by reaction with methanol. The propagation steps are shown in Figure 10.

Co-polymerization is favoured (over methyl propionate formation) for most bidentate ligands since the chelating diphosphine imposes a *cis* stereochemistry on the two other coordination sites, occupied by the growing polymer chain and the incoming monomer units. This *cis* stereochemistry is ideal for the sequential ligand insertions required for co-polymerization. By contrast, two monodentate phosphine ligands, tend to coordinate mutually *trans* due to steric considerations, so termination becomes relatively more competitive. Despite this, it is interesting that Lucite's methyl propionate process (Section 4.3.2) actually employs a bidentate diphosphine ligand (Figure 9). It is thought that the steric bulk of the t-Bu substituents in that case is important in changing the selectivity. Although their properties promised so much, and despite at least two major companies (Shell and BP) developing the manufacture of the polyketone polymers, the projects were dropped, and no commercial process operates.

Discussion Point DP4: *The palladium-catalyzed carbonylation of ethylene both to methyl propionate and to polyketones use weakly coordinating anions to achieve high activity. Discuss the features of the catalytic mechanism which give rise to this effect.*

4.5 Oxidative Carbonylation of Methanol to Dimethyl Carbonate and Dimethyl Oxalate

Organic carbonates especially dimethyl carbonate, $(MeO)_2CO$, DMC, are becoming increasingly important compounds, both to make polycarbonate materials and as intermediates in chemical synthesis. Polycarbonates are used in making (unbreakable) CDs and in other electronic devices and, because of their excellent optical properties, in window glazing. Their synthesis originally involved the use of the very toxic phosgene ($COCl_2$). However oxidative carbonylation reactions have essentially made the phosgene processes redundant. The basic reaction (Equation 28) makes DMC by a methoxycarbonylation,

$$2MeOH + CO + \tfrac{1}{2}O_2 \rightarrow (MeO)_2CO + H_2O \qquad (28)$$

which is then converted into diphenyl carbonate (the preferred intermediate to polycarbonates) by transesterification with phenol. World production is around 200 kt/a.

Two oxidative carbonylation processes to DMC have been commercialized, one by EniChem uses a copper chloride catalyst; the other, developed by Ube and Bayer uses methyl nitrite (from methanol, NO and oxygen).

The ENI copper chloride process can be represented by the redox reactions (Equations 29 and 30),

$$2MeOH + 2CuCl + \tfrac{1}{2}O_2 \rightarrow 2Cu(OMe)Cl + H_2O \qquad (29)$$

$$2Cu(OMe)Cl + CO \rightarrow (MeO)_2CO + 2CuCl \qquad (30)$$

$PdCl_2$ in the presence of a base, and cobalt-Schiff base complexes have also been proposed as catalysts.

The second process occurs in the gas phase; in the first step methanol reacts with NO and oxygen at about 50°C to give methyl nitrite (Equations 31 and 32):

$$2NO + \tfrac{1}{2}O_2 \rightarrow N_2O_3 \qquad (31)$$

$$N_2O_3 + 2MeOH \rightarrow 2MeONO + H_2O \qquad (32)$$

The methyl nitrite then reacts with CO over a supported palladium chloride catalyst (at 100–120°C) to give the DMC product and to regenerate the NO (Equation 33):

$$2MeONO + CO \rightarrow (MeO)_2CO + 2NO \qquad (33)$$

In the methoxycarbonylation with a palladium(II) acetate catalyst dimethyl oxalate is formed together with DMC (Equation 34):

$$2MeOH + 2CO + \tfrac{1}{2}O_2 \rightarrow MeO_2C\text{-}CO_2Me + H_2O \qquad (34)$$

Oxalate formation appears to be promoted by base and it is believed that Pd(0) is the active catalyst whereas DMC formation is facilitated by Pd(II). The nitrite/NO is believed to promote the re-oxidation of Pd(0) to Pd(II) and a possible mechanistic sequence (Equations 35a–d) has been proposed,

$$Pd(0) + RONO \rightarrow Pd(OR)(NO) \tag{35a}$$

$$Pd(OR)(NO) + CO \rightarrow Pd(CO_2R)(NO) \tag{35b}$$

$$Pd(CO_2R)(NO) + RONO \rightarrow Pd(CO_2R)(OR) + 2NO \tag{35c}$$

$$Pd(CO_2R)(OR) + CO \rightarrow Pd(CO_2R)_2 \rightarrow RO_2CCO_2R + Pd(0) \tag{35d}$$

[mechanism: D. Delledonne, F. Rivetti, U. Romano, *Applied Catalysis A: General* 2001, **221**, 241; S.-I. Uchiumi, K. Ataka, T. Matsuzaki, *J. Organomet. Chem.* 1999, **576**, 279].

4.6 Hydroformylation of Olefins

The hydroformylation (or "oxo") reaction was discovered in 1938 by Roelen who was working on the formation of oxygenates as by-products of the Fischer-Tropsch reaction over cobalt catalysts. It soon became clear that the aldehydes and alcohols found were the products of secondary reactions undergone by the 1-alkenes (which are the primary products of the Fischer-Tropsch reaction, Section 4.7.2) with syngas. Further work showed that Roelen had discovered a new reaction, in which the elements of H and CHO were added to an olefin (hence *hydroformylation*), and which was catalyzed by cobalt. It was later found that the true precatalyst was not cobalt metal but derivatives of dicobalt octacarbonyl, such as the hydride, $CoH(CO)_4$.

α-Olefins are hydroformylated to give both linear (normal, *n*-) and branched chain (iso, *i*-) aldehydes (Equation 2)

$$R\diagup\!\!\!\!\diagdown + CO + H_2 \longrightarrow R\diagup\!\!\!\diagdown\!\!\!\diagup CHO + \underset{R}{\diagup}\overset{CHO}{\diagdown} \tag{2}$$

The process was rapidly commercialized by BASF and has been developed around the world especially to make C_4 oxygenates (butyraldehde and butanol) from propylene. The process is very important and many catalyst variations are known.

4.6.1 Manufacture of *n*-Butyraldehyde and *n*-Butanol

n-Butyraldehyde and *n*-butanol were originally made from acetaldehyde via an aldol condensation to crotonaldehyde (MeCH=CHCHO) followed by hydrogenation. The many uses for the C_4 chemicals have sparked the development of the metal catalyzed hydroformylation of propylene.

The desired product is normally the *n*-butyraldehyde, which is generally favoured if a metal catalyst bearing a very bulky ligand is used. The first catalysts were cobalt salts which generated cobalt carbonyls under the reaction conditions (syngas, $CO + H_2$, 130–175°C, 250 bar). Rhodium complexes bound to specific organic ligands, frequently triphenylphosphine, such as the Wilkinson hydride complex, $Rh(H)(PPh_3)_3(CO)$, are now used for propylene hydroformylation, as these catalysts allow operation under much milder conditions (~ 100°C, 10–20 bar). However some modified cobalt catalysts are preferred for longer chain 1-alkenes.

The largest volume hydroformylation reaction converts propylene into *n*-butyraldehyde, from which is made 1-butanol for solvents, or 2-ethylhexanol (the phthalate ester of which has been widely used as a plasticizer for PVC) via an aldol condensation. Estimated world production of butanol is approaching 2 Mt/a.

Higher α-olefins react similarly, and reactions of industrial interest that are of increasing importance include the hydroformylation of 1-hexene to heptaldehyde, of 1-octene to nonaldehyde (pelargonic aldehyde) and of decene to the C_{11} aldehyde, in each case the target being detergent alcohols. This development is driven by the increasing availability of the appropriate α-olefins either from Fischer-Tropsch product mixtures (Section 4.7.2) or from the catalyzed oligomerization of ethylene (Section 5.5). Internal olefins can also be hydroformylated but such reactions have been less exploited commercially, as double bond isomerization there competes strongly.

4.6.2 "Unmodified" Cobalt Catalysts

Since the desired product from propylene hydroformylation is *n*-butyraldehyde, considerable attention has been devoted to increasing the selectivity; this focussed attention on the mechanism, especially the step where the propylene inserts into the Co-H bond, since this can be either Markovnikov or anti-Markovnikov.

Simplified general cycles for 1-alkene hydroformylation, based on $CoH(CO)_4$ acting as the pre-catalyst, and taking account of the observed kinetics are shown in Figures 11 and 12.

$$d[\text{aldehyde}]/dt = k_{\text{obs}}[\text{alkene}][H_2][Co][CO]^{-1} \tag{36}$$

The rate (above a minimum threshold of CO pressure) is given by Equation 36, where the inverse dependence on CO pressure suggests a key step involving CO dissociation from the catalyst. Loss of CO from $CoH(CO)_4$ frees a coordination site for binding alkene (see Figure 11). If a 1:1 ratio of H_2/CO is maintained, the rate will be essentially independent of total pressure, since the rate is proportional to pH_2 but inversely proportional to pCO. However, a certain minimum CO partial pressure is required to maintain the stability of

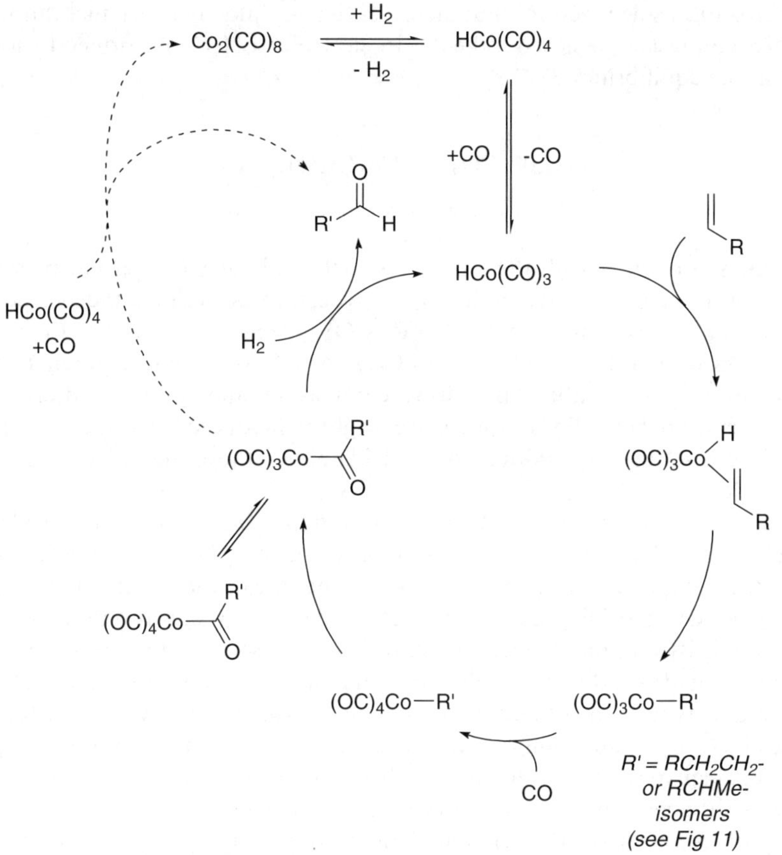

Figure 11 *Scheme representing reactions proposed to occur during the unpromoted-cobalt catalyzed hydroformylation of α-olefins.*

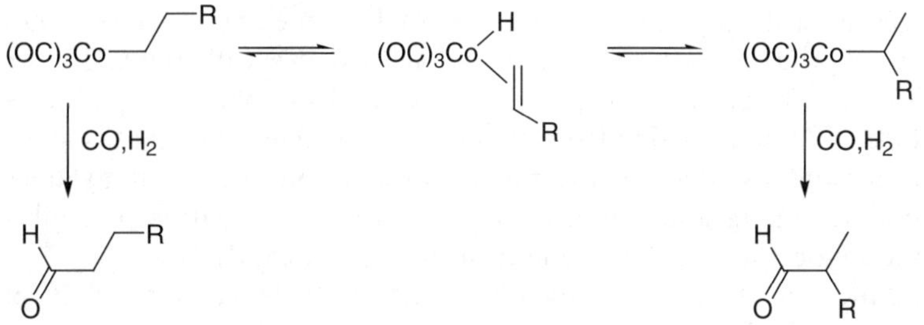

Figure 12 *Isomerization of cobalt alkyls leading to linear and branched hydroformylation products.*

$CoH(CO)_4$ which decomposes to cobalt metal at low p_{CO}. Thus, reasonable reaction rates in the range 110–180°C require rather high p_{CO}, and total H_2/CO pressures of 200–300 bar. Higher CO partial pressure *decreases* the hydroformylation reaction rate and also decreases the amount of alkene isomerization side reactions, but increases the *n-:i-* aldehyde product ratio. The "active catalyst" is believed to be the 16-electron $CoH(CO)_3$, formed by loss of CO in the equilibrium 37.

$$CoH(CO)_4 \rightleftharpoons CoH(CO)_3 + CO \qquad (37)$$

This is consistent with the reduced activity at higher CO partial pressures. The lower regioselectivity at lower CO pressure is because the alkene iso-merization is easier for the 16e-$Co(R)(CO)_3$ species as shown in Figure 12. The unsaturated 16-electron $Co(R)(CO)_3$ will have a long enough lifetime at a lower p_{CO} to allow β-hydride elimination and alkene reinsertion to give the branched alkyls, which are slightly favoured thermodynamically. CO insertion and H_2 addition then yields either the branched or the linear aldehyde.

There is a compromise between rate and regioselectivity; thus while the reaction of $CoH(CO)_4$ with 1-alkenes typically gives *n-/i-* aldehyde ratios of 3-4 to 1, the *n-/i-* product ratio is increased but the rate of the reaction is decreased at higher CO partial pressure. Alkene isomerization side reactions are also decreased. By contrast higher temperatures increase the rate, but decrease the linear aldehyde product regioselectivity and increase undesirable side reactions. Some aldehyde hydrogenation to alcohols is usually observed (5–12%), but little alkene hydrogenation to alkane occurs ($\sim 1\%$), particularly at higher p_{CO}. As the aldehydes are usually later hydrogenated to alcohols in industry alde-hyde hydrogenation is not an unwelcome side reaction.

Exxon uses its hydroformylation plant in Louisiana to produce some 400 kt/a of a mixture of C_7 to C_{13} aldehydes and alcohols from a feedstock based on propylene dimerization/oligomerization to a mixture of C_6 to C_{12} branched internal alkenes. $CoH(CO)_4$ is also a powerful isomerization catalyst: thus the catalyzed hydroformylation of a mixture of 1-octene and 4-octene (150°C, 200 bar; 1:1 H_2:CO) gives a mixture of nonylaldehydes with the CHO group in positions 1- (50–60%); 2- (22%); 3- (7%) and 4- (6%). The rather similar distributions found starting from either 1- or 4-octene show that isomerization does occur and also that the linear hydroformylation to give *n*-nonylaldehyde is favoured. Labelling studies indicated that alkene isomerization generally occurs without dissociation of the alkene from the cobalt catalyst.

In addition to the hydroformylation reactions, side reactions of the product alcohols and aldehydes occur to form *heavy ends*, particularly at higher reaction temperatures, and usually account for $\sim 9\%$ of the product distribution. Industrial reactors usually start using high boiling solvents, but after a while these heavy ends become the solvents.

To recover the catalyst, BASF oxidizes $CoH(CO)_4$ with oxygen to form water soluble Co^{2+} salts that are extracted and reduced under syngas to reform $CoH(CO)_4$, while Exxon deprotonates $CoH(CO)_4$ with aqueous NaOH to make $Na[Co(CO)_4]$.

4.6.3 Phosphine-Modified Cobalt Catalysts

In the 1970s Shell developed a process in which trialkylphosphine ligands were added to the cobalt carbonyl catalyzed hydroformylation reactions. The pre-catalyst was now a phosphine cobalt carbonyl in equilibrium with the hydride $CoH(CO)_3(PR_3)$. This caused major changes in the rate and regioselectivity of the reaction; for example, since the Co-CO bonding in $CoH(CO)_3(PR_3)$ is stronger than in $CoH(CO)_4$, a much lower CO partial pressure is required to stabilize the catalyst and prevent formation of Co metal. Thus the reaction with $CoH(CO)_3(PR_3)$ only needs 50–100 bar of pressure, and can be run at higher temperatures without significant decomposition to metallic cobalt.

The electron-donating phosphine also increases the hydridic nature of the Co–H bond ($CoH(CO)_4$ is quite acidic) and increases the hydrogenation ability of the $CoH(CO)_3(PR_3)$ catalyst. Thus the aldehydes produced are hydrogenated to make alcohols *in situ*. Less electron-rich phosphines, such as PPh_3, give less hydrogenation to the alcohol, and lower linear regioselectivities. The better hydrogenation ability, however, also results in increased alkene hydrogenation side-reactions producing alkanes; which can account for 10–20% of the product. Because of the aldehyde hydrogenation step more H_2 is needed, so H_2:CO ratios of 2:1 (or slightly higher) are typically used.

Another, largely electronic, effect of phosphine substitution is that the higher stability of the $CoH(CO)_3(PR_3)$ catalyst, arising from stronger Co–CO bonding, means that this catalyst is about 5–10 times less active than $CoH(CO)_4$. Just as with the unmodified cobalt catalyst, CO dissociation from the saturated 18 e-species is needed to open up an empty coordination site on the cobalt to allow coordination of alkene and H_2. However this now requires higher reaction temperatures together with longer reaction times and larger reactor volumes. A schematic representation of the trialkylphosphine promoted olefin hydroformylation is in Figure 13. The lower part of the scheme illustrates the hydroformylation itself (together with some isomerization processes) while the upper part of the scheme illustrates the cycle in which aldehyde is hydrogenated to the alcohol. The key intermediate common to both cycles is the coordinatively unsaturated 16-electron $CoH(PR_3)(CO)_2$.

Simple trialkylphosphines do not appear to be effective but patents have been issued to Shell [R. F. Mason, J. L. van Winkle, US Patent 3 400 163; J. L. van Winkle, R. C. Morris and R. F. Mason US Patent 3 420 898], describing bicyclic phosphine ligands with hydrocarbon tails, such as 9-substituted-9-phosphabicyclononanes (commonly known as *phobanes*), which have been successfully used in these reactions. Regioselectivities of 7-8:1 for $CoH(CO)_3(PR_3)$, by comparison with *n-/i-* ratios of 2-3:1 found for $CoH(CO)_4$, are obtained. It is found that the selectivity is optimized at a PR_3 cone angle of about 132°, after which the bulk of the phosphine ligand does not increase the product linear regioselectivity any further. Phosphine modified cobalt hydroformylation is used by Shell to convert a C_4–C_{20} mixture of linear, internal alkenes (formed in its Shell Higher Olefins Process (SHOP)) into detergent grade alcohols.

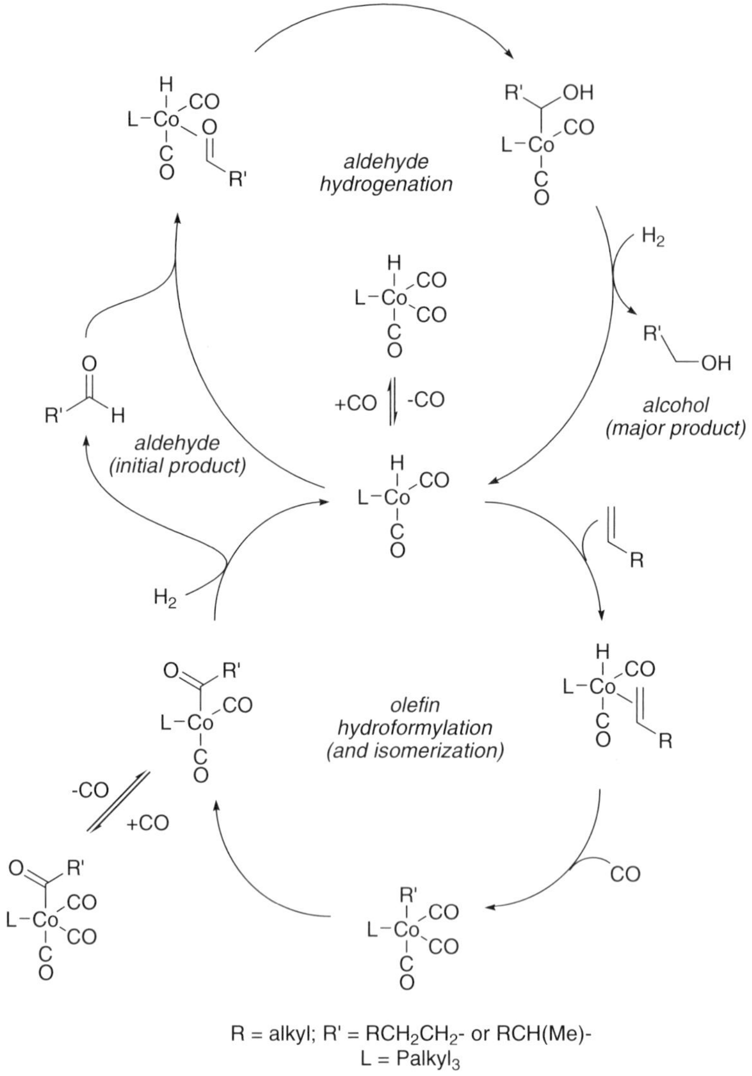

R = alkyl; R' = RCH₂CH₂- or RCH(Me)-
L = Palkyl₃

Figure 13   *Scheme representing reactions proposed to occur during the trialkylphosphine-cobalt catalyzed hydroformylation of α-olefins (R = alkyl; R' = RCH₂CH₂–or RCH(Me)–).*

4.6.4   Rhodium-Catalyzed Hydroformylation

The move from a cobalt- to a rhodium-based process that was seen in methanol carbonylation (Section 4.2.4) is echoed in hydroformylation, thus the late 1960s saw the development of rhodium catalysts here too. Wilkinson and his colleagues found that RhH(CO)(PPh₃)₂ was an outstanding catalyst as it was very selective to aldehyde products (no alcohol formation, no alkene hydrogenation or iso-merization occurred) and that very high *n-/i-* aldehyde selectivities of 20 : 1 for a variety of 1-alkenes could be obtained under mild conditions (25°C, 1 bar; 1:1 H_2/CO). The rate increased at higher temperatures but the regioselectivity dropped (to 9:1 at 50°C); the regioselectivity also decreased when the reaction was run at higher pressure (3:1 at 80–100 bar of $H_2 + CO$). As was found for the methanol carbonylation, the milder conditions and the better selectivity more than made up for the cost of using the more expensive rhodium-based catalyst.

Currently some three-quarters of all industrial hydroformylation processes are based on rhodium triarylphosphine catalysts, especially for the lower alkenes where high regioselectivity to linear aldehydes is critical. The initial catalyst system was derived from Wilkinson's hydrogenation catalyst, $RhCl(PPh_3)_3$, but it was rapidly discovered that halides inhibited hydroformylation. Commonly used precatalysts include $Rh(acac)(CO)_2$ which, under process conditions and in the presence of excess PPh_3, gives $RhH(CO)(PPh_3)_3$ *in situ*. The active species is the 4-coordinate $RhH(CO)(PPh_3)_2$.

The technology was then developed, in the early 1970s, at Union Carbide in conjunction with Davy Powergas and Johnson Matthey to give the first commercial low pressure hydroformylation process using rhodium and excess PPh_3. The excess phosphine is needed because dissociation of PPh_3 results in conversion of $RhH(CO)(PPh_3)_3$ into a series of complexes, $RhH(CO)_x(PPh_3)_{4-x}$ (Equations 38 and 39) which are more active, but less selective pre-catalysts for hydroformylation.

$$RhH(CO)(PPh_3)_3 + CO \rightleftharpoons RhH(CO)_2(PPh_3)_2 + PPh_3 \qquad (38)$$

$$RhH(CO)_2(PPh_3)_2 + CO \rightleftharpoons RhH(CO)_3(PPh_3) + PPh_3 \qquad (39)$$

The excess phosphine ligand shifts the equilibria back towards the more selective $RhH(CO)(PPh_3)_3$ pre-catalyst. The regioselectivity of Rh/PPh_3 catalyzed hydroformylation is related to both $[PPh_3]$ and the H_2/CO ratio used. If propylene hydroformylation is run in molten PPh_3 a 16:1 *n-/i-* aldehyde ratio is obtained. Commercially, propylene is hydroformylated with a PPh_3 concentration of about 0.4 M and [Rh] of about 1 mM to give a *n-/i-* selectivity of $\sim$8-9:1. Higher CO partial pressures lower the product regioselectivity, in contrast to the situation for a $CoH(CO)_4$-catalyst.

The currently accepted mechanism for Rh/PPh_3 catalyzed hydroformylation is shown in Figure 14. Again the first step is loss of a ligand from the starting pre-catalyst ($RhH(CO)(PPh_3)_3$) and its replacement by the olefin; hydride migration then occurs to give either the *n-* or the *i-*alkyl. This then takes up CO and undergoes migratory insertion to give a Rh-acyl, which then reacts with hydrogen to regenerate $RhH(CO)(PPh_3)_2$ and releases the aldehyde product. While the basic form of the cycle is generally agreed, there is still dispute about the rate determining step, and it has been suggested that several of the fundamental steps have similar rate constants, making it difficult to specify one overall rate determining step. The presence of excess PPh_3 ligand also minimizes ligand fragmentation reactions such as Ph-P cleavage that lead to catalyst deactivation.

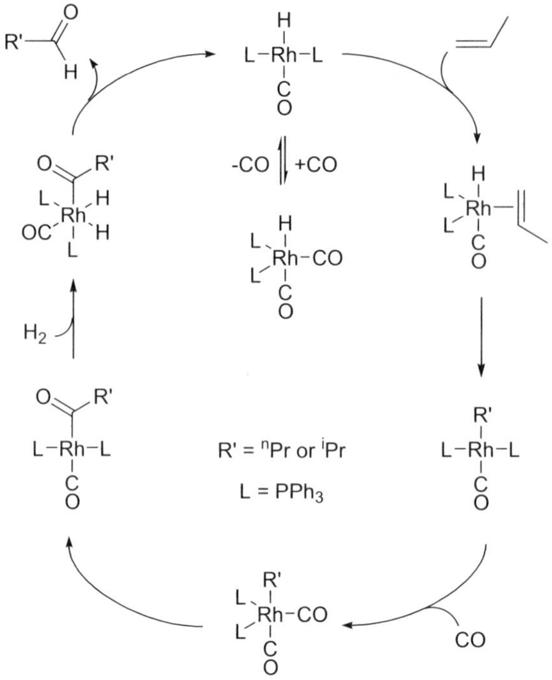

Figure 14   *Representation of the chief steps during the Rh/PPh$_3$ catalyzed hydroformylation of propylene.*

4.6.5   Two-Phase (Water-Soluble) Rhodium Hydroformylation Catalysts

One important variant of Rh/PPh$_3$ catalysis is the two-phase catalyst system developed by Kuntz at Rhône-Poulenc in 1981, using a sulfonated triphenylphosphine ligand, P(C$_6$H$_4$-*m*-SO$_3$Na)$_3$ (TPPTS) to generate the water soluble catalyst: RhH(CO)[P(C$_6$H$_4$-*m*-SO$_3$Na)$_3$]$_3$.

  Since the catalyst has a very high (9+) formal charge it is totally insoluble in all but the most polar solvents. The resultant two-phase catalyst system has the advantage over the completely homogeneous system in that all the organic product, butyraldehyde, is essentially in the organic phase and can easily be separated. Similarly, recovery of the catalyst is straightforward as it all stays in the aqueous phase. An excess of the phosphine ligand is required for good *n-/i-* selectivities, as with conventional Rh/PPh$_3$ catalysts, but lower concentrations are required because the TPPTS phosphine dissociation equilibrium in water is shifted towards the Rh-TPPTS coordinated complexes. Shorter chain alkenes (C$_2$–C$_4$) are water soluble enough that sufficient migrates into the aqueous catalyst phase to allow hydroformylation. Rather high linear to branched regioselectivities of 16–18:1 for propylene can be obtained using this water soluble catalyst, but rates are slower than with conventional Rh/PPh$_3$ catalysts due to lower alkene concentrations in the water phase. Alkenes higher than

1-pentene are not soluble enough in water, and the process is limited to shorter chain alkenes that have some degree of water solubility. Celanese-Ruhrchemie currently operates several hydroformylation plants based on this water soluble rhodium catalyst technology. [C. W. Kohlpaintner, R. W. Fischer, B. Cornils, *Applied Catalysis A: General* 2001, **221**, 219.]

4.6.6 Rhodium Hydroformylation Catalysts with Bidentate Ligands

Several research groups, notably at Eastman and Union Carbide, have developed catalysts containing chelating bis-phosphine or bis-phosphite rhodium compounds that show remarkably high product regioselectivities and good activity. Some of the best ligands form 9-membered chelate rings with the Rh centre, and catalysts based on these ligands are highly regioselective, giving *n-/i-* aldehyde product ratios of $> 30:1$ for propylene with rates about twice that of Rh/PPh_3 for the diphosphines and even higher for the diphosphites. As the phosphite ligands are poorly σ-donating they dissociate readily and make the rhodium centre extremely active; such catalysts are also good enough isomerization catalysts that they can even hydroformylate some internal alkenes (such as 2-butene) and could form the basis of a new generation hydroformylation technology.

4.6.7 Enantioselective Hydroformylation

Although there has been a great deal of research activity to find good hydroformylation procedures to make aldehydes and alcohols enantioselectively, none seems as yet to have been commercialized. Nevertheless the potential is so high that it will certainly not be long before this methodology is applied to problems in the pharmaceutical and agrochemical industry. Much of the interest has focussed on Pt(II) complexes containing chiral diphosphines, which have been used elsewhere for enantioselective hydrogenations. However, it has been suggested that as the coordination requirements for the critical hydroformylation intermediates are likely to be 5-coordinate they may well be different from those for hydrogenation. Rhodium complexes based on very bulky chiral diphosphites, for example some derived from homochiral (2R,4R) pentane-2,4-diol, *L*-diethyl tartrate, or 1,2:5,5-di-isopropylidene-*D*-mannitol give good *ee's*, as do complexes derived from the mixed phosphine-phosphite BINAPHOS. For example, the complex $Rh(H)(CO)_2(R,S\text{-BINAPHOS})$ can hydroformylate styrene to an 88:12 mixture of $PhCHMe(CHO)$ and $PhCH_2CH_2CHO$ at 99% conversion, the former having an ee of 94%, in 43h at 60°C with $CO + H_2$ (100 bar; 1:1). [A comprehensive discussion of hydroformylation and related reactions is by P. W. N. M. van Leeuwen, *Homogeneous Catalysis*, Kluwer, Academic Publishers, Dordrecht, 2004.]

Discussion Point DP5*: Methanol carbonylation processes have used all three of the group 9 metals, whereas only cobalt and rhodium based hydroformylation processes are significant. Is there any potential for iridium as a useful hydroformylation catalyst?*

Discussion Point DP6: *High pressure, in situ spectroscopy has played an important role in identifying the complexes present during homogeneous catalytic carbonylations. What are the merits and drawbacks of a) IR and b) NMR spectroscopy for probing the mechanisms of these processes?*

Discussion Point DP7: *Select a chiral pharmaceutical or agrochemical which could usefully be synthesized using an enantioselective hydroformylation procedure, and show how this might be accomplished.*

4.7 CO Hydrogenation

In addition to the many reactions in which CO or syngas is reacted with an alkene, an alcohol or other oxygenate to give new products of substantially higher value, there are also two major processes in which CO itself is hydrogenated directly. One is the synthesis of methanol, the other the Fischer-Tropsch reaction to make linear hydrocarbons, especially 1-alkenes. In the first both the C and the O from carbon monoxide are retained, while in the Fischer-Tropsch reaction only the carbon is retained in the product, the oxygen being used to form water, which can be regarded as driving the reaction thermodynamically (See Box 7). The two processes occur under rather different conditions and over different heterogeneous catalysts and there is usually only a small overlap in the products. Key to both these reactions is syngas ($CO + H_2$) Section 4.1.2. Much research time has been expended in trying to develop homogeneously catalyzed in solution versions of methanol synthesis or the Fischer-Tropsch reaction but no commercial process has yet emerged (see Annex 1.3).

Box 7 Thermodynamics of CO Hydrogenation

Thermodynamics as well as kinetics play key roles in determining which of the various possible CO hydrogenation reactions take place. Table 1 shows that the formation of water plus hydrocarbons is very favourable, showing negative $\Delta G^\circ_{(500K)}$. Oxygenate formation where no water is co-produced is unfavourable to varying degrees, thus the formation of methanol, formaldehyde, and ethylene glycol all show positive $\Delta G^\circ_{(500K)}$. Only when the oxygenate has a reasonable hydrocarbon tail and water is again formed, as with ethanol, do the thermodynamics improve again.

Table 1

reaction	$\Delta G^\circ_{(500K)}$, kJ/mol
$3H_2 + 1\ CO = H_2O + CH_4$	−94
$2H_2 + 1\ CO = H_2O + 1/3\ (C_3H_6)$	−31
$2H_2 + 1\ CO = CH_3OH$	+21
$H_2 + 1\ CO = HCHO$	+51
$3H_2 + 2\ CO = HOCH_2CH_2OH$	+66
$4H_2 + 2CO = C_2H_5OH + H_2O$	−27

4.7.1 Methanol Synthesis

Methanol is very important both as a product and as a feedstock in chemical industry; the total world capacity is currently over 30 Mt/a and is rising at about 3% p.a. Syngas is now the only realistic feedstock for making methanol. Major plants to make it from very cheap raw materials have been built in Trinidad, Saudi Arabia and elsewhere; these use methane from oil wells, which was previously flared to waste, but can easily be converted into syngas. Such methanol plants use what is termed "stranded gas" which is natural gas in a remote area where it cannot be economically used for any other purpose. Very large "mega-scale" methanol plants $\geq$ 1.5 Mt/a are now built. For example in Trinidad, which is now the world's largest exporter of methanol, with a total production capacity of 6.5 Mt/a.

The major uses of methanol are to make formaldehyde (ca. 30%); the fuel additive methyl *t*-butyl ether (MTBE, ca. 30%, though this may now be phased out on environmental grounds); acetic acid (ca. 10%); acetate esters as solvents, methyl methacrylate, chloromethane (to make silicones) and dimethyl terephthalate (DMT, fibres) (ca. 30%).

Commercially, methanol is produced from the hydrogenation of CO (syngas) over heterogeneous CuZn oxide based catalysts using fixed bed reactors (Equation 11).

$$CO + 2H_2 \rightarrow CH_3OH \ (\Delta H_{298K} \ -91 \ kJ/mol; \ \Delta G_{298K} \ -25 \ kJ/mol) \quad (11)$$

Other reactions that play a significant role include the hydrogenation of carbon dioxide (Equation 40),

$$CO_2 + 3H_2 \rightarrow CH_3OH + H_2O$$
$$(\Delta H_{298K} \ -49 \ kJ/mol; \ \Delta G_{298K} \ 3 \ kJ/mol) \quad (40)$$

and the WGSR (Section 4.1.3, Equation 7),

$$CO + H_2O \rightleftharpoons CO_2 + H_2 \quad (7)$$

The first industrial methanol "high pressure" process, operating at 340 bar and 320–380°C, was developed by BASF in the 1920s. It remained the dominant technology for over 45 years. Since the synthesis gas employed at that time was based on low grade German coal (lignite, which contains chlorine and sulfur impurities), a relatively poison resistant zinc- and chromium-oxide based catalyst (ZnO/Cr_2O_3) was developed. In fact the thermodynamic equilibrium is most favourable at high pressure and low temperature and considerable efforts were made to improve the process. This came about in the 1960s when ICI developed the use of a copper catalyst, and found ZnO to be a perfect dispersant for the copper. The new catalyst enhanced the reactivity, allowed milder operating conditions, and facilitated the development of the "low pressure" process, over a Cu–ZnO–Al_2O_3 catalyst (50–100 bar, 240–260°C) which is today the main process used to make methanol. Just as in the other large scale heterogeneous catalytic reactions, the engineering is critical, especially the mode by which the heat of reaction is removed.

The form of the catalyst, its morphology, and its method of preparation vary widely and this is reflected in its catalytic properties. The actual mechanism of the CO hydrogenation is complicated and still the subject of debate, and the roles played by each of the metals in the catalysis is not agreed. However it appears that the copper is responsible for the CO coordination and activation, while the heterolytic splitting of H_2 into H^+ and H^- occurs on the zinc oxide; most important is the synergy between the Zn and Cu. The alumina plays several roles as promoter.

One cycle that has been proposed involves the intermediacy of surface formate formed from surface OH and CO (Equations 41a–c)

$$OH_{(surf)} + CO \rightarrow HCOO_{(surf)} \tag{41a}$$

$$HCOO_{(surf)} + 2H_{(surf)} \rightarrow CH_3O_{(surf)} + O_{(surf)} \tag{41b}$$

$$CH_3O_{(surf)} + H_2O \rightarrow CH_3OH + OH_{(surf)} \tag{41c}$$

It has also been found that mixtures of CO and CO_2 undergo hydrogenation faster than either component individually. The role of CO_2 here is not clear but some studies have concluded that the actual CO hydrogenation catalytic cycle may involve carbon *di*oxide which is first converted into formate on the catalyst surface before reduction to surface methoxy and methanol (Equations 42a–c).

$$CO_2 + {}^{1}\!/{}_{2}H_2 \rightarrow HCOO_{(surf)} \tag{42a}$$

$$HCOO_{(surf)} + 2H_{(surf)} \rightarrow CH_3O_{(surf)} + O_{(surf)} \tag{42b}$$

$$CH_3O_{(surf)} + H_{(surf)} \rightarrow CH_3OH \tag{42c}$$

There is great interest in using carbon dioxide as the raw material for methanol synthesis. Unfortunately the catalysts that are effective for CO rich feeds are not as effective for CO_2 rich feeds. Thus, much effort is going into the search for an efficient catalyst for the methanol synthesis reaction from CO_2 rich feeds. ZrO_2 containing catalysts have shown promise but the technology still seems confined to the laboratory.

An interesting variation on the methanol formation is that in some cases higher oxygenates can be formed (e.g., ethanol, acetic acid or isobutanol), over mixed oxides (such as $ZrO_2/ZnO/MnO/K_2O/Pd$) or promoted copper catalysts. These are probably secondary products derived from methanol and formate by more standard organic reactions.

4.7.2 Hydrocarbons from the Hydrogenation of CO: the Fischer-Tropsch (F-T) Reaction

Methanol is one important commodity chemical obtained from the heterogeneously catalyzed hydrogenation of CO. Changing conditions and catalyst changes the product distribution to make a quite different set of very useful chemicals: linear hydrocarbons, especially 1-alkenes (together with internal alkenes, methane and other alkanes). This reaction, named after Fischer and Tropsch who discovered it in 1923, is again attracting widespread attention for its commercial potential as well as for the engineering and the chemistry involved.

As previously described (Section 4.1.1) the first CO hydrogenation reaction was the nickel catalyzed methanation. However when other transition metal catalysts were used (notably Fe, Co or Ru) carbon-carbon bond forming reactions occurred and the products were composed of longer chain alkenes (Equation 43) and alkanes (Equation 44); methane was always a major component, especially when higher temperatures were used. The thermodynamics strongly favour hydrocarbon formation (see Box 7) since water is the other product of the Fischer-Tropsch (F-T) reaction. By contrast, direct oxygenate formation is more difficult and only becomes thermodynamically favoured for longer chain oxygenates. Labelling studies have indicated that alcohols are formed in quite separate processes, and that they do not arise from the hydrocarbons or vice-versa.

$$nCO + 2nH_2 \rightarrow CH_3(CH_2)_{n-3}CH{=}CH_2 + nH_2O \qquad (43)$$

$$nCO + (2n + 1)H_2 \rightarrow CH_3(CH_2)_{n-2}CH_3 + nH_2O \qquad (44)$$

The original F–T reaction was an iron- or cobalt-catalyzed hydrogenation of CO and these metals, with appropriate promoters, have remained the catalysts of choice. Ruthenium, although more active than Fe or Co, does not appear to be used in F–T plants due to its higher cost. Rhodium is another metal with F–T potential, especially towards oxygenate formation, but the activity is low and the cost is again high. Although there are significant differences in detail between the activities of the four metals (for example Co promotes hydrogenation more easily, and hence can give more alkanes than alkenes), the primary product distributions can be remarkably similar, and indicate at least some mechanistic similarities.

When the mass fraction of the long-chain hydrocarbon products of the F–T synthesis (W) is plotted against the carbon number (N) it is found that W decreases approximately monotonically with molecular size. Thus the major product is the C_1, methane, followed by the C_2 hydrocarbons (ethylene and ethane), the C_3 hydrocarbons, and so forth, as shown in Figure 15. This distribution follows Schultz-Flory statistics for a polymerization involving the sequential addition of C_1 units to a chain, given by the dotted line in Figure 15. Further and more detailed consideration of the mechanisms is in Annex 1.

In practice the primary F–T hydrocarbon products then go on to react further in secondary processes. These include isomerization (both by H-migration and skeletal isomerization), hydrogenation and dehydrogenation, and aromatization reactions, as well as in the incorporation of CO, of other oxygenates, and even nitrogenous compounds. A typical F–T product

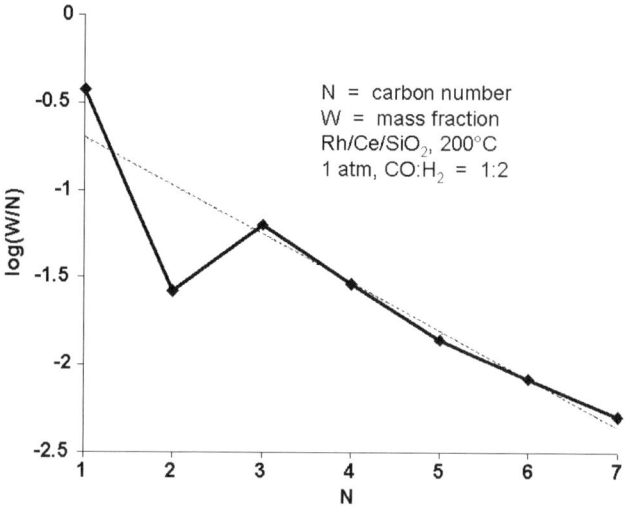

Figure 15   *Anderson-Schulz-Flory plot for CO hydrogenation (dashed line theoretical plot for polymerization of C_1 species; solid line, experimentally observed values. See M. L. Turner, H. C. Long, A. Shenton, P. K. Byers, P. M. Maitlis, Chem. Eur. J., 1995, **1**, 8).*

distribution is therefore extremely complex and can include various types of alkenes, alkanes, aromatics, and oxygenates (such as alcohols, esters and ketones). However for the purposes of the present discussion we will only consider the primary products here.

F–T is a major part of the processes known collectively as *gas-to-liquids* (GTL) or *coal-to-liquids* (CTL) technology. GTL allows the conversion of natural gas (chiefly methane) *via* reforming to syngas, into liquid hydrocarbons for use as automobile fuels. Since the F–T hydrocarbons are largely linear they are well suited as diesel fuel (high cetane number). By contrast, petrol engines need branched chain (high octane number) fuels. Other feedstocks that can be turned into syngas include coal, which is used in South Africa and will also be used in the CTL Chinese F–T plants. Technology is also now being developed to allow the conversion of biomass into syngas, and this should prove an important way forward in the future since biomass, a readily renewable material, could become a sustainable feedstock, which none of the others are.

Very significantly, so far as we are concerned here, the α-olefins produced by F–T reactions are being used as feedstocks for further elaboration in the petrochemicals industry. An example is the hydroformylation of α-olefins to aldehydes and alcohols. F–T technology is now seen as a simple way of utilizing stranded reserves of natural gas and a valuable means for turning natural gas into commodity chemicals. Since the syngas is relatively easily purified from contaminants such as aromatics and sulfur- and nitrogen- containing compounds, F–T products derived from it are very high quality feedstocks for the chemicals industry; the purity of the products also offer more environmentally acceptable liquid hydrocarbon fuels such as ultra-low sulfur diesel.

CTL F–T technology was first used in Germany during World War 2 to turn lignite (poor quality coal) into motor fuel. Although that work ended with the collapse of the German economy in 1945, the information on the details of the F–T processes proved invaluable to industries in the US, UK and other countries, and allowed them to extend the technology. With the ready availability of oil supplies over the following decades further development of F–T technology slowed down but South Africa, which was subjected to an oil embargo during the 1970s and later, developed an indigenous CTL F–T industry (Sasol) to supplement its fuel supplies. Large GTL F–T plants are now being operated in South Africa by Sasol, and are being built (or are in advance planning, in 2005) by major multinationals, in Malaysia, Qatar, Nigeria, Egypt, Indonesia, China, and Australia. It has been estimated that F–T technology becomes competitive when the oil price is significantly above *ca.* US $25/barrel.

Prewar (1938) German production of F–T fuel was around 660 kt/a; a trial plant built in Brownsville Texas in the 1950s produced 360 kt/a, and Sasol production in South Africa is currently (2005) around 7 Mt/a. Shell has a plant in Malaysia with a capacity of 500 kt/a to make waxes (long chain alkanes) by F–T from syngas using a hydrogenating Co catalyst.

The F–T GTL or CTL processes generate complex mixtures, some of which are large molecules. Just as for crude oil, these need to be broken down to be useful either as fuels or as building blocks for the chemicals industry. Thus the final stage in GTL or CTL technology is a series of refining operations very similar to those used in conventional petroleum upgrading.

4.7.3 Fischer-Tropsch Technology

Much of the more recent development of F–T technology has come from the work of scientists at Sasol, but with important contributions from workers at Shell and others.

There are two modes of commercial operation, the Low Temperature Fischer-Tropsch process (LTFT) which aims to make high molecular weight hydrocarbon waxes and the high temperature (HTFT) process which mainly produces alkenes and gasoline. Currently the most efficient reactor type for LTFT is the slurry reactor in which syngas is passed up through a slurry consisting of catalyst particles suspended in molten F–T waxes. The operating temperature depends on whether the catalyst is cobalt or iron based, but is usually below 250°C in order to minimize unwanted methane production and to maximize wax selectivity. A lower temperature operation also minimizes wax hydrocracking.

The HTFT process operates with an iron based catalyst at about 350°C with the syngas passing through a fluidized bed of finely divided catalyst. Low temperatures cannot be used as the two phase system (gas and catalyst) would become "defluidized" by the formation of liquid waxes. At the higher temperature of the HTFT process, the catalyst is much more active than it is in the LTFT process and the hydrocarbon production rate is much higher.

The hydrocarbons produced in the FT process are predominantly linear, and the alkenes are predominantly 1-alkenes. The high 1-alkene content makes it advantageous to use F–T hydrocarbons as feedstocks for the production of higher value added chemicals.

An interesting variation on this is practised by Shell. In recent years, probably also helped by the sharp rise in the world oil price, there has been renewed interest in the F–T route to high quality diesel fuel. The Shell F–T plant in Malaysia which came on stream in 1993 uses multitubular fixed bed reactors to produce long chain hydrocarbon waxes by a LTFT process (over a promoted cobalt catalyst). These waxes are then selectively cracked over zeolites to give the desired shorter chain molecules suitable as diesel fuel. Since F–T hydrocarbons are predominantly linear they are not suitable for petrol engines, but are ideal for diesel.

The diesel produced in this way has several good features: it is biodegradable, and has a cetane number of 74 as against 40 for conventional US diesel fuel. The aromatic content is about 2 mass% as against about 32% in US diesel fuel. As well as having low levels of aromatics, F–T fuels have extremely low levels of sulfur, making them very attractive since they comply with stringent new regulations on sulfur levels laid down by the EC and the state of California, without needing the difficult and expensive deep HDS treatment.

Annex 1 Concerning the Mechanism of the Fischer-Tropsch Reaction

Annex 1.1 How are 1-alkenes formed from syngas?

As we mentioned in Section 4.1.2 and in Chapter 1, the ready availability of syngas and the increasingly short supply, or difficulty of access, of other carbon-based feedstocks, implies that syngas based reactions will increase in significance in the chemical industry. Amongst the most important is the Fischer-Tropsch (F–T) process, and there has therefore been considerable interest in its mechanism.

F–T technology hydrogenates CO over promoted transition metal catalysts into a mixture of organic products including alkenes, alkanes and methane. Some aromatics and oxygenates (including alcohols, aldehydes, ketones and acids) are also formed. Careful analysis of the reaction products indicates that the alkanes and internal olefins as well as the aromatics and some of the oxygenates are largely products of secondary reactions undergone by the 1-alkenes. It is now generally agreed that the primary F–T reaction involves the formation of 1-alkenes (under largely kinetic control) and that the other products are derived from them in a wide variety of subsequent reactions, some of which are catalyzed by the metals and/or the supports.

Interestingly, the free energies (Box 7) and the enthalpies of formation of the products of CO hydrogenation are also reflected in the rates (kinetics) of the various reactions; thus hydrocarbons are formed relatively easily. While there is still some controversy about the precise mechanism, over the years a consensus has emerged that the F-T reaction is a polymerization of surface CH_2 species derived from the hydrogenation of adsorbed CO. This is for example shown by the product hydrocarbon distribution where there is a monotonic decrease in yield with increase in molecular size. Plotting $\log(W/N)$ against W (W = mass fraction; N = carbon number) gives a characteristic graph (dotted line, Figure 15) which is typical for polymerization kinetics.

The experimentally obtained Anderson-Schulz-Flory (ASF) distribution (solid line) follows the theoretical values closely and was an early indication that the reaction to form the hydrocarbons was a type of polymerization, and indeed of C_1 species. An interesting feature of the ASF plot is that it is not quite smooth but has a kink at $N = 2$ which comes below the curve (see Figure 15). The reason why substantially less ethane and ethylene than expected is formed has been widely debated: it can occur if fewer free C_2 species are produced or if the C_2 fraction preferentially undergoes further reaction. The former explanation seems to be the more accepted one, in other words the rate at which surface-attached C_2 undergoes further polymerization is faster than the rate of liberation of the free C_2 hydrocarbons from the surface.

Mechanistic studies have mostly focussed on the 1-alkene formation by polymerization of surface CH_2 (methylenes). The formation of the CH_2 species by a deoxygenation/hydrogenation sequence of adsorbed CO is not well understood as there are few convincing organometallic models, but it is usually depicted as shown in Figure 16. The path involves stepwise de-oxygenation to a surface carbide followed by sequential H transfer to make various intermediate $C_1H_{x(ad)}$ (x = 0 – 4) species and finally methane which is liberated.

One interesting and revealing aspect of the more recent studies is the importance of a C_2 intermediate species (which may be, for example, a vinyl, $-CH{=}CH_{2(ad)}$, or vinylidene, ${=}C{=}CH_{2(ad)}$), and which has been confirmed by a number of independent investigations. This is also suggested by the kink at [C_2]

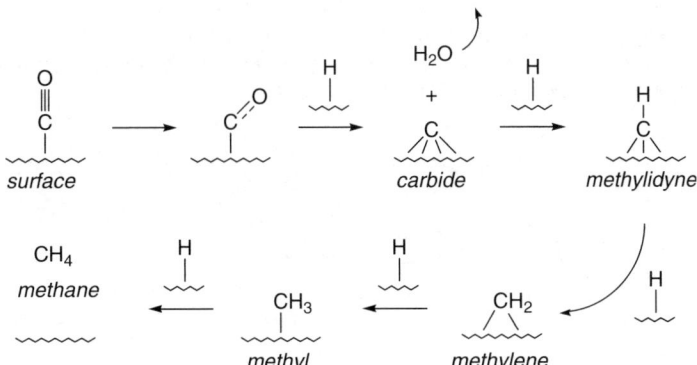

Figure 16   *Scheme representing reactions proposed to occur during the initial stages of the Fischer-Tropsch hydrogenation of CO on a metal surface. Various steps involving deoxygenation and hydrogenation are shown.*

in the ASF plot (solid line, Figure 15). Thus the rate determining (slow) step of the Fischer-Tropsch reaction is not, as might be supposed, the formation of the CH_2 surface species, but of the C_2 species that initiates the subsequent polymerization. There seems substantial agreement that the formation of the deoxygenated $C_{1(ad)}$ species is in fact very fast over metal surfaces. The $C_{2(ad)}$ intermediate could arise from the coupling of a methylene and a methylidyne species in the surface (Figure 17a) or from the dimerization of a surface carbide to a dicarbide ($C_{2(ad)}$), followed by a hydrogenation (Figure 17b). Both of these processes have organometallic analogues.

The subsequent steps (illustrated in Figure 18) have been proposed on the basis of detailed labelling studies and organometallic reactions which model them. The basic reaction is the combination of a C_1 fragment (methylene, $CH_{2(ad)}$) with a surface vinyl (or alkenyl, $RCH=CH_{(ad)}$), to give an allylic species ($RCH=CHCH_{2(ad)}$), which then isomerizes to a new surface alkenyl ($RCH_2CH=CH_{(ad)}$), which can again combine with further methylene in the same manner. Termination of the polymerization occurs by transfer of surface hydride to the alkenyl (reductive elimination) to give the alkene directly.

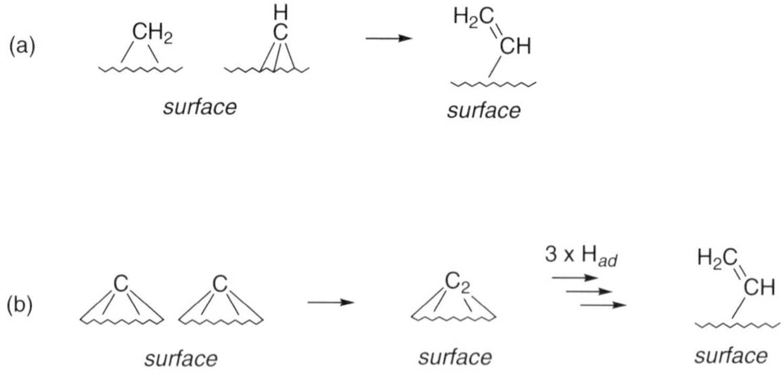

Figure 17   *Scheme representing possible surface reactions that occur to form the surface C_2 species (probably $CH_2=CH_{(ad)}$) that initiate Fischer-Tropsch polymerization: (a) by reaction of a methylidyne and a methylidene; (b) via formation of a dicarbide which is then hydrogenated in the surface.*

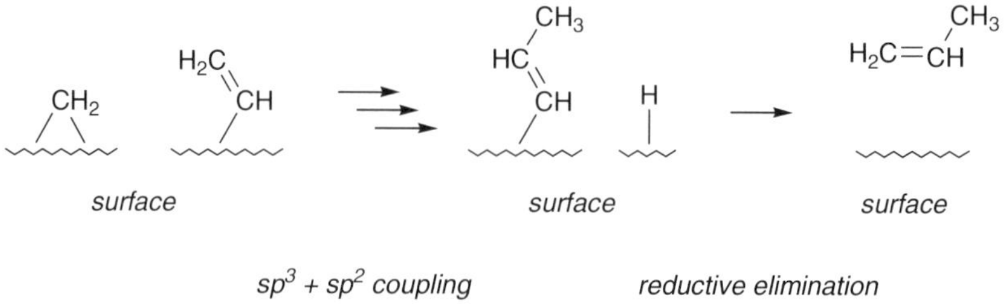

Figure 18   *Scheme representing reactions proposed to occur during the propagation steps of the Fischer-Tropsch hydrogenation of CO to propylene on a metal surface.*

Annex 1.2 Other mechanistic proposals

The F-T reaction has been the subject of innumerable investigations that have given rise to many proposals for the mechanism of 1-alkene formation; unfortunately many attractive ideas have been ruled out by subsequent detailed studies, especially using isotope labelling. Many of the earlier mechanistic suggestions are now mostly seen to predict the wrong products, or to involve steps that theoretical studies indicate are likely to have very high energy barriers.

Among mechanisms which now have to be discounted are the following:

(i) The $C_{1(ad)}$ species is a hydroxymethylene ($CH(OH)_{(ad)}$) which polymerizes by condensation with loss of water giving ($CHC(OH)_{ad}$) which is then hydrogenated and reacts further [This and other earlier ideas were summarized by R. B. Anderson, *The Fischer-Tropsch Synthesis*, Academic Press, New York, 1984, p. 175]. The direct spectroscopic observation of surface species derived from CO such as $CH_{(ad)}$, $CH_{2(ad)}$, etc now makes this unlikely.

(ii) Sequential additions of methylene to an *alkyl* chain in the surface ($CH_{2(ad)} + RCH_{2(ad)} \rightarrow RCH_2CH_{2(ad)} \rightarrow \rightarrow \rightarrow$ alkene), see R. C. Brady, R. Pettit, *J. Am. Chem. Soc.*, 1980, **102**, 6181; 1981, **103**, 1287. In fact, the $Csp^3 - Csp^3$ coupling reactions that would be required are now recognized to be at very high energy and the products do not agree with labelling studies either.

(iii) Paths involving cationic surface vinylidenes have been proposed; however they are expected to give a high proportion of branched chain hydrocarbons, which is contrary to the experimental facts; see E. Hoel, *Organometallics* 1986, **5**, 587.

(iv) Metathesis-type mechanisms involving the direct reaction of terminal methylenes ($M=CH_2$) and an olefin via metallacyclobutanes (see, for example, F. Hughes *et al.*, *Nouveau J. Chimie.*, 1981, **5**, 207). This would link to the metal catalyzed polymerization of ethylene on single site catalysts (see Chapter 7, Section 7.4), and explain the need for a C_2 initiator. However it predicts isobutene to be a major product which it is not; detailed isotope labelling studies also rule this out.

(v) Paths involving surface methylidyne ($CH_{(ad)}$) rather than methylene ($CH_{2(ad)}$) have been suggested by I. M. Ciobica *et al.*, *Stud. Surf. Sci. Catal.*, 2001, **133**, 221, based on the greater stability of $CH_{(ad)}$ over $CH_{2(ad)}$. Chain growth then occurs by sequential addition of first $CH_{(ad)}$, followed by $H_{(ad)}$, to a growing alkyl or alkylidene chain. This is a serious contender mechanism and since methylene and methylidyne are related by a H-transfer process (Equation A1) the active species will be determined by the hydrogen concentration. On a H-covered surface during the F-T reaction it is however more likely to be $>CH_{2(ad)}$ rather than $\equiv CH_{(ad)}$ and therefore a methylidyne intermediate seems less probable.

$$\equiv CH_{(ad)} + H_{(ad)} \rightleftarrows \; >CH_{2(ad)} \quad\quad\quad (A1)$$

Annex 1.3 Homogeneous CO hydrogenation

CO hydrogenation homogeneously processes can also occur in solution, especially catalyzed by Ru or Rh complexes, but is quite different in all respects from the heterogeneously catalysed reactions. While the latter give largely linear hydrocarbons, the homogeneous reactions give mainly C_1 and C_2 oxygenates and virtually no hydrocarbons. Details of researches using various Group 8 and 9 metal catalysts in solution have been collected [G. Braca, *Oxygenates by homologation or CO hydrogenation with metal complexes*, Kluwer Academic Publishers, Dordrecht, 1994]. Solution reactions usually require higher temperatures ($>220°C$) and higher syngas pressures than the heterogeneous ones, as well as certain oxy-solvents (glymes, N-methylpyrrolidone, sulfolane, acetic acid) and iodide or occasionally bromide, as co-promoters. Because of the temperatures and pressures required and the lack of selectivity so far achieved no commercial process has been adopted. However a methanol producing reaction using a nickel carbonyl and base catalyst under milder conditions ($80–120°C$, 10–50 bar) and believed to proceed via the intermediacy of methyl formate has also been described. There may be an analogy to the formation of methanol on a surface, for example, Equation 41a–c.

Discussion Point DP8: *While the formation of long chain alcohols is thermodynamically favourable, a reaction in which syngas is converted into a long-chain alcohol in a single step is not easily realised. Thus a more realistic plan is to carry out the conversion in two stages, benefitting from the favourable thermodynamics of forming the alkyl chain and then coupling it to an oxyfunction. Plan a sequence along these lines to make 1-heptanol from syngas.*

Annex 2 Some Hints for Discussion Points

Our understanding of the mechanisms underlying much of the chemistry developed in this book is still at quite an early stage and there are often no "right" answers. The Discussion Points we propose are therefore deliberately rather unstructured and are intended to encourage you, the reader, in thinking as widely as possible. We hope you will use the problems to research the original literature and increase your understanding of metal-catalyzed reactions and of current practice in the chemical industry. Do talk to your colleagues and your teachers and involve them in the discussions. Use as starting source the references cited here; if you do not have easy access to books or review papers, try using Google on the web; even quite an unsophisticated entry such as "hydroformylation propylene", will yield some useful leads.

Below we offer some further hints on the Discussion Points to start you off.

Discussion Point DP1 hint You may wish to consider the conditions under which the reactions are conducted, and what one might do to facilitate a reaction at lower temperature, for example.

Discussion Point DP2 hint Instead of water, what other solvent(s) could you suggest for the reaction?

Discussion Point DP3 hint A good way to begin considering a synthetic procedure is to carry out a retroanalysis, as exemplified in Chapter 5, Box 11.

You could for example, analyse the structure of isobutyric acid into a series of smaller units and find ways of making and linking them.

Discussion Point DP4 hint What characteristics of the counter-anions give rise to the two types of product? van Leeuwen and his colleagues have developed the concept of the *bite angle* of a chelate ligand to explain such differences in behaviour; on that basis what sort of ligand would be ideal to get each of the two products? See also the websites, *http://pubs.acs.org/cgi-bin/abstract.cgi/orgnd7/2003/22/i25/abs/om034012f.html* and *http://www.chem.wisc.edu/ ~ casey/hydroformylation.html*

Discussion Point DP5 hint Using the concepts behind the successful iridium catalyzed methanol carbonylation as guide; which would you expect the difficult steps in a hydroformylation to be.

Discussion Point DP6 hint Timescales and sensitivities are at the heart of this problem.

Discussion Point DP7 hint This will require a knowledge of the structures of some (relatively simple) biologically active chiral alcohols and/or aldehydes where a starting olefin is readily available.

Discussion Point DP8 hint As for DP3 good way to begin considering a synthetic process is to carry out a retroanalysis; see, for example Chapter 5, Annex 1, page 195.

References

General: Eds., J. A. Moulijn, P. W. N. M. van Leeuwen, R. A. van Santen, *Catalysis, an integrated approach*, NIOK, Elsevier, Amsterdam, 1993.

P. W. N. M. van Leeuwen, *Homogeneous catalysis*, Kluwer Academic Publisher, Dordrecht, 2004.

Methanol carbonylation: T. W. Dekleva, D. Forster, *Adv. Catal.*, 1986, **34**, 81.

Rhodium catalyzed methanol carbonylation: P. M. Maitlis, A. Haynes, G. J. Sunley, M. J. Howard, *J. Chem. Soc., Dalton Trans.*, **1996**, 2187.

Cobalt-catalyzed hydroformylation: M. F. Mirbach, *J. Organomet. Chem.*, 1984, **265**, 205.

Rhodium catalyzed hydroformylation mechanism: P. W. N. M. van Leeuwen, C. Claver, *Rhodium catalyzed hydroformylation*, Kluwer Academic Publishers, Dordrecht, 2002.

Web reference to hydroformylation: G. G. Stanley; http://chemistry.lsu.edu/outreach/stanley/Chem-4571-Notes.htm.

Fischer-Tropsch technology: A. P. Steynberg, *Studies in Surface Science and Catalysis* 2004, **152**, 1; and M. E. Dry, *Catalysis Today*, 2002, **71**, 227.

Fischer-Tropsch Mechanisms: P. M. Maitlis, *J. Mol. Catal (A, chemical)*, 2003, **204–5**, 55; more detailed analysis and discussion, are given in M. L. Turner, P. M. Maitlis, et al., *J. Am. Chem. Soc.*, 2002, **124**, 10456, and Supplementary material thereto.

Gas to liquids technology: K. Aasberg-Petersen, *et al.*, *Applied Catalysis A: General* 2001, **221**, 379.

Biomass as feedstock to syngas: K. Tomishige, M. Asadullah, and K. Kuni-
mori, *Catalysis Today* 2004, 89, 389; M. Rohde, D. Unruh, P. Pias, K-W.
Lee, and G. Schaub, *Studies in Surface Science and Catalysis* 2004, **153**
(Carbon Dioxide Utilization for Global Sustainability), 97.

Methanol synthesis: X.-M. Liu, G. Q. Lu, Z.-F. Yan, and J. Beltramini, *Ind
Eng. Chem. Res.*, 2003, **42**, 6518; J. M. Thomas and W. J. Thomas, *Principles
and practice of heterogeneous catalysis*, VCH, 1997, p. 515.

CHAPTER 5
Carbon–Carbon Bond Formation

FAUSTO CALDERAZZO,[a] MARTA CATELLANI[b] AND
GIAN PAOLO CHIUSOLI[b]

[a] *Dipartimento di Chimica e Chimica Industriale, Università di Pisa, Via Risorgimento 35, Pisa I-56126*
[b] *Dipartimento di Chimica Organica e Industriale, Università di Parma, Parco Area delle Scienze 17/A, Parma I-43100*

5.1 Introduction

Carbon-carbon bond formation is one of the oldest and most important topics of organic chemistry, and for a long time has been dominated by Friedel–Crafts and Grignard reactions. The former are based on stabilized carbocations as reagents and are exemplified by benzene alkylation (Equation 1), while the latter involve stabilized carbanions and are exemplified by acetone alkylation (Equation 2).

Thus, benzene can be alkylated by a carbocation R^+, stabilized by $AlCl_4^-$ and acetone can be alkylated by a carbanion R^-, stabilized by $MgCl^+$:

$$Me_2C{=}O + [R^-MgCl^+] \rightarrow Me_2C(R){-}O{-}MgCl \qquad (2)$$

It should be noted that the species written as $[R^+ \ AlCl_4^-]$ and $[R^- \ MgCl^+]$ belong to the category of polar organometallics, in which the metal and carbon charges tend to come together to form an ion-pair. Their structures range from monomers to complex aggregates. The formulae given here are shorthand representations often used in discussing the organic chemistry of such species. In fact such reagents and intermediates really contain polar covalent bonds and truly ionic species are rare.

To convert these reactions into efficient industrial processes an important requisite is that they are catalytic. Friedel–Crafts alkylations meet this requirement and in some cases are still carried out today with conventional catalysts such as $AlCl_3$. Many acid- and base-catalyzed C–C bond-forming reactions are practised on a small scale. We shall not treat these reactions except to introduce

163

more recent developments involving the use of solid metal oxides or transition metals. As we shall see, the advent of these classes of catalysts (Appendices 1 and 2) has indeed changed the scene of industrial organic chemistry substantially because of their greater efficiency combined with a number of advantages that make their use preferable to conventional processes. The data quoted here are usually taken from patents but do not necessarily correspond to actual practice, which in many cases is kept secret by the manufacturing companies.

This Chapter is subdivided into five Sections dealing with: alkylation and related reactions; activation of aryl and vinyl halides by transition metals to manufacture fine chemicals; butadiene chemistry; olefin oligomerization; carbene chemistry and asymmetric cyclopropanation.

C–N and C–O (in addition to C–C) coupling processes are now becoming significant in industrial metal-catalyzed organic syntheses; however the paucity of available information restricts our ability to deal adequately with these topics.

5.2 Alkylation and Related Reactions

Conventional acid- or base-catalyzed alkylation and related reactions have evolved towards the use of solid metal compounds with acid or base properties, as explained below.

5.2.1 Ethylbenzene by Alkylation of Benzene with Ethylene

Alkylation of benzene with ethylene to give ethylbenzene according to Equation 3, an example of the classical Friedel–Crafts chemistry, is one of the largest tonnage C–C bond forming processes.

$$\text{C}_6\text{H}_6 \; + \; \text{H}_2\text{C}{=}\text{CH}_2 \quad \xrightarrow[\text{90 °C}]{\text{AlCl}_3,\ \text{HCl}} \quad \text{C}_6\text{H}_5\text{CH}_2\text{CH}_3 \tag{3}$$

This is one of the few metal catalyzed processes involving a *non-transition* metal where conventional chemistry has been transferred to a large commercial scale. The world production of this intermediate to styrene (Section 3.9.3) amounts to *ca.* 27 Mt/a.

Classical Friedel–Crafts chemistry is at work here, ethylene being converted by protonation with HCl and AlCl$_3$ into an ion-pair (Figure 1).

An ion pair of the type, H$^+$ AlCl$_4^-$, is believed to be the actual catalyst in the liquid-phase process operating at *ca.* 90 °C. As is well known in electrophilic aromatic substitution, electron-releasing substituents favour alkylation; thus ethylbenzene can undergo further alkylation to diethylbenzenes. To limit overalkylation, benzene must be used in large excess. The peralkylated benzenes, however, can be recycled to obtain ethylbenzene by exchange with

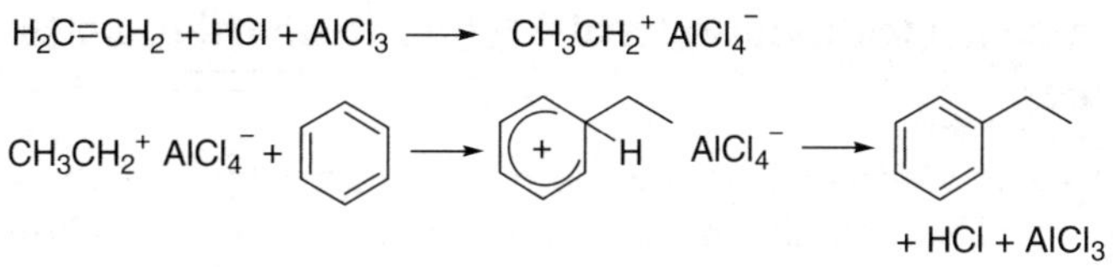

$$H_2C=CH_2 + HCl + AlCl_3 \longrightarrow CH_3CH_2^+ \, AlCl_4^-$$

Figure 1 *Schematic representation of ethylbenzene formation.*

benzene, since Friedel–Crafts alkylation is a reversible process. It should be added that *p*-diethylbenzene has its own market as a desorbent in adsorptive separation processes (≥ 1 kt/a world production).

The process has some unattractive aspects essentially connected with plant corrosion and waste disposal associated with the presence of chloride. These difficulties have spurred worldwide research for more suitable catalysts. These efforts have been successful largely due to the use of zeolites and other metal oxides possessing strong acid character (Box 1 and Appendix 2), which are now used for both liquid- and gas-phase reactions to a much larger extent than the AlCl₃-based process. The zeolite pore size and morphology determine the selectivity in alkylation and transalkylation reactions, and research in this area to overcome drawbacks connected with selectivity and catalyst life is still ongoing. Conditions are relatively mild and this has led to the development of environmentally more benign processes. The Lummus/UOP EBone process (liquid phase proprietary zeolite catalyst), commercialized in 1996, offers 99.99% pure ethylbenzene with a yield of 99.6%. Other producers are Mobil-Badger (gas phase process, zeolite ZSM-5 catalyst), Mobil-Raytheon (EBMax liquid phase process, MCM zeolite), EniChem (liquid phase process, zeolite beta-based PBE-1 catalyst). The proton in the acid zeolite structure plays a role similar to that of the Brønsted acid in the traditional process. The weakly coordinated ethylene gives rise to a stabilized zeolite-segregated carbocation able to attack benzene. How stabilization occurs still is a matter of discussion. The carbonium ion may form a bond with the zeolite structure before benzene alkylation occurs or may be directly transferred to benzene in a concerted process, *i.e.* ethylene protonation is simultaneous with benzene ethylation.

Box 1 Zeolites as Alkylation Catalysts

The requirements of industrial processes (easy separation of catalysts from products, low levels of corrosion and of polluting fluids) have led to the development of heterogeneous catalysts that avoid the presence of corrosive or undesirable species in solution and in the products, in contrast to homogeneous processes. Several types of zeolites can be used for both the liquid- and the gas-phase processes, such as ZSM–5, MCM–22, Y and Beta (Appendix 2).

5.2.2 Toluene Dealkylation and Methyl Redistribution to Benzene and Xylenes

Benzene, toluene and *p*-xylene are the most industrially important aromatic hydrocarbons. Their relative proportions must be adjusted according to the needs of the market, which presently requires more benzene and xylene. Dealkylation of toluene can be carried out by hydrogenolysis with H_2 using Cr, Mo or Pt oxides as catalyst precursors (Equation 4). Methyl redistribution in toluene is best carried out around 420°C on ZSM–5 zeolite. Under these conditions, xylene is mainly present as the *para* isomer (90% selectivity, Equation 5) due to the shape-selectivity of the zeolite.

$$\text{Toluene} + H_2 \xrightarrow[\text{600 °C, 25 bar}]{\text{Cr, Mo, Pt}} \text{Benzene} + CH_4 \qquad (4)$$

$$\text{Toluene} \xrightarrow[\text{420 °C}]{\text{ZSM-5 zeolite}} \text{Benzene} + \text{xylene} \qquad (5)$$

5.2.3 Cumene from Benzene and Propylene

Cumene (isopropylbenzene) is the major intermediate (10.6 Mt/a was the world capacity in 2004) for the manufacture of phenol. It is formed by alkylation of benzene with propylene according to Equation 6. Propylene is believed to react with hydrogen chloride forming the more stable secondary carbocation, which is associated with the tetrachloroaluminate anion. The ion pair reacts further with benzene to give the product.

$$\text{Benzene} + H_2C{=}CHCH_3 \xrightarrow{AlCl_3, HCl} \text{cumene} \qquad (6)$$

As for ethylbenzene production, acid- and $AlCl_3$-based catalysts were superseded by solid catalysts, especially zeolites. The pore size problem is particularly critical for cumene. If the pores are too small, as in $ZSM{-}5$ zeolite (rings formed by 10 SiO_4/AlO_4 tetrahedral units), utilized in the methyl redistribution of toluene (Equation 5), product diffusion becomes difficult: a higher temperature is required and extensive isomerization of the *i*-propyl to the *n*-propyl group may occur. By contrast, Y, Beta or MCM-22 zeolites (CD-Tech catalytic distillation technology; EniChem, fixed bed, liquid phase; Mobil-Raytheon, fixed bed, liquid phase) with pores of 12 SiO_4/AlO_4 units give a high selectivity (99.95% at 95% conversion at about 230°C). Several proprietary catalysts are used, but their detailed structural properties have not been disclosed.

Commercial zeolites are usually mixed with a binder such as γ-alumina and shaped into pellets. The shaping procedure causes changes in binder porosity to give zeolites with different physical properties.

Box 2 Research for Alternative Routes to Aromatic Alkylation

No industrial process can be considered as the ultimate, and research continuously provides new alternatives for economic assessment.

We have seen that in Friedel–Crafts type reactions the olefin is activated towards electrophilic attack. One could inquire however whether activation might be caused to occur in the opposite sense by forming a metal–coordinated aryl reagent which would then migrate onto the unactivated olefin. This is what happens in certain transition metal–catalyzed reactions involving arene activation either by means of an electrophilic metal species or by oxidative addition to a metal. Indeed, as shown in Figure 2, ethylbenzene has been obtained via benzene C–H activation by the electrophile $IrCl_3$ or by oxidative addition of benzene to Ir(I) [J. Oxgaard, R. P. Muller, W. A. Goddard, III and R. A. Periana, *J. Am. Chem. Soc.*, 2004, **126**, 352]. The catalytic efficiency of this reaction is low and at present no industrial application can be envisaged. However, the concept is important and further developments should be sought.

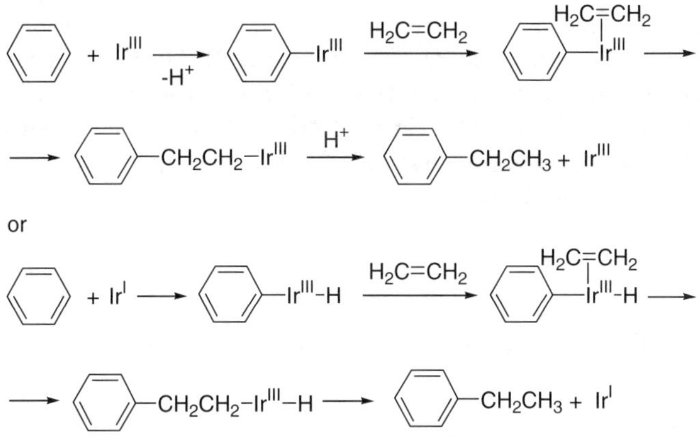

Figure 2 *Suggested Scheme for the Ir–catalyzed synthesis of ethylbenzene.*

5.2.4 2,6-Di-isopropylnaphthalene

An analogous process has been used industrially for the synthesis of 2,6-di-isopropylnaphthalene from naphthalene and propylene in the presence of H-mordenite, another shape-selective zeolite (Equation 7). Oxidation of

2,6-di-isopropylnaphthalene gives 2,6-naphthalenedicarboxylic acid, an important monomer for condensation polymers with ethylene glycol.

$$\text{naphthalene} \ + \ 2 \ \text{CH}_2{=}\text{CH-CH}_3 \ \xrightarrow{\text{cat.}} \ \text{2,6-di-isopropylnaphthalene} \tag{7}$$

5.2.5 Other Alkylations of Aromatics

Friedel–Crafts technology and zeolite- or other solid catalyst-based processes
are currently used for other aromatic alkylations, in particular for the manufacture of linear alkylbenzenes (LABs) made from C_{10}–C_{14} olefins (Equation 8),
or from the corresponding chloroparaffins and benzene, and also to make
m– and *p*–cymene (isopropyltoluene; Equation 9). LABs are used for the
production of sulfonate detergents, while cymenes lead to *m*– and *p*–cresols
through a procedure analogous to that used for the cumene-to-phenol process.

   Discussion Point DP1: *What problems might you expect if you wanted to carry
out research on the alkylation of biphenyl?*

   Friedel–Crafts alkylation processes were traditionally operated at 65–70°C
with $AlCl_3$ and at 40–60°C with HF. A variety of solid acid catalysts have been
developed at the laboratory level, mainly based on zeolites, heteropolyacids or
sulfated zirconia (zirconia treated with sulfuric acid). The most recent industrial
achievement is the Detal process (UOP–CEPSA) which is based on silica–
alumina impregnated with HF. The selectivity towards linear alkylbenzenes
exceeds 95%. The cymene processes use $AlCl_3$ in the liquid phase or supported
phosphoric acid as catalysts.

$$\text{benzene} \ + \ \text{R-HC}{=}\text{CH-R}' \ \longrightarrow \ \text{alkylbenzene} \tag{8}$$

$$\text{toluene} \ + \ \text{H}_2\text{C}{=}\text{CHCH}_3 \ \longrightarrow \ \text{cymene} \tag{9}$$

5.2.6 Alkane Cracking and Isomerization on Solid Acid Catalysts

The basic concept underlying alkylation reactions of aromatics is the formation of a stabilized carbocation able to attack nucleophilic substrates. Hydrocarbon cracking and hydrocracking, alkane isomerization, and olefin alkylation are important processes based on related alkane carbocation chemistry in the production of various types of hydrocarbons such as the branched ones for high octane gasolines. Zeolites and metal oxides are the preferred catalysts.

It is widely accepted that in hydrocarbon cracking a stabilized carbocation is first formed by hydride removal from the hydrocarbon through protonation effected by zeolite- or metal oxide-bonded protons. The protons can also be generated by metal-catalyzed dissociation of dihydrogen, which is added to the reaction mixture under pressure. As is well known, the carbocation thus formed may rearrange (probably through a cyclopropane intermediate) to the more stable tertiary carbocation. It can accept hydrogen (skeletal isomerization), lose hydrogen to give olefins (dehydrogenation, also before rearrangement), or alkylate olefins. Figure 3 depicts the behaviour of *n*-butane in a zeolite (ZO$^-$ $^+$H = protonated zeolite) in a simplified way; see references to Section 5.2.

We shall not treat cracking processes here due to the complexity of these high-temperature (usually around 500°C) reactions. However, cycloalkane dehydrogenation to aromatics (Appendix A2.4.4), alkane isomerization and olefin alkylation (leading to branched alkanes from linear ones) occur *via* such carbocation rearrangements.

Alkane isomerization is carried out under hydrogen pressure on zeolite-type catalysts, which are particularly robust, being insensitive to poisons such as sulfur-containing compounds and water at 200–280°C. The use of so-called dual catalysts, containing both a solid acid species and a transition metal such as

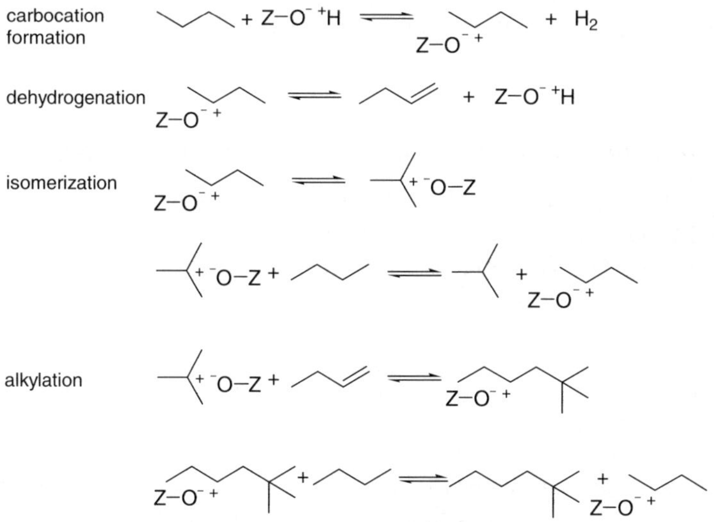

Figure 3 *Schematic representation of zeolite-catalyzed reactions of n-butane.*

palladium or platinum, *e.g.* chlorided platinum on alumina at 120–170°C, has produced good results, but the sensitivity to poisons may offset the advantage of working at lower temperatures. When these metals are present they are thought to activate H_2 for protonation and also for hydrogenation of the olefins that originate from the carbocations. Thus, a Pt/beta-zeolite catalyst isomerizes *n*-octane with 77% selectivity (at 28% conversion) to methylheptanes at 200°C and 15 bar of H_2. More recently, some metal oxides (sulfated zirconia) have been found to be active at 180–200°C, as well as resistant to many contaminants.

Olefin alkylation, long carried out in solution with Brønsted acids, is now beginning to turn to the use of zeolites or other solid acid reagents under mild conditions. The main problem is removal of both the heat generated by the reaction and the carbon deposits. While the former problem can be solved by slurry reactor techniques, the latter requires a chemical treatment which is best carried out by metal–catalyzed hydrogenation at high temperature.

Box 3 Gasoline from Methanol

The production of gasoline from methanol is a parallel process to the Fischer–Tropsch synthesis of hydrocarbons from syngas (Section 4.7.2). A shape–selective zeolite (ZSM–5) was the catalyst of choice in the process put on stream in 1987 by Mobil in New Zealand; however the plant was later closed. The zeolite was used at *ca.* 400°C in a fluid catalyst reactor, which allows prompt removal of the heat of reaction.

In this process, methanol deprotonation is believed to afford a methoxy intermediate, which is stabilized in the zeolite cavity; the carbon–to–oxygen bond is then cleaved to form a methylene which oligomerizes to gasoline. Like the Fischer–Tropsch, this process becomes economically interesting in a regime of high oil prices. It has the additional drawback, however, of leading to a gasoline high in undesirable aromatics.

5.2.7 *o*–Pentenyltoluene from *o*–Xylene and Butadiene

Analogously to acid–catalyzed alkylations, which are now mostly carried out with zeolites or solid acids, base–catalyzed alkylations can also be conducted over solid catalysts. The usage is much smaller, however.

$$(10)$$

The production of *o*–pentenyltoluene (world production 40 kt/a) is based on the alkylation of *o*–xylene with butadiene at *ca.* 145°C in the presence of supported alkali metals, for example K on K_2CO_3 [D. L. Sikkenga, US Patent 4

990 717 (1991), to Amoco] (Equation 10). The selectivity of butadiene conversion is 65–70%. *o*–Pentenyltoluene is a useful intermediate for the manufacture of 2,6–naphthalenedicarboxylic acid (see also Section 5.2.4). The probable reaction course is shown in Figure 4.

Two potassium atoms transfer an electron each to butadiene forming a dianion; transmetallation with *o*–xylene then gives the potassium–bonded carbanion, which inserts butadiene. A second transmetallation with *o*–xylene liberates the potassium–stabilized benzylcarbanion, which is the actual catalytic species and generates *o*–pentenyltoluene. This can then be cyclized to 1,5–dimethyltetralin, which, after dehydrogenation to the corresponding naphthalene and isomerization to the 2,6–isomer, affords 2,6–naphthalenedicarboxylic acid by oxidation.

5.2.8 C–C Bond Formation Through Multifunctional Catalysis By Mixed Metal Oxides

Another example where metal oxides have replaced old procedures is the *n*–butyraldehyde condensation–dehydration leading to an intermediate which is hydrogenated *in situ* to the industrially important 2–ethylhexanol, used for the manufacture of the di–2–ethylhexyl phthalate plasticizer. This can now be achieved with mixed Mg–Al–Pt oxides, the first acting as a base for condensation, the second as an acid for dehydration while the platinum oxide, which is readily reduced to metallic Pt, is a hydrogenation catalyst both for the double bond and the formyl group (Figure 5). This illustrates how industrial processes involving the use of solid acid–base catalysts are constantly expanding.

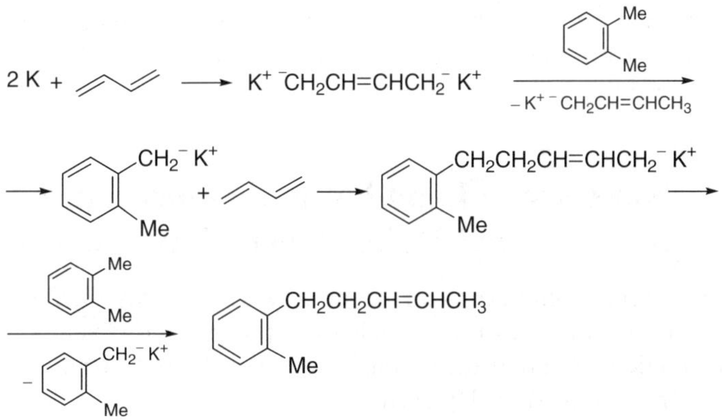

Figure 4 *Scheme showing the proposed base–catalyzed formation of o–pentenyltoluene.*

$$2\,CH_3(CH_2)_2CHO \xrightarrow{\text{base}} CH_3(CH_2)_2CHOHCH(CH_2CH_3)CHO \xrightarrow{\text{acid}}$$

$$\longrightarrow CH_3(CH_2)_2CH=C(CH_2CH_3)CHO \xrightarrow{H_2} CH_3(CH_2)_3CH(CH_2CH_3)CH_2OH$$

Figure 5 *Scheme of a sequence of reactions with a multifunctional catalyst.*

Box 4 Transition Metal-Catalyzed Reactions

Metal oxide species with acid or basic properties as efficient catalysts for alkylation and related reactions have been discussed in Section 5.2. An alternative approach is based on reactions of covalent metal-to-carbon (M–C) bonds. Transition metals are well-suited for this task, as they form directional bonds using hybrid orbitals, and undergo low-energy electron promotion and transfer processes. There are now many industrial processes involving transition metal-catalyzed carbon-carbon bond formation (for example, carbonylation, metathesis, and polymerization reactions, see Chapters 4, 6 and 7, respectively). In sections 5.3-5.4 we deal with other C–C bond forming reactions that can lead to fine chemicals (see Chapter 1).

Although C–C coupling reactions of aryl bromides with copper were first reported by Ullmann in 1901, research on transition-metal catalyzed C–C coupling reactions really gained momentum from Kharasch's discovery in 1941 that these metals profoundly alter the course of Grignard reactions (C-halide bond activation). Subsequent studies established protocols for C–C bond formation based on direct activation of C-halide or C–O bonds by transition metals to give M–C σ-bonds, which, in turn react with various substrates to give the organic product and to regenerate the metal in its original oxidation state. Another root from which research on C–C coupling developed can be found in hydroformylation and hydrocarboxylation reactions (Chapter 4). M–C σ-bonds are also generated by insertion (of olefins, CO, *etc*) into metal-heteroatom (M–O, M–N) bonds, and by external attack by water or other nucleophiles on the coordinated olefin, as in the Wacker and related processes (Chapter 2). M–C bonds are also present in metalla-cycles formed from unsaturated species and low-valent metals.

5.3   Carbon-Carbon Bond Formation through Activation of Aryl- or Vinyl-Halide bonds: Fine Chemicals

Low valent metal complexes can generate aryl- or vinyl-metal bonds by oxidative addition of aryl or vinyl halides (Appendix 1), which can undergo various catalytic C–C bond formation reactions. The industrially most significant ones are described in this Section.

5.3.1   Vinylarenes by Vinylation of Aromatics

Some intermediates to fine chemicals are prepared industrially by vinylation of aromatics. We have selected three well-known products as representative cases, namely an intermediate to Naproxen, a non-steroidal antirheumatic drug (Albermarle) (Equation 11); an intermediate to Prosulfuron, a potent herbicide (Ciba Geigy-Novartis) (Equation 12); and a monomer for polymerization to Cyclotene electronic resins (Dow Chemicals) (Equation 13).

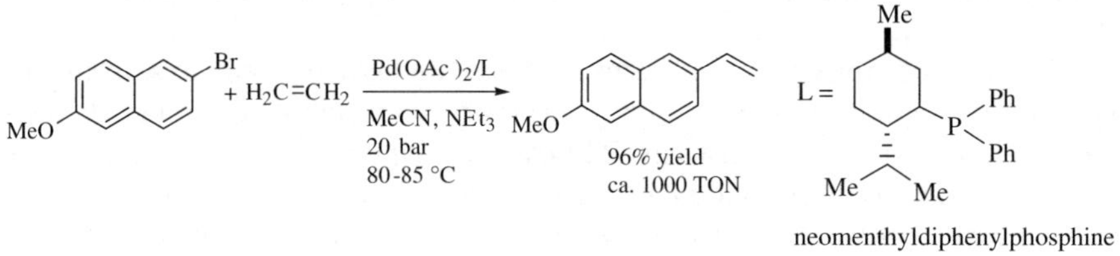

(11)

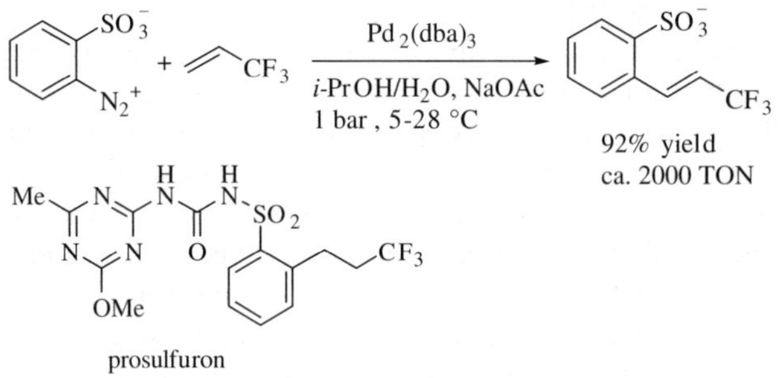

prosulfuron

(12)

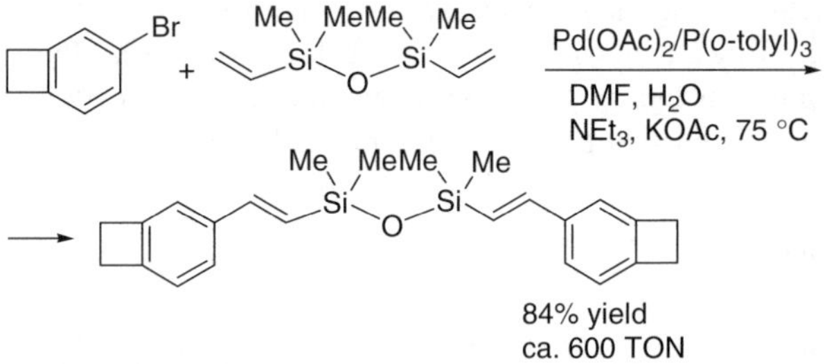

(13)

84% yield
ca. 600 TON

The intermediate to Naproxen, 6-methoxy-2-vinylnaphthalene, is prepared from 2-bromo-6-methoxynaphthalene and ethylene in the presence of Pd(OAc)$_2$, a convenient precursor to palladium(0) complexes, and a base as neutralising agent, in acetonitrile solution (Equation 11). According to a patent [T.-C. Wu, US Patent 5 536 870 (1996), to Albermarle] the catalyst is formed *in situ* starting from Pd(OAc)$_2$ and using neomenthyldiphenylphosphine as ligand. Several dozen tonnes of this product are currently prepared in a one-pot process for the carbonylation step leading to Naproxen (Section 4.3.2). The process can also be used for the preparation of vinyl intermediates to other non steroidal antirheumatic "profen" compounds.

The sodium salt of 2-(3,3,3-trifluoropropenyl)benzenesulfonic acid, the intermediate to Prosulfuron, is manufactured from 2-diazobenzenesulfonate and 3,3,3-trifluoropropene in the presence of palladium(0) and NaOAc (Equation 12). According to a patented procedure [P. Baumeister, G. Seifert and H. Steiner, EP Patent 584 043 A1 (1994), to Ciba-Geigy], aniline-2-sulfonic acid is diazotised with isopropyl nitrite in an isopropanol/water solution at 15–20°C. The diazo suspension is transferred to a pressure vessel where it is reacted with 3,3,3-trifluoropropene in the presence of a palladium(0) dibenzylideneacetone (dba) complex, and an excess of NaOAc as buffer at 1 bar, initially at 5–10°C, then at 27–28°C.

The monomer for Cyclotene electronic resin, divinyltetramethyldisiloxanebis-benzocyclobutene, derives from 2-bromobenzocyclobutene and divinyl-tetramethyldisiloxane, again using $Pd(OAc)_2$ as catalytic precursor and a base as neutralizing agent (Equation 13). The reaction is carried out, in line with a recent general protocol for Heck-type reactions, in a mixture of DMF and water preferably using tris-*o*-tolylphosphine as ligand, and triethylamine and/or KOAc as neutralizing agent [R. A. DeVries and H. R. Frick, US Patent 5 136 069 (1992), to Dow Chemicals].

These reactions are commonly interpreted to be composed of three main steps, namely: *a*) oxidative addition of an aryl-X species to palladium(0) with formation of an arylpalladium(II) bond; *b*) insertion of a terminal olefin; and *c*) reductive elimination regenerating palladium(0). To achieve a catalytic cycle, the rates of these steps have to match each other. The basic process was discovered by Heck in 1968. The mechanism has not yet been well defined and several variants have been proposed. A widely accepted scheme is reported in Figure 6.

With aryl halides oxidative addition is often the rate-determining step; this is promoted by appropriate ligands such as triarylphosphines. Finely divided palladium metal or ligand-free palladium, prepared *in situ* by reduction of palladium(II) species, can also be used in weak donor media. Palladium(0) with a triarylphosphine as ligand reacts with aryl halides in the order ArI > ArBr > ArCl. Electron-withdrawing groups on the aryl halide increase the reactivity

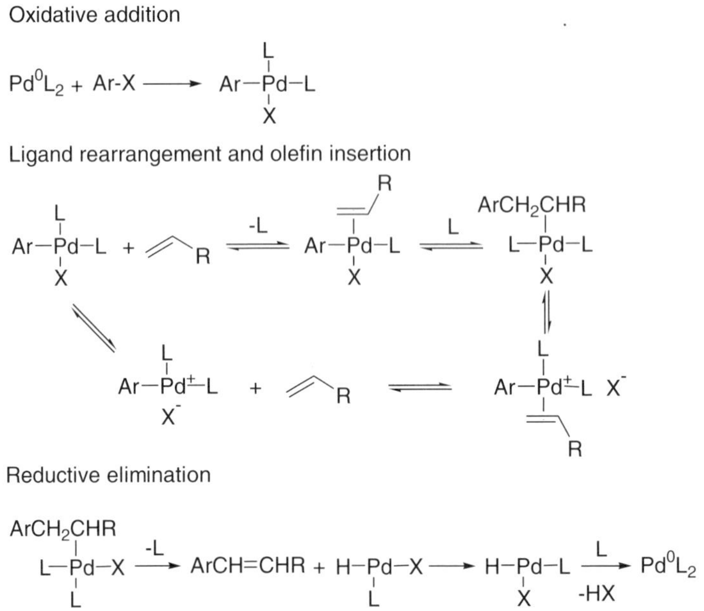

Figure 6   *Scheme illustrating the Heck arylation of olefins (R = alkyl, aryl; X = halide or other leaving groups; L = ligand).*

and aryl bromides or even chlorides containing such substituents can be made sufficiently reactive for an industrial process. A *cis* arylpalladium complex is likely to be first formed (in equilibrium with the *trans*, which is the usually isolated one) [A. L. Casado and P. Espinet, *Organometallics* 1998, **17**, 954].

The insertion step is believed to require dissociation of either the ligand or the halide to allow the olefin to coordinate to the arylpalladium species; a further requirement for a facile reaction is for the olefin to be *cis* to the aryl-Pd bond. DMF used as solvent stabilizes the ionic form by virtue of its high polarity, hence favouring halide dissociation. Insertion (or migration of the aryl onto the coordinated olefin) then occurs; for α-olefins bearing electron-withdrawing groups, the insertion is more than 99% regioselective for the terminal position. The final step is reductive elimination to H-Pd-X, which decomposes to Pd(0) + HX. If the palladium species is kept in solution by sufficiently coordinating ligands [*e.g.* P(aryl)$_3$] the catalytic cycle continues. If, however, ligands are altered, removed or no longer able to coordinate effectively, metallic palladium separates out and the reaction is inhibited and slowed down or stopped.

One of these inhibiting reactions has been recognized to be palladium-promoted triarylphosphine quaternization with aryl halides to [P(aryl)$_3$(aryl′)]$^+$, accompanied by separation of palladium metal. Triarylphosphines may also undergo C–P bond exchange with a Pd-bonded aryl group. A possible pathway, involving C–P oxidative addition to palladium is shown in Equation 14. The aryl group thus cleaved may then undergo an undesired Heck reaction. Other mechanisms involving aryl substitution at phosphorus by nucleophiles, rather than oxidative addition, have also been proposed, however. A variety of nucleophiles can act in this way including organo-palladium species as well as acetato- and similar anions.

$$
\begin{array}{ccccc}
\underset{\underset{PAr_3}{|}}{\overset{\overset{PAr_3}{|}}{Ar^*\!-\!Pd\!-\!X}} & \xrightarrow{-PAr_3} & \underset{\underset{X}{|}}{\overset{\overset{PAr_2}{|}}{Ar^*\!-\!Pd\!-\!Ar}} & \xrightarrow{PAr_3} & \underset{\underset{PAr_3}{|}}{\overset{\overset{PAr_2Ar^*}{|}}{Ar\!-\!Pd\!-\!X}}
\end{array} \qquad (14)
$$

Inhibition also occurs when products, substrates, or reagents acting as ligands stabilize the palladium intermediates too much, thus blocking the reaction sequence. Ligand excess may produce a similar effect. Thus many palladium-catalyzed reactions are inhibited by excess ligand whose electronic and steric properties must therefore be taken into account.

There are several ways by which a palladium(II) species can be reduced to palladium(0). In the reaction to form 6–methoxy-2-vinylnaphthalene (Equation 11) reduction of the palladium(II) species to palladium(0) occurs with concomitant oxidation of the tertiary phosphine (Equation 15).

$$
Pd(PR_3)_2(OAc)_2 + PR_3 \rightarrow O{=}PR_3 + AcOAc + Pd(PR_3)_2 \qquad (15)
$$

The electron-releasing phosphine promotes oxidative addition of the bromo derivative to Pd(0) and, because of its bulkiness, readily generates free coordination sites by dissociation. Ethylene coordination and insertion then occur, followed by reductive elimination, triethylamine acting as a base to neutralize hydrogen bromide. As in most cases of transition metal-catalyzed reactions the fine details of the mechanism are still under investigation. Thus recent studies by Amatore's group suggest that the palladium(0) species formed by reduction of palladium acetate is an anionic acetato complex.

The synthesis of sodium trifluoropropenylbenzenesulfonate, (Equation 12) uses the diazo compound obtained from aniline-2-sulfonic acid and isopropyl nitrite in isopropanol for the oxidative addition to palladium(0)/dba, in the absence of phosphine ligands, with NaOAc as buffer. The substrate itself contributes to palladium stabilization (Equation 16).

$$\text{(structure: benzene ring with } SO_3^- \text{ and } N_2^+ \text{)} + Pd^0L_2 \longrightarrow \text{(structure: benzene ring with } SO_3^- \text{ and } PdL_2^+ \text{)} + N_2 \qquad (16)$$

By contrast to the normally unstable diazonium salts, the one generated *in situ* under the conditions of this process is stable and so there is no serious safety problem and the use of anilinesulfonic acid as starting material turns out to be cost effective.

The synthesis of the monomer for Cyclotene resins (Equation 13) is accompanied by the formation of by-products that derive from internal rather than terminal aryl attack on the vinyl group. Further decrease of selectivity is caused by aryl attack on only one of the terminal carbon atoms of the two siloxane vinyl groups and by hydrogenolysis at the C-Si bond carbons, followed by aryl group attack on the terminal or the internal carbon atom of the vinyl group thus formed.

When tris-*o*-tolylphosphine was used as ligand the main product (Equation 13) was formed in *ca.* 84% yield by GC area analysis. No correlation of regioselectivity with electronic or steric properties was found for several different tertiary phosphines. The final step consisting of *cis* β-hydrogen elimination leading to the *E* product occurs stereoselectively because only one of the two possible conformations leading to elimination is favoured.

This product must be purified from phosphorus before it can be used as an adhesion promoter for electronic devices. This is done by oxidising the phosphine to phosphine oxide with hydrogen peroxide or organic peroxides and by separating the products chromatographically.

5.3.2 Alkynylarenes by Vinylation of Triple Bonds

Vinylation of the triple bond is used industrially for the synthesis of Terbinafin, an anticholesterol drug possessing selective inhibiting activity on squalene epoxidase, which has been commercialized by Sandoz under the trade name of Lamisil.

According to the procedure described in a patent [S. Nakagawa, A. Asai, S. Kuroyanagi, M. Ishihara, Y. Tanaka, US Patent 5 231 183 (1993), to Banyu Pharmaceutical Co.] the final step in the synthesis is carried out by reacting *E*-N-(3-chloroprop-2-enyl)-N-methyl-1-naphthalenemethanamine with *tert*-butylacetylene at 20°C under the conditions shown in Equation 17.

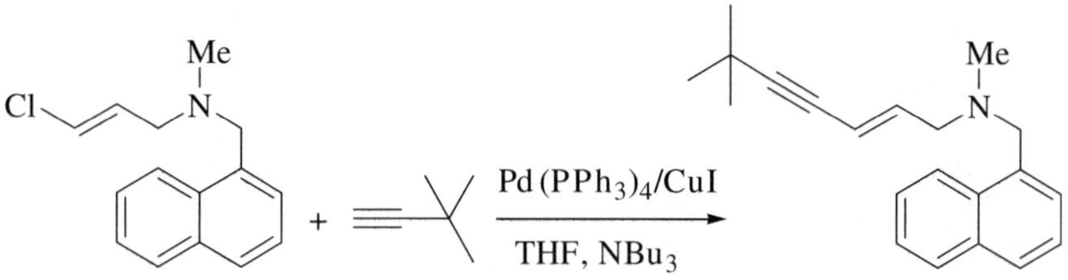

89% yield
ca. 5000 TON

terbinafin

(17)

The synthesis, based on the Cassar-Sonogashira protocol for arylation of alkynes, probably involves the oxidative addition of the organic halide (or an equivalent with a good leaving group) and differs from Heck-type reactions in that halide replacement followed by C–C coupling occurs instead of insertion followed by β-hydrogen elimination. In fact the formation of an alkynyl-palladium bond is strongly preferred in the case of terminal acetylenes. In order for the coupling to occur, the vinyl- and the alkynyl-palladium bonds must be *cis*. This may require a halide or ligand dissociation to generate the needed three- coordinate T-shaped intermediate. A largely accepted reaction pathway is shown in Figure 7.

5.3.3 Biaryls by Aryl Coupling

We consider here the syntheses of 2-cyano-4'-methylbiphenyl and of 4-chloro-2'-nitrobiphenyl as the best known examples. The former product is an important intermediate to Valsartan (Clariant) and analogous "sartan" compounds used to control blood pressure by inhibiting angiotensin, while 4-chloro-2'-nitrobiphenyl is an intermediate to the fungicide Boscalid (BASF-Novartis).

Examples from patents [S. Haber and N. Egger, US Patent 6 140 265 (2000), to Clariant; S. Haber and H.-J. Kleiner, US Patent 5 756 804 (1998), to Hoechst-Clariant] describe the synthesis of 2-cyano-4'-methylbiphenyl by reaction of 2-chlorobenzonitrile and *p*-tolueneboronic acid, in the presence of Na_2CO_3, $PdCl_2$, and triphenylphosphinetrisulfonic acid (4 mol) in a water-toluene system, with a co-solvent miscible with water such as DMSO or

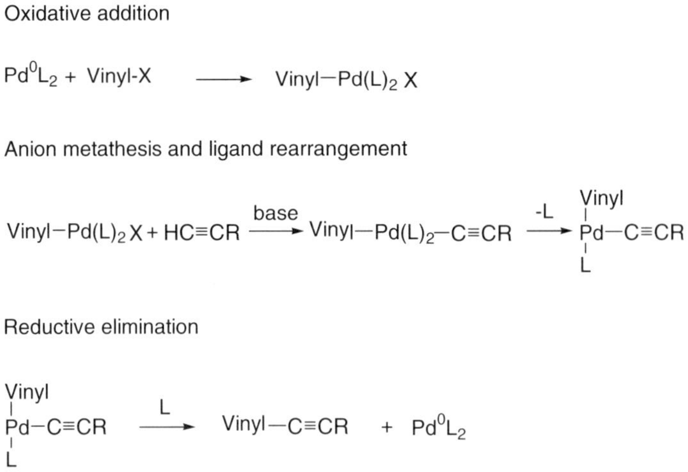

Oxidative addition

$$Pd^0L_2 + Vinyl\text{-}X \longrightarrow Vinyl\text{-}Pd(L)_2\,X$$

Anion metathesis and ligand rearrangement

$$Vinyl\text{-}Pd(L)_2X + HC\equiv CR \xrightarrow{\text{base}} Vinyl\text{-}Pd(L)_2\text{-}C\equiv CR \xrightarrow{\text{-}L} \underset{\underset{L}{|}}{\overset{\overset{Vinyl}{|}}{Pd}}\text{-}C\equiv CR$$

Reductive elimination

$$\underset{\underset{L}{|}}{\overset{\overset{Vinyl}{|}}{Pd}}\text{-}C\equiv CR \xrightarrow{\text{L}} Vinyl\text{-}C\equiv CR + Pd^0L_2$$

Figure 7   *Probable reaction path for vinyl-alkynyl coupling on palladium.*

triethanolamine (Equation 18). The correct proportion of the three solvents leads to a homogeneous system at the reaction temperature (120°C) while at 20°C separation of the organic layer, containing the majority of the product occurs. The catalyst system is recovered from the water-rich layer.

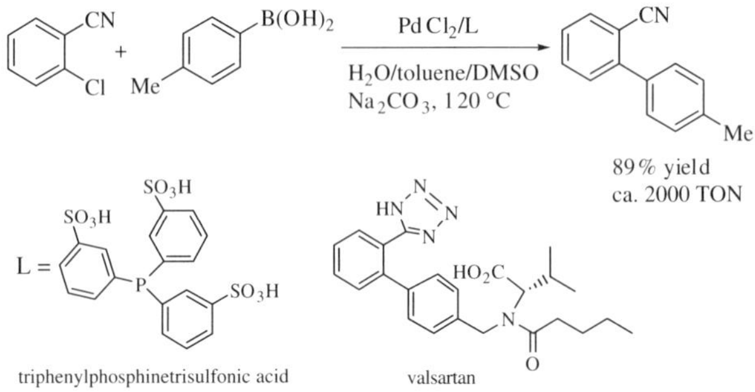

$$(18)$$

This process has the advantage of giving the product in only one step in place of the five previously needed.

   Difficulties connected with the disposal of boron waste and also with its cost have prompted Sanofi to build a plant in France for the production of 2-cyano-4'-methylbiphenyl based on the palladium-catalyzed Grignard reaction of *p*-tolylmagnesium chloride with *o*-bromobenzonitrile. According to a patent [M. Alami, G. Cahiez, B. Castro and E. Riguet, US Patent 6 392 080 (2002), to Sanofi] the process is carried out at 20°C in THF and dimethoxyethane in the presence of *ca.* half the stoichiometric amount of $MnCl_2$ (presumably to form a manganese organometallic species that does not attack the cyano group as the Grignard reagent does) + 2LiCl and a very small amount of the catalyst,

PdCl$_2$-dppp. The yield is 95%. Magnesium and manganese salts are discarded after the work up and incorporated in building materials. This is an ecological strategy that may be generally useful for industrial effluents.

Boscalid was shown to be an effective fungicide and its industrial production is expected to begin soon. The biphenyl intermediate, 4′-chloro-2-nitrobiphenyl, is prepared by palladium-catalyzed coupling of 2-nitrochlorobenzene with 4-chloroboronic acid according to Equation 19. Other details are not available. Boscalid is the 2-chloronicotinic acid amide derivative of the amine obtained by reduction of the nitrobiphenyl product.

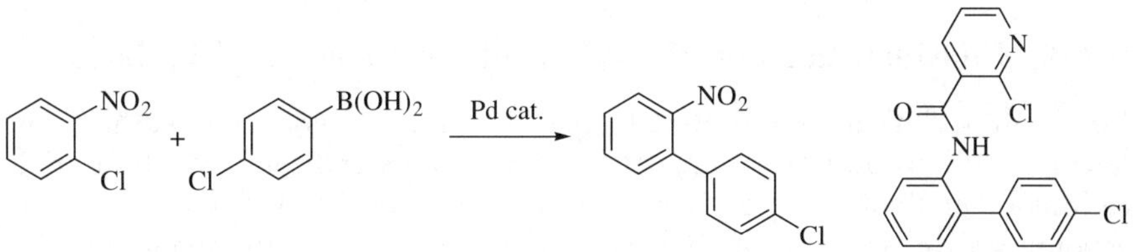

boscalid

(19)

Reactions 18 and 19 are based on the Suzuki procedure, which consists of the oxidative addition to palladium(0) of the aryl derivative used for the Heck and Cassar-Sonogashira reactions, followed by anion exchange with an arylboronic acid and aryl coupling according to the general Scheme shown in Figure 8. The coupling step implies two *cis* aryl groups. Although analogous sequences of elementary steps take place in the procedures by Stille, Kumada and Negishi,

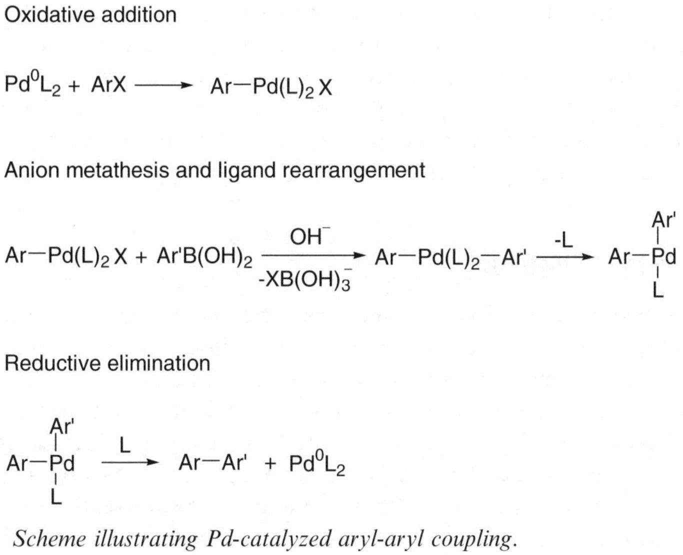

Oxidative addition

$$Pd^0L_2 + ArX \longrightarrow Ar\!-\!Pd(L)_2X$$

Anion metathesis and ligand rearrangement

$$Ar\!-\!Pd(L)_2X + Ar'B(OH)_2 \xrightarrow[-XB(OH)_3^-]{OH^-} Ar\!-\!Pd(L)_2\!-\!Ar' \xrightarrow{-L} Ar\!-\!\underset{L}{\overset{Ar'}{Pd}}$$

Reductive elimination

$$Ar\!-\!\underset{L}{\overset{Ar'}{Pd}} \xrightarrow{L} Ar\!-\!Ar' + Pd^0L_2$$

Figure 8 *Scheme illustrating Pd-catalyzed aryl-aryl coupling.*

where the anion exchanges occur on other organometallics, the Suzuki procedure is the one largely utilized industrially.

The synthesis of the intermediate to Valsartan offers a valuable illustration of the two-phase technology for aryl coupling reactions. The activating CN group favours C–Cl bond cleavage, which is the rate-determining step. This allows the use of an aryl chloride, which is an important advantage in view of the higher cost of the corresponding bromide and iodide. Similar comments apply to the synthesis of the intermediate to Boscalid where the electron-withdrawing nitro group activates the C–Cl bond.

5.3.4 Considerations on Reactions of Aryl and Vinyl Halides

For both technical and economic reasons Heck, Cassar-Sonogashira and Suzuki reactions do not yet appear to have found significant industrial applications other than those described in Sections 5.3.1–5.3.3. Serious problems encountered in this type of chemistry are connected with the requirement to remove even traces of palladium from the products for pharmaceutical, cosmetic and fine chemicals applications, and with the need to prevent catalyst deactivation (Box 5) to ensure catalytic efficiency. Removal of traces of palladium can be achieved by using strong coordinating compounds such as 2,4,6-trimercaptotriazine or solid supports [G. W. Kabalka, R. M. Pagni and C. M. Hair, *Org. Lett.* 1999, **1**, 1423]. In addition, the use of aryl halides and of special ligands is rather expensive, may require product purification, and may cause corrosion of reactor materials and waste disposal problems. Moreover, as pointed out in Chapter 1, the added value of a pharmaceutical or speciality product is mainly based on its properties, with the manufacturing process being a minor part of the total cost, thus improving the process may not influence the final price significantly.

In spite of these drawbacks however the ability to use functionalized substrates directly in highly efficient and selective catalyses under mild conditions, without the need for laborious protection and deprotection of sensitive groups, makes these simpler processes very attractive. They also benefit from lower costs and fewer adverse environmental impacts than more traditional routes, and are likely to be much further expanded in future.

5.3.5 Why Palladium?

On going through the industrial processes described above the question arises of the reasons for the extensive use of palladium. To drive such catalytic processes it is necessary to find a compromise, for example a balance between the tendency of a transition metal to undergo oxidative addition on the one hand and reductive elimination to liberate the organic product on the other. This means that the metal must be sufficiently "noble" to favour reductive elimination without impairing its ability to undergo a sufficiently fast oxidative addition. In certain cases, as in the Suzuki reaction, the less expensive nickel could be used, but the slower reductive elimination reduces the catalytic efficiency and favours competitive pathways leading to undesired by-products. Intermediate steps such as migratory insertion may also be rate determining. It is therefore necessary that all the steps proceed at comparable rates in order to avoid accumulation of intermediates. Thus metals forming very inert σ-bonds cannot be used without extra activation. Other practical reasons (the need for metal recovery, toxicity problems, *etc*) also restrict the choice of metal. The discovery of new ligands able to cause a metal to work more efficiently under suitable reaction conditions could change this picture however (See also Box 5). Some hints for improvements are offered in Annex 1.

Discussion Point DP2: *Devise a palladium-catalyzed C–C bond-forming process for the synthesis of an industrially significant compound, for example a stilbene (1,2-diphenylethylene) derivative starting from an aromatic bromo– derivative.*

Box 5 Metal Catalysts: Liganded Complexes versus "Naked" Nanoparticulates?

Catalytic efficiency is a crucial issue in considering industrial applications. Catalytic activity is often inhibited by substances that coordinate metals too strongly. Designing an ideal catalyst requires a precise knowledge of the behaviour of the actual catalytic species, which is not an easy task. For example a common inhibition of palladium-catalyzed reactions arises from the formation of inert metal ("palladium black"). However in some cases very active colloidal palladium nanoparticles are present. Such nanoparticles which may be stabilized by weakly coordinated ligands, such as dipolar aprotic solvents or quaternary ammonium halides, are in equilibrium with the mono– or di–nuclear species. High dilution can favour this process, a hypothesis that seems to be confirmed by a recent example concerning the Suzuki reaction to form biphenyl, where it has been found that at very low loading of the catalyst precursor (concentrations of 0.01–0.1mol%) palladium does not separate. Even traces of palladium (less than 50 ppb), contained in the Na_2CO_3 used as a base in the process, can efficiently catalyze the reaction without any further addition of palladium. TONs around 50×10^6 have been reached with a palladium loading of 2.5 ppm [R. K. Arvela, N. E. Leadbeater, M. S. Sangi, V. A. Williams, P. Granados and R. D. Singer, *J. Org. Chem.*, 2005, **70**, 161].

Other examples are now recognized where a mononuclear compound that is catalytically active under "homogenous" conditions at one temperature can also act as a source of a highly active nanoparticulate metal catalyst at higher temperature. This was recently demonstrated for a rhodium complex in hydrogenation reactions [C. M. Hagen, J. A. Widegren, P. M. Maitlis and R. G. Finke, *J. Am. Chem. Soc.*, 2005, **127**, 4423].

5.4   Chemistry of Allylic Compounds: Butadiene as Substrate

A series of transition metal-catalyzed reactions, based on the formation of allylic-metal complexes from butadiene is described in this Section. The formation of allylic-metal complexes may occur by hydride transfer from the metal to coordinated butadiene (Section 5.4.1), by oxidative addition of a protonic acid to the metal, followed by hydride transfer to butadiene (Section 5.4.4) or by oxidative addition of two butadiene molecules to the metal (Sections 5.4.2 and 5.4.3).

5.4.1   1,4-Hexadiene from Butadiene and Ethylene

1,4-Hexadiene is utilized as a third co-monomer in ethylene/propylene elastomers. While the terminal double bond is incorporated into the polymeric chain, the internal one is not and can be cross-linked (vulcanized) to confer elasticity to the material.

1,4-Hexadiene in its *E* form is manufactured from butadiene and ethylene (Equation 20). The *Z* isomer must be kept to a minimum owing to its adverse effect on polymerization. The reaction, first reported in 1961, was commercialized by DuPont. The present world production is estimated around 2.5–3 kt/a. Various metal salts (Ni, Co, Fe) were tested, but the best results were obtained with $RhCl_3$.hydrate.

$$\text{(butadiene)} + \text{(ethylene)} \xrightarrow[\text{EtOH, 50 °C}]{RhCl_3} \text{(1,4-hexadiene)} \qquad (20)$$

ca. 70% yield
ca. 1000 TON

Optimal conversion to 1,4-hexadiene is favoured by a large excess of ethanol over $RhCl_3$. 3-Methyl-1,4-pentadiene and 2,4-hexadiene are also formed in low yields. Butadiene and ethylene are used in excess to restrict isomerization to 2,4-hexadiene as they compete with the product (1,4-hexadiene) for the coordination sites involved.

The reaction is stereoselective, 1,4-hexadiene being mainly obtained in the *E* configuration; the Z-isomer is present only to a small extent and its formation can be further decreased by donor ligands such as $Bu_3P{=}O$ and $(Me_2N)_3P{=}O$. The reaction is also regioselective: the positional isomer $CH_2{=}CHCH(CH_3)CH{=}CH_2$, derived from ethylene insertion into the more substituted carbon-rhodium bond of the allylic system, is present to less than 1% when ethanol is in large excess.

As in the other cases of C–C bond formation catalyzed by transition metal complexes, the critical step is to form the initial metal-carbon bond, which is butadiene insertion into a Rh–H bond to form a Rh-allylic species.

More detailed studies showed that what is supposed to be the actual catalyst is formed from the $RhCl_3$ precatalyst by replacement of Cl by ethoxy (from

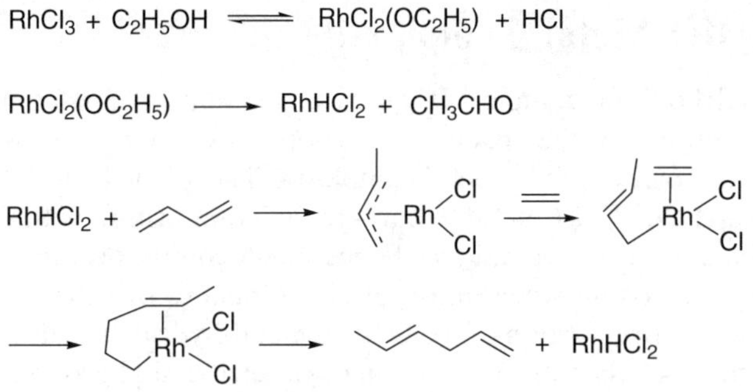

$$RhCl_3 + C_2H_5OH \rightleftharpoons RhCl_2(OC_2H_5) + HCl$$

$$RhCl_2(OC_2H_5) \longrightarrow RhHCl_2 + CH_3CHO$$

Figure 9 *Representation of the Rh-catalyzed formation of E 1,4-hexadiene from but-adiene and ethylene (other ligands on Rh omitted for simplicity).*

ethanol), followed by β-H elimination and the formation of a (solvated) $RhHCl_2$ species and CH_3CHO (Figure 9). Butadiene insertion into the Rh–H bond is then proposed to give a butenylrhodium bond, which inserts ethylene. β-H-elimination reforms the initial hydride catalyst, and liberates the hexadiene product. Since some Rh(III) is reduced to Rh(I) during the reaction, HCl can be added to convert it again into $RhHCl_2$ by oxidative addition.

It is interesting to note that the same catalyst can dimerize ethylene at a comparable rate, suggesting that the preference for hexadiene is due to the relative stability of the η^3-1-methylallylrhodium complex.

5.4.2 Cyclooctadiene and Cyclododecatriene from Butadiene

1,5–Cyclooctadiene (COD) is an intermediate for the manufacture of poly-octenamers, while 1,5,9–cyclododecatriene (CDT) is the starting material for the manufacture of dodecanoic acid and lauryl lactam. The latter is converted to the polyamide fiber Vestamide (DuPont). Both COD and CDT are made from butadiene (Equations 21 and 22).

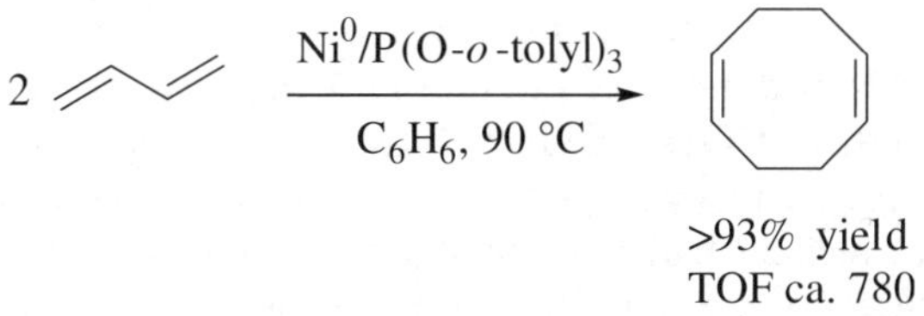

$$(21)$$

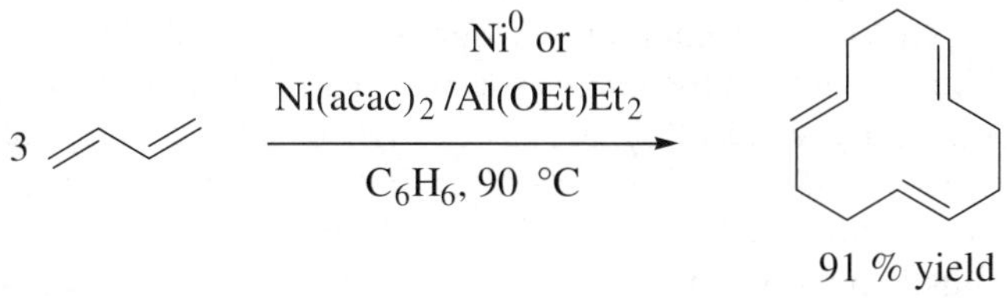

>93% yield
TOF ca. 780

$$(22)$$

91 % yield

Box 6 Allyl Metal Complexes

The butenylrhodium complex (Figure 9), as well as related nickel and palladium complexes that catalyze reactions described in Sections 5.4.1–5.4.4, are examples of allylic metal complexes. The allylic ligand binds to the metal through η^1-C–M bonds that are in equilibrium with η^3-bonded ligands which exist in *syn* and *anti* forms. Such complexes are prepared by two main paths: oxidative addition of allylic halides or other allylic derivatives to metals, or insertion of dienes into metal-hydride bonds (Figure 10). In the absence of other strongly binding ligands the allyl group is in the η^3-form, but addition of ligands, including appropriate substrates, causes it to convert into the more reactive η^1-form.

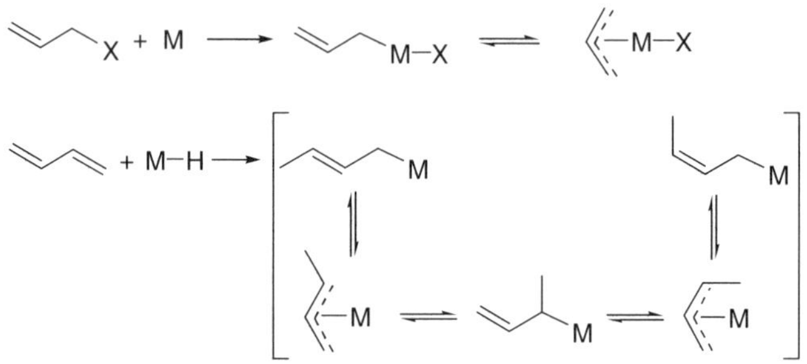

Figure 10   *Schematic representation of the formation and equilibria between η^1- and η^3-allyl and -butenyl ligands.*

According to Wilke the Ni(0) complex of tris(*o*-tolyl)phosphite (formed *in situ* in benzene at 90°C), converts butadiene into COD in high yield, small amounts of vinylcyclohexene (VCH) and CDT also being formed as by-products. CDT, predominantly in the *E,E,E* form, is produced selectively in the absence of phosphorus ligands. COD formation can be contrasted with the (uncatalyzed, electrocyclic) Diels–Alder dimerization of butadiene, which gives mainly VCH plus small amounts of COD. The nickel-catalyzed dimerization is no longer a concerted electrocyclic reaction, but a stepwise reaction which is controlled by the formation of nickel-carbon σ–bonds from oxidative addition of two molecules of butadiene to nickel. Owing to the allylic nature of the nickel-carbon bonds, several species are in equilibrium (Figure 11). Ligands exert a key role in shifting these equilibria and in causing the formation of the cyclic organic products by reductive elimination. Phosphite ligands are effective, especially those containing bulky substituents such as tris[(*o*-phenyl)phenyl]phosphite, which gives even better results than tris–*o*-tolyl phosphite. The electron-withdrawing aryl phosphites stabilize nickel(0), while the steric hindrance caused by the *ortho* substituent limits access to the metal centre of only two molecules of butadiene thus avoiding formation of CDT (Figure 11) which would require

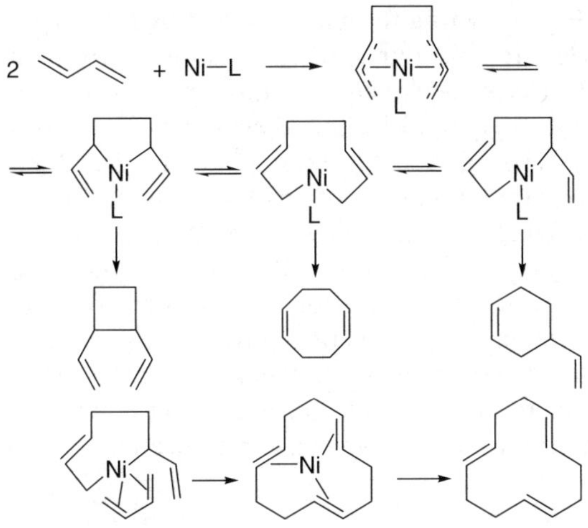

Figure 11 *Representation of di- and trimerization of butadiene on nickel(0) to give bis-allylic intermediates and cyclic olefins.*

Box 7 Metallacycles

Metallacycles, such as those obtained from butadiene and nickel(0), represent an important class of compounds containing two metal-carbon σ-bonds and are useful in transition metal-controlled organic syntheses.

Metallacycles have also been prepared from two molecules of alkenes or alkynes or from palladated *o*-alkyl- or aryl- substituted aryls by C–H activation. Metallacycles usually have five- or six-membered ring structures, and four-membered metallacycles have been shown to be intermediates in metathesis (Chapter 6).

The reactivities of metallacycles obtained from the same substrate with different metals can vary dramatically; for example, in contrast to nickel, the palladacyles obtained from butadiene mainly convert into open-chain dimers as a consequence of the greater tendency of palladium to undergo β-hydrogen elimination.

---

coordination of a third butadiene molecule. Conversely, strongly binding ligands should not be used if the aim is to make CDT; thus reduction of Ni(acac)$_2$ by Al(OEt)Et$_2$ in the presence of butadiene readily gives CDT with only a minor amount of VCH, while in the presence of an electron-releasing ligand such as tricyclohexylphosphine a 1:1 ratio of VCH and COD is formed. Several allylic species related to the observed organic products have been identified by NMR and some have been crystallographically established.

As shown in Figure 11 ring formation occurs by reductive elimination from these nickelacycles. It is favoured when L is an electron-withdrawing bulky phosphite, while electron-releasing ligands such as tricyclohexylphosphine coordinate more strongly to nickel and their steric effect favours the formation of the smaller VCH ring to a substantial extent. The absence of the added ligand L allows coordination of the third butadiene to form CDT.

Interestingly a Ziegler Natta catalyst (Chapter 7) consisting of $TiCl_4$ associated with an aluminium alkyl co-catalyst in a 1:1 molar ratio, polymerizes butadiene, but it is also able to afford cyclododecatriene provided that the $TiCl_4$/co-catalyst molar ratio is kept approximately to 0.1 instead of 1 [M. Rapoport and D. L. Sullivan, Can. Patent 1 055 052 (1979), to DuPont]. Although the mechanism is unclear, this is a good example of how delicate can be the modulation of a catalytic system for a selective synthesis.

5.4.3   Octadienol from Butadiene

Although 2,7-octadien-1-ol itself has a variety of uses, its main market involves hydrogenation to *n*-octanol, whose phthalate diester is utilized as a plasticizer with superior properties to the currently used branched octanol (2-ethyl-hexanol).

The synthesis from butadiene and water occurs according to Equation 23.

$$(23)$$

According to a patent [Y. Tokitoh, T. Higashi, K. Hino, M. Murosawa and N. Yoshimura, US Patent 5 057 631 (1991), to Kuraray Industries] the reaction is conducted with butadiene in sulfolane / water in the presence of $Pd(OAc)_2$ as catalyst precursor and a soluble triarylphosphine (or its phosphonium bicarbonate, which is formed from octadienol itself and carbon dioxide) as ligand. The selectivity to 2,7-octadien-1-ol is 92–94% (TOF > 1000), while the isomeric 1,7-octadien-3-ol accounts for another 3–5%. The product is extracted with hexane, while the aqueous sulfolane solution, containing the catalyst (*ca.* 1 mmol/l) and triethylamine, is recycled. In the absence of carbon dioxide, the main product is 1,3,7-octatriene, an open-chain butadiene dimer.

As shown in Figure 12, the reaction can be viewed as involving the attack by $[NHEt_3]^+$ HCO_3^- on the palladium(II)-coordinated butadiene-derived allylic

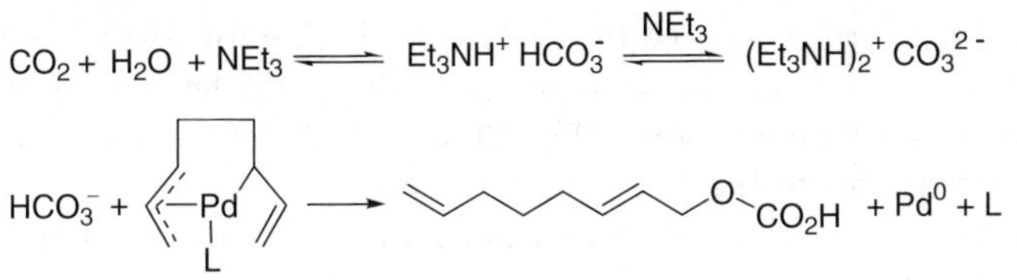

$$CO_2 + H_2O + NEt_3 \rightleftharpoons Et_3NH^+ \, HCO_3^- \xrightarrow{\text{NEt}_3} (Et_3NH)_2{}^+ \, CO_3{}^{2-}$$

$$HCO_3^- + \left\langle\!\!\left\langle -Pd \atop L \right\rangle\!\!\right\rangle \longrightarrow \text{\char`\~}\text{\char`\~}\text{O}\text{-}CO_2H \; + Pd^0 + L$$

Figure 12 *Schematic representation of the formation of 2,7-octadienol acid carbonate.*

species. This reaction is reminiscent of the attack of water on palladium-coordinated ethylene, in the Wacker process (Section 2.15).

The process reported here uses a clever combination of the factors that promote catalyst life and efficiency. The soluble phosphine or its phosphonium salt, used in a molar excess of about 50 over palladium, stabilizes the palladium complex in aqueous solution; the sulfolane-water solution ensures the solubility of the reactants, while extraction with hexane under CO_2 pressure recovers the product with only small contamination by palladium, phosphorus or nitrogen. The phosphine or its phosphonium salt and the ammonium bicarbonate remain in the aqueous solution. Since the TON is good and the solution can be recycled, consumption of palladium is very low.

5.4.4 Adiponitrile by HCN Addition to Butadiene

Adiponitrile is an important intermediate for the manufacture of hexa-methylenediamine (Section 3.4), which, together with adipic acid, is used to produce nylon 6,6. Although adiponitrile is still largely produced from adipic acid, obtained by vapour phase oxidation of cyclohexane (Section 2.2), it is also manufactured from butadiene by DuPont on the basis of the process first patented in 1970 (Equations 24–26). Conversion is 99% with >90% selectivity to adiponitrile.

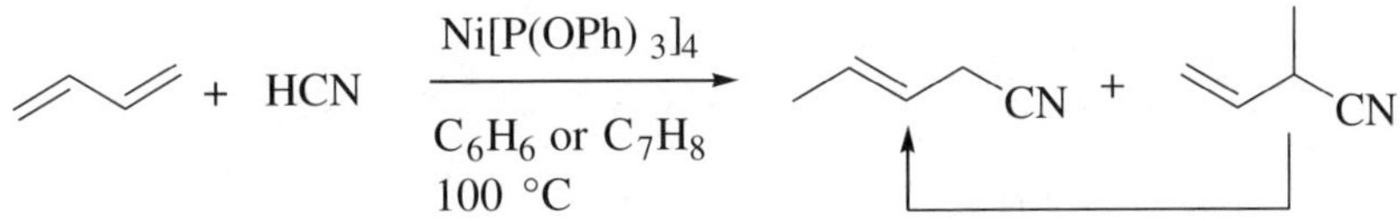

$$\text{\char`\~}\text{\char`\~} + \text{HCN} \xrightarrow[\substack{C_6H_6 \text{ or } C_7H_8 \\ 100\ ^\circ C}]{Ni[P(OPh)_3]_4} \text{\char`\~}\text{\char`\~}\text{CN} + \text{\char`\~}\text{CN} \quad (24)$$

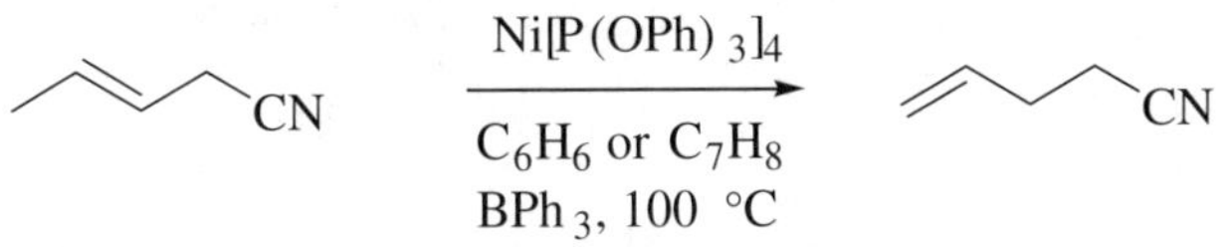

$$\text{\char`\~}\text{\char`\~}\text{CN} \xrightarrow[\substack{C_6H_6 \text{ or } C_7H_8 \\ BPh_3, \, 100\ ^\circ C}]{Ni[P(OPh)_3]_4} \text{\char`\~}\text{\char`\~}\text{CN} \quad (25)$$

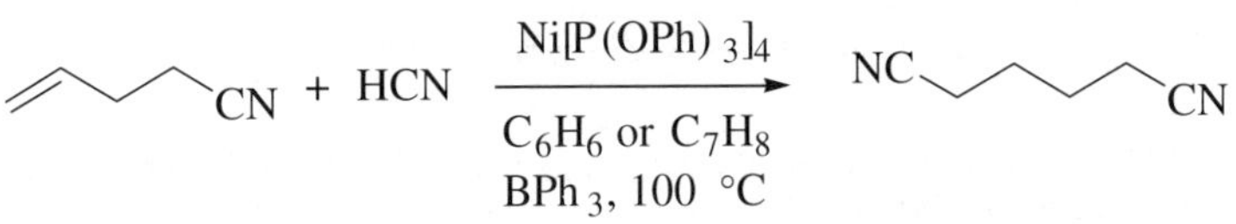

$$\text{\char`\~}\text{\char`\~}\text{CN} + \text{HCN} \xrightarrow[\substack{C_6H_6 \text{ or } C_7H_8 \\ BPh_3, \, 100\ ^\circ C}]{Ni[P(OPh)_3]_4} \text{NC}\text{\char`\~}\text{\char`\~}\text{CN} \quad (26)$$

Equation 24 describes the Ni(0)-catalyzed addition of HCN to butadiene, which leads to 3-pentenenitrile together with its allylic isomer 2-methyl-3-butenenitrile (1.5:1 molar ratio). The branched allylic isomer, however, progressively isomerizes to the linear 3-pentenenitrile.

Double-bond isomerization of 3-pentenenitrile (Equation 25) gives an equilibrium mixture containing only *ca.*7% of 4-pentenenitrile with respect to the 3-isomer. This isomerization occurs much faster than that leading to the thermodynamically favoured 2-pentenenitrile. 4-Pentenenitrile is preferentially hydrocyanated to adiponitrile because it reacts faster than the 3-isomer for steric reasons (Equation 26). In addition to adiponitrile, the final reaction mixture contains 2-methylglutaronitrile (from hydrocyanation of 2-methyl-3-butenenitrile), ethylsuccinonitrile (from cyanation of 3-pentenenitrile before isomerization) and 2-pentenenitrile (from isomerization of 3-pentenenitrile, which does not undergo further hydrocyanation).

The process is paradigmatic of the problems that have to be faced by scientists studying catalysis of organic reactions involving several steps. Oxidative addition of HCN to nickel(0) and butadiene insertion yield an allylmetal intermediate, which undergoes C–C coupling with CN. As allylnickel complexes exist in the η^3 and η^1 forms (in equilibrium with each other), coupling with CN gives both the linear and the branched nitriles. The branched nitrile progressively converts into the linear via equilibria as shown in Figure 13. At this point 3-pentenenitrile undergoes a Ni(II)- and Lewis acid (BPh$_3$ is preferred)–promoted double bond isomerization to 4-pentenenitrile, followed by a second HCN oxidative addition to Ni(0) and C–C coupling [Figure 13; L = PR$_3$ or P(OR)$_3$, n = 1 or 2].

Although not all the fine details of the DuPont process have been disclosed, enough is known to show the fascinating chemistry underlying it. For example nickel(0) is probably a better catalyst than palladium(0) (coordinated by the

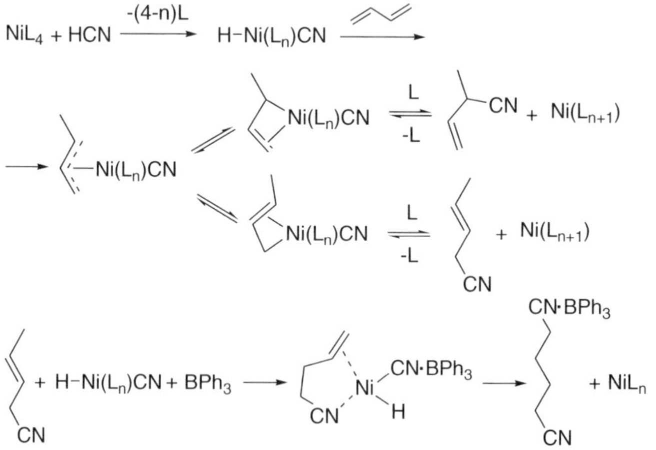

Figure 13   *Schematic representation of the steps presumably involved in the Ni-catalyzed hydrocyanation of butadiene to adiponitrile.*

same ligands), because, being more basic, it favours oxidative addition of HCN (which may be formally regarded as a protonation of the basic metal).

The chosen ligands have large cone angles and therefore have a strong tendency to dissociate from the metal, thus providing coordination sites for entering substrate and reagents. In this reaction ligand dissociation has also been experimentally found to be favoured by oxidative addition of HCN to nickel(0), especially if the latter is bonded through nitrogen to a Lewis acid which enhances its strength. That a Lewis acid such as BPh_3 favours double bond isomerization is probably due to its coordination to the Ni-bonded CN group, which increases the concentration of the catalytically relevant species. This is also useful for the subsequent addition of the second HCN molecule.

The use of tris(o-tolyl) phosphite in place of triphenylphosphine has been found to favour the reductive coupling of the metal-coordinated methylallyl (or of the cyanobutyl group in the second step) with the CN group, which liberates the metal in its original (0) oxidation state.

A drawback connected with the tendency of Ni(0) to undergo oxidative addition of HCN is the formation of inactive nickel cyanide complexes. Such inhibition can be overcome, however, by keeping the HCN concentration low. This also has the further advantage of preventing sudden temperature increases arising from the high exothermicity of the reactions with HCN.

5.5 Oligomerization of Olefins

The reiteration of the insertion step of unsaturated compounds into metal-alkyl bonds leads to polymerizations, dealt with in Chapter 7. However a series of important commodities, mainly used as monomers for polymerization, are manufactured by catalytic processes based on early termination steps of sequential olefin additions.

Shell manufactures α-olefins from ethylene by oligomerization with a nickel catalyst in a polar solvent such as ethylene glycol, under the conditions specified in Equation 27. This corresponds to the first part of the SHOP process (Shell Higher Olefin Process) described in Section 6.2.2. The world production is estimated to be over 1 Mt/a.

$$nH_2C{=}CH_2 \xrightarrow[\substack{HOCH_2CH_2OH \\ 80\text{-}100\ ^\circ C\ ,\ 70\text{-}140\ bar}]{Ni} CH_3CH_2(CH_2CH_2)_{n\text{-}2}CH{=}CH_2 \quad (27)$$

99% yield
>6000 TON

C_{12}–C_{18} α-olefins are distilled out and used directly for the manufacture of detergents, while C_4–C_{10} olefins and those > C_{20} undergo isomerization and metathesis as described in Chapters 3 and 6, respectively, to obtain suitable C_{10}–C_{14} fractions.

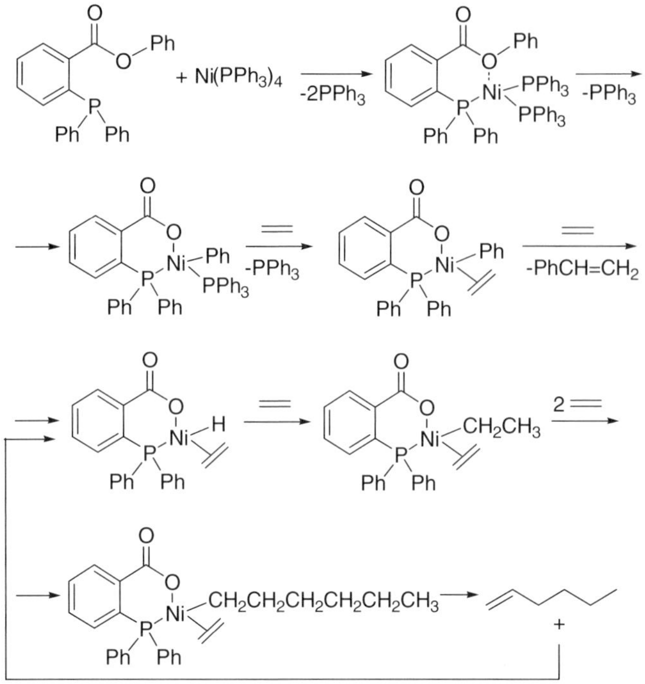

Figure 14 *A proposed mechanism for the Ni–catalyzed trimerization of ethylene to 1–hexene using a hemilabile P⌢O ligand.*

The oligomerization reaction is believed to involve a nickel hydride, which is generated *in situ* from nickel(0), triphenylphosphine, and the phenyl ester of *o*-diphenylphosphinobenzoic acid. The trimerization of ethylene to 1–hexene is described in Figure 14.

The first step corresponds to coordination of the *ortho*-phosphinated benzoic acid phenyl ester to the pre-catalyst, a triphenylphosphinenickel(0) complex. This is followed by the oxidative addition of the phenyl ester group, then ethylene coordinates and inserts into the phenyl–nickel bond. This sequence presumably requires ligand (PPh$_3$) dissociation and provides a good example of how complex the formation of the "true" catalyst from the initially added pre–catalyst can be. The initial hydride that starts the catalytic cycle is probably formed by reductive elimination of styrene from the nickel-bonded phenylethyl group. At this point ethylene coordinates, probably *trans* to oxygen, and a migratory insertion into the nickel hydride bond occurs, the coordination sites of the species involved being dictated by the P⌢O system. The hemilabile P⌢O ligand (Box 8) turned out to be best suited for catalysis; a related P⌢P ligand was not effective. Repetition of these steps leads to the formation of a nickel–bonded hexyl group. Reductive elimination occurs by hydride transfer to nickel, the latter reverting to its initial state and starting a new catalytic cycle. The presence of the chelating ligand allows hydride transfer to be faster than further ethylene insertion, which would lead to polymerization (Chapter 7). A possible alternative to hydride transfer to nickel is hydride transfer to coordinated ethylene, which generates a nickel-ethyl bond able to start a new oligomer chain. This point is still the subject of debate.

Box 8 Mixed and Hemilabile Ligands

Research on C–C bond formation and polymerization has identified a series of effective mixed ligands. *Mixed ligands* offer at least two coordination sites with different properties, and can thus exert differing influences on the coordination of other ligands. For unsymmetrical chelate ligands with two different donor ends (eg., -O and PR_3) one end can dissociate more easily from the metal than the other: such ligands are called *hemilabile*. These properties can be exploited by a substrate to gain access to the reaction centre either by selective opening of one coordinating site or by taking advantage of a coordination site made available selectively after dissociation of a stabilizing ligand. Well established examples are offered by $P^\frown O$ systems: the one shown in Figure 14 undergoes oxidative addition to the metal at the labile ester bond and also favours ethylene coordination *trans* to the resulting metal to oxygen bond. Benzylcyclopentadienyl-type ligands, extensively used in polymerization (Chapter 7), also show hemilabile behaviour, being able to coordinate to metals through either the cyclopentadienyl or the arene moieties [A. Bader and E. Lindner, *Coord. Chem. Rev.* 1991, **108**, 27].

The oligomerization product thus formed is a mixture of olefins with a Schulz-Flory distribution of molecular weights (see also Section 4.7.2), whose composition can be modified, for example by adding an excess of tertiary phosphine or by changing the ancillary ligand.

A similar catalytic process (Dimersol, Institut Français du Petrole), based on a nickel hydride formed *in situ* from a nickel complex and an aluminium alkyl, has been applied industrially to oligomerize ethylene, propylene, butenes or mixtures of the three.

Ethylene dimerization forming 1-butene can also be carried out selectively with a titanium(IV) derivative which is reduced *in situ* to titanium(II) (Alphabutol process, Institut Français du Petrole). The result has been attributed to the formation of a titanacyclopentane, which decomposes to 1-butene. The absence of a hydride species active in oligomerization would account for the high selectivity (Figure 15). No additional solvent is required, as 1-butene also acts as solvent. The total world butene production capacity by this process is estimated to be > 300 kt/a.

Ethylene has also been trimerized selectively to 1-hexene, used as a monomer for copolymerisation with ethylene (Chapter 7).

Chevron-Phillips operates a plant in Qatar, based on the catalytic trimerization with a chromium complex prepared from chromium(III) 2-ethylhexanoate,

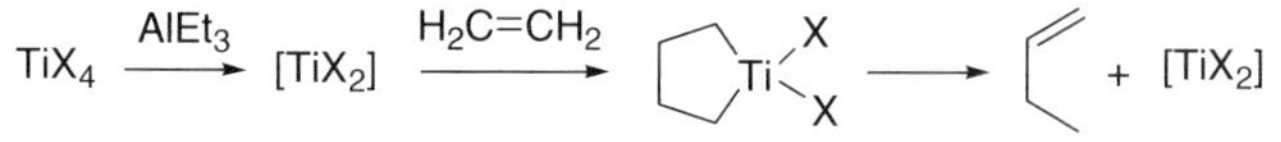

Figure 15 *Schematic representation of a possible mechanism for the titanium-catalyzed dimerization of ethylene to 1-butene via a titanacyclopentane intermediate.*

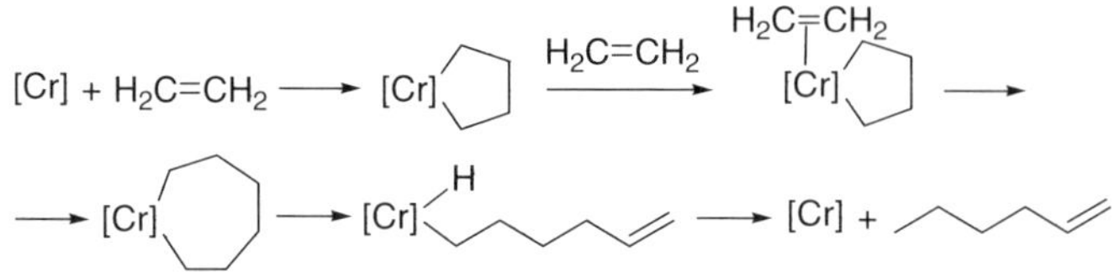

Figure 16 *Schematic representation of a possible path for ethylene trimerization involving 5- and 7-membered chromacycles.*

2,5-dimethylpyrrole, diethylaluminium chloride and triethylamine in toluene at 115°C and 100 bar. Oligomers, containing *ca.* 94% of hexenes are obtained, 1-hexene being present to the extent of ca 99%. The TOF is *ca.* 156 kg /g Cr/h. The aromatic solvent presumably stabilizes the chromium complex through coordination [W. Freeman, J. L. Buster and R. D. Knudsen, US 5 856 257 (1999); R. D Knudsen, J. W. Freeman and M. E. Lashler US 5 563 312 (1996), to Phillips Petroleum]. The real nature of the catalytic species is not known, and although an isolated polynuclear tris(pyrrolyl)chromium chloride complex was found to be active, its catalytic efficiency was lower than that obtained by preparing the catalyst *in situ*. This is a not unusual observation as the true catalytic species are often labile, present in low concentration and not easily identifiable.

Although the actual oxidation state of chromium in the active catalyst is unclear, the reaction has again been interpreted as an ethylene dimerization leading to a metallacycle, in this case followed by ethylene addition and β-hydrogen elimination (Figure 16).

The effect of the pyrrolyl ligand may be due to a type of hemilability (see Box 8) arising from an η^5 to η^1 coordination shift, the former facilitating metallacycle formation and the latter favouring further ethylene insertion and formation of the chromium-bonded 5-hexenyl species.

5.6 Carbene Chemistry and Asymmetric Synthesis: Chrysanthemic Esters

Carbene species can be stabilized by complexation to metals and transferred to olefinic substrates in catalytic reactions. Although the main industrial application of carbenes is in metathesis (Chapter 6), an important application in the area of fine chemicals is to asymmetric cyclopropanation.

Chrysanthemic acid esters, prepared according to Equation 28, belong to the family of pyrethroids, an important class of insecticides. Both the *trans-* and the *cis*-isomers possess insecticidal activity. The other two of the four possible isomers are inactive. The process by Sumitomo Chemical Co. (100–200 t/a, estimated) consists of the reaction of 2,5-dimethyl-2,4-hexadiene with an optically active diazoacetic ester in the presence of a catalytic amount of a copper(I) complex (Equation 28 and Figure 18). The reaction, carried out at 40°C, gives an *ee* of 94% for the *trans* isomer and 46% for the *cis* isomer with a *cis/trans* ratio of 7/93.

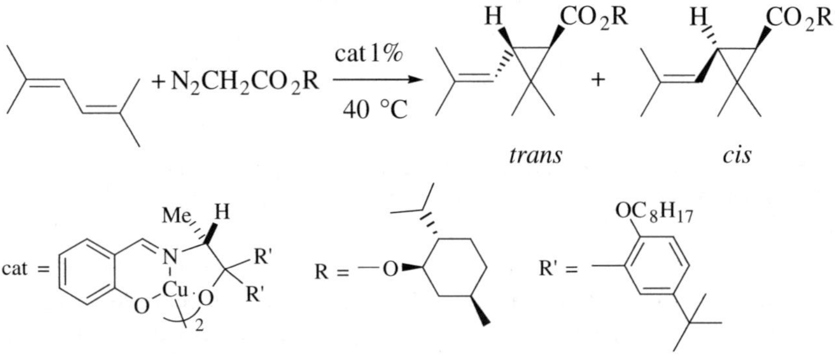

(28)

Box 9 Use of Ionic Liquids

Modifications have recently been proposed for the Dimersol and the Phillips processes by Institut Français du Pétrol and Sasol Technologies, respectively, based on the use of an ionic liquid such as imidazolium tetrachloroaluminate (Figure 17), which is liquid at room temperature and an excellent solvent for the organometallic catalyst.

The ionic liquid can, for example, be added to the butene effluent from the Dimersol process to obtain octenes by butene dimerization; the octene can be carbonylated (Section 4.6) and hydrogenated to *iso*-nonanol, used to make phthalate plasticizers. In the case of the Phillips trimerization process the use of an ionic liquid allows an easy separation of the trimers and the catalyst for recycling (see also reviews to Section 5.5). However, the industrial use as solvents of ionic liquids, containing halide species (especially anions such as BF_4^-, PF_6^-, or $AlCl_4^-$) has the disadvantage that they readily break down to give HX, which can adversely affect the reaction. New types of non-halide containing ionic liquids are being actively researched.

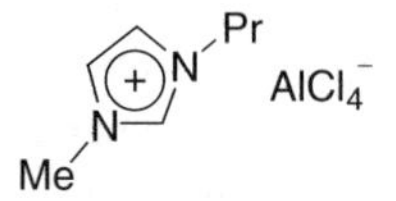

Figure 17 *An example of an ionic liquid.*

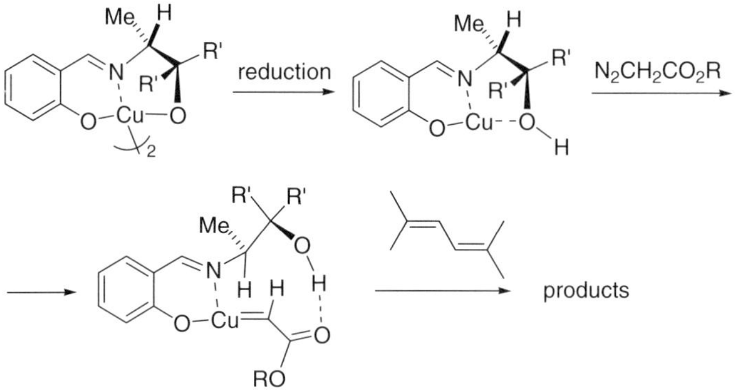

Figure 18   *A schematic representation of the proposed copper intermediates in cyclo-propanation (R and R' as in Equation 28).*

The pre-catalyst, a salicylaldiminato complex of copper(II), is reduced *in situ* to the corresponding copper(I) complex by a hydrazine derivative or by the diazoacetic ester. The latter then reacts with the activated complex evolving N_2 and forming a carbene complex, which in turn reacts with the diene substrate to give a cyclopropane ring. The stereo- and enantio-selectivities are induced when the diene substrate approaches the carbene moiety at the face not screened by the N-bonded substituted hydroxyalkyl group. Computational models have given a satisfactory picture of the preferred pathway. The chelat-ing ligand is bonded to copper through the O and N atoms forming a system of C_1 symmetry. C_2 symmetric ligands are reported in Sections 2.7 for oxidation and 3.5–3.8 for hydrogenation.

One double bond of the diene attacks the copper-bonded carbene on the side of the carbene chain opposite to the N–C–C–OH group to form the cyclopro-pane ring. The carbene chain is intrinsically chiral, its plane being orthogonal to that of the N–Cu–O plane (with the C–H bond pointing upwards) and rigid, due to charge transfer from the chelated copper towards the ester carbonyl group; a further contribution to rigidity comes from a hydrogen bond to the hydroxyl group of the N–C–C–OH chain. The R and R' groups effect enantio- and stereo-selectivity: R' is usually a 2-substituted aryl group such as 5-*t*-butyl-2-octyloxyphenyl and its bulkiness is essential both to induce stereochemistry towards the *trans* isomer and to increase enantioselectivity. The stereogenic centre, which is in a diastereoisomeric relationship with the chiral carbene, usually bears a methyl group; bulkier groups have an adverse effect on enantioselectivity. R in the ester can be an alkyl group: since the carbonyl oxygen is hydrogen-bonded to the oxygen of the side chain, R is forced to occupy the region where the diene must travel to attack the carbene ligand, thus influencing the diastereoselectivity. Normally bulky ligands are the most effec-tive (Equation 28) and *l*-menthyl is preferred as it enhances both the enantio- and the diastereo-selectivity. Thus a combination of factors that ensure pref-erential access of the diene substrate to one face of the carbene ligand are responsible for the high stereo- and enantio-selectivity obtained [K. Suenobu, M. Itagaki and E. Nakamura, *J. Am. Chem. Soc.* 2004, **126**, 7271].

Discussion Point DP3: *Propose an application of the techniques reported in this chapter to the synthesis of the potentially interesting industrial compound methyl p-cyanocinnamate (which after hydrogenation to the aminoester, could be a possible monomer for polymerization). What will be the advantage over the conventional procedures and what difficulties need to be overcome?*

Box 10 Recent Trends in the Asymmetric Catalysis of C–C Bond Formation

Metal-catalyzed industrial processes for asymmetric C–C bond formation have not yet fully developed in spite of the increasing demand of single enantiomers for pharmaceuticals, agrochemicals and other applications of fine chemicals. As already mentioned this arises from problems connected with the use of metal complexes in fine chemical processes, and the strong competition by enzymatic catalysis. However, as more enzyme structures are elucidated and computational methods are refined, the design of synthetic catalyst precursors will become more accurate and the transfer of these procedures to industrial processes will be more likely to occur. Research has already worked out a number of new procedures, including Heck and Suzuki reactions, nucleophilic addition to carbonyl compounds, Diels-Alder type reactions, aldol reactions, cyclopropanation, and metathesis. These procedures are based on the design of new metal complexes (*e.g.* of Rh) of C_2 or C_1 symmetry, on the use of new reaction media, on exploiting chirality transfer through achiral ligands, and on the amplification of chirality from a low to a very high enantiomeric excess.

The cyclopropanation procedure shown in Section 5.6 has been studied with many other metals (Rh, Ru, Co, Fe, Os, Pd, Pt, Cr) and with a variety of new asymmetric ligands such as the bis-oxazolines.

Annex 1 Devising New Synthetic Pathways

To solve problems such as the one proposed in *DP3* try to apply a retroanalysis of the molecule to be synthesized, cutting its structure into readily available or easy to prepare sub-units. At this point look at the possibility to synthesize the target molecule from these sub-units using organometallic procedures that have found successful industrial applications.

A simple example is offered by adiponitrile: the sub-units used can derive from cutting the adiponitrile molecule into a four-carbon unit (butadiene) and two one-carbon units (HCN) as in the process shown in Section 5.4. It is also possible, however, to cut the same molecule into two three-carbon units (acrylonitrile) and a coupling procedure of the latter can be devised. A rhodium(I) chloride species in the presence of H_2 has indeed proved to effect this coupling, but an industrial application has not yet been achieved, however [D. J. Milner and R. Whelan, *J. Organomet. Chem.* 1978, **152**, 193].

An electrochemical procedure for acrylonitrile coupling has been industrially successful. Here a reductive acrylonitrile coupling is effected by the electrons coming from the cathode of an electrolytic cell. The first formed radical carbanion attacks the second acrylonitrile molecule. Uptake of a new electron from the cathode and of two protons gives adiponitrile in high yield [M. Baizer and D. E. Danly, *Chem. Ind.* (London), 1979, **435**, 439].

Finally in carrying out the retroanalysis it is possible to retain the six-carbon skeleton of adiponitrile by starting from cyclohexane and ammonia. The former can be readily converted to adipic acid by catalytic oxidation (Chapter 2).

Annex 2 Hints to Improve or to Develop Alternative Processes for the Synthesis of Aromatics Catalyzed by Transition Metals

While processes based on Pd-catalyzed C–C coupling reactions dependent on aromatic carbon-halogen activation are used in the laboratory, they have so far been little used in industry. Strategies that could be employed to expand their use include:

–using bulky and electron-releasing phosphines such as tris-*t*-butylphosphine or biphenyl-containing tertiary phosphines to facilitate the oxidative addition of aromatic C–Cl bonds to palladium;
–attaining high catalytic efficiency with cyclometalated complexes as catalyst precursors or with phosphite or carbene ligands containing bulky substituents in the *ortho* position;
–replacing palladium with nickel; this should favour the oxidative addition step, as Ni(0) shows a higher basicity, other factors (ligands, stereochemistry) being equal.

In addition other strategies include:

–replacing the aromatic C–Cl bond with the C–O in anhydrides, thus diminishing halide use and waste disposal problems;
–replacing organoboron with organosilicon compounds;
–exploring highly diluted palladium catalysts in the presence of weakly coordinating ligands (or solvents other than phosphines), to prevent metal precipitation and to avoid phosphine ligands at the same time.

In parallel with the studies on C–C coupling, C–N and C–O coupling reactions catalyzed by palladium or other transition metals, have recently been investigated and important results have been obtained by the Buchwald and Hartwig groups in coupling bromo or chloroarenes with amine and congeners or alcohols. Industrial processes are still in their infancy but the prospects are promising.

Annex 3 Perspectives in C–C Bond Forming Organic Syntheses

Annex 3.1 Catalytic Efficiency and Selectivity

The design of more efficient catalysts approaching and even surpassing the TON and selectivity of enzymes is an actively pursued goal. An ideal catalyst will match the substrate selectively and ensure comparable rates of the reaction steps.

Some new homogeneous catalysts, containing sophisticated ligands, have already been considered (Annex 2) and applications to industrially interesting procedures can be expected.

Annex 3.2 Reaction Media for Catalysis

C–C bond forming catalytic reactions have been carried out in a variety of media including water and several organic solvents. Many reactions are promoted by polar solvents. Aqueous-organic biphases (Section 5.3.3), and phase transfer have been used. Micellar catalysis allows reactions to be carried out in water using long-chain ammonium or phosphonium salts in the absence of organic solvents. For example 2,7-octadien-1-ol (Section 5.4.3) has been successfully synthesized on a laboratory scale from butadiene and water [E. Monflier, P. Bourdauducq, J.Couturier, J. Kervennal and A. Mortreux, *Appl. Cat A: General* 1995, **131**, 167].

Another biphasic technique uses a fluorinated substance (*fluorous phase*), which is immiscible with water and most non-fluorinated solvents at room temperature, but which can form a single phase at higher temperatures. Products can thus be separated at room temperature and the soluble catalyst recycled. One such catalyst is made by adding two perfluorinated chains to dba, a classic palladium(0) ligand.

Ionic liquids have also been used as reaction media (Box 9). The high polarity of quaternary ammonium salts is exploited to favour polar reaction steps. For example, with 1-*n*-butyl-3-methylimidazolium tetrafluoborate, 2,7-octadien-1-ol (Section 5.4.3) was obtained in good yield from butadiene and water and was readily separated as it was immiscible with the ionic liquid below 5°C [J. E. L. Dullius, P. A. Z. Suarez, S. Einloft, R. F. deSouza, J. Dupont, J. Fisher and A. DeCian, *Organometallics* 1998, **17**, 815].

Supercritical fluids and particularly carbon dioxide can also be useful media for C–C bond forming reactions. It should also be pointed out that many of the above procedures, although attractive on a laboratory scale, do involve expensive reagents, ligands, or equipment, and thus are most likely to be applied only in making products of very significantly high value.

Heterogeneous or heterogenised catalysts (Appendix 2) are extensively used in the form of solid oxides for large industrial production (as shown in Section 5.2), but so far have not found significant uses in fine chemicals manufacture, although some attractive routes have been worked out. For example 2,7-octadien-1-ol (Section 5.4.3) can be readily obtained from but-adiene and water through an alternative process by using supported palladium on montmorillonite (a layered silicate clay) [B. I. Lee, K. H. Lee and J. S. Lee, *J. Mol. Catal. A, Chem.*, 2000, **156**, 283]. Metal or metal complex nanoparticles prepared in media of various kinds also hold promise.

Annex 3.3 C-H Activation

The activation of C-H bonds for direct C–C bond formation reactions has the potential to become very important especially if it can be accomplished for sp^3 C–H bonds, in methane or alkanes as these are the major feedstocks available. In addition, C–H bond activation of functionalized organic compounds for selective C–C bond formation has been and will continue to be a very impor-tant goal of organometallic catalysis. So far the use of transition metal complexes has led to interesting results which however are not yet industrially relevant.

Annex 3.4 Multistep Reactions

Another rapidly progressing field is that of multistep reactions which occur in ordered sequences chemo-, regio- and stereo-selectively on a transition metal species. To this end, it is necessary to delay release of the desired product until the whole series of steps has been completed; competitive terminations (such as hydride elimination) must be prevented or must only occur at low rates compared to the main sequence. An example, reported by Chiusoli in the late 50s, is offered by the nickel-catalyzed synthesis of methyl 2,5-heptadienoate from 2-butenyl chloride, acetylene, CO and methanol. The reaction is chemo-, regio- and stereo-selective: the four molecules react in the order shown in Equation A3.4 (chemoselectivity); the butenyl group attacks the terminal allylic carbon rather than the internal one (regioselectivity) and acetylene insertion leads to a *Z* double bond (stereoselectivity).

$$\text{CH}_3\text{CH}=\text{CHCH}_2\text{Cl} + \text{HC}\equiv\text{CH} + \text{CO} + \text{MeOH} \xrightarrow{\text{Ni}^0, \text{MgO}} \text{CO}_2\text{Me}$$

$$(A3.4)$$

Remarkable results were recently achieved with the catalytic use of metallacy-cles, which form in the course of the reaction sequence, direct a key step selectively and are cleaved spontaneously [M. Catellani, F. Frignani and A. Rangoni, *Angew. Chem. Int. Ed. Engl.* 1997, **36**, 119]. The use of these techniques can reduce the number of separate steps usually needed in the manufacture of pharmaceutical compounds, agrochemicals and other speciality chemicals.

Another important research trend refers to cooperative catalysis, in which two or more metals cooperate in different steps of a process.

References

Section 5.2: aromatic Friedel–Crafts-type alkylations: C. Perego and P. Ingallina, *Catal. Today* 2002, **73**, 3; A. Corma, *Chem. Rev.* 1995, **95**, 559; alkane cracking and isomerization: Y. Ono, *Catal. Today* 2003, **81**, 3; A. Feller and J. A. Lercher, *Adv. Catal.* 2004, **48**, 229; isoparaffin-olefin alkylation: A. Corma and A. Martinez, *Catal. Rev. Sci. Eng.* 1993, **35**, 483; acid–base catalysis with metal oxides: K. Tanabe and W. F. Hoelderich, *Appl. Catal. A: Gen.* 1999, **181**, 399.

Section 5.3.1: Heck reaction: R. F. Heck, *J. Am. Chem. Soc.* 1968, **90**, 5518; R. F. Heck, *Org. React.* 1982, **27**, 345; W. Cabri and I. Candiani, *Acc. Chem. Res.* 1995, **28**, 2; I. Beletskaya and A. V. Ceprakov, *Chem. Rev.* 2000, **100**, 3009; inhibiting effects: P. W. N. M. van Leeuwen, *Appl.Cat. A: Gen.* 2001, **212**, 61; P. E. Garrou, *Chem. Rev.* 1985, **85**, 171; nature of palladium intermediates: C. Amatore and A. Jutand, *Acc. Chem. Res.* 2000, **33**, 314; applications to pharmaceuticals: M. Prashad, *Topics Organomet. Chem.* 2004, **6**, 181; C. E. Garret and K. Prasad, *Adv. Synth. Catal.* 2004, **346**, 889.

Section 5.3.2: Synthesis of terbinafin: U. Beutler, J. Mazacek, G. Penn, B. Schenkel, and D. Wasmuth, *Chimia* 1996, **50**, 154; E.-I. Negishi and L. Anastasia, *Chem. Rev.* 2003, **103**, 1979.

Section 5.3.3 Cross coupling: *J. Organomet. Chem.*, special issue, 2002, 653; Suzuki reactions: A. M. Rouhi, *Chem. Eng. News* 2004, **82**, (36) 49; A. Suzuki in *Modern Arene Chemistry*, ed. D. Astruc, Wiley, New York, 2002, p. 53.

Section 5.4.1: Reaction of butadiene with ethylene: A. C. L. Su, *Adv. Organometal. Chem.* 1979, **17**, 269.

Section 5.4.2 Nickel-catalyzed butadiene chemistry: G. Wilke, *Angew. Chem. Int. Ed. Engl.* 1988, 27, 185; S. Tobish, *Adv. Organomet. Chem.* 2003, **49**, 167.

Metallacycle chemistry: J. P Collman, L. S. Hegedus, J. R. Norton and R. G. Finke, *Organometallic Chemistry of Transition Metals: Principles and Use*, University Science Book, Mill Valley, Calfornia, 1989, Chapter 10.

Section 5.4.3 Octadienol from butadiene: E. Monflier, P. Bourdauducq, J.-L. Couturier, J. Kervennal and A. Mortreux, *J. Mol. Cat A* 1995, **97**, 29; *Appl. Cat. A* 1995, **131**, 167; B. I. Lee, K. H. Lee and J. S. Lee, *J. Mol. Cat. A: Chem.* 2001, **166**, 233.

Section 5.4.4. Butadiene hydrocyanation: C. A. Tolman, W. C. Seidel, J. D. Druliner and P. J. Domaille, *Adv. Catal.* 1985, **33**, 1, *idem*, Organometallics, 1984, 3, 33.

Section 5.5: Olefin oligomerization: Y. Chauvin and H. Olivier-Bourbigou, *ChemTech* 1995, **25**, 26; Y. Chauvin, S. Einloft and H. Olivier-Bourbigou, *Ind. Eng. Chem. Res.* 1995, **34**, 1149. J. T. Dixon, M. J. Green, F. M. Hess and D. H. Morgan, *J. Organomet. Chem.* 2004, **689**, 3641; F. Speiser, L. Saussine, and P. Braunstein, *Acc. Chem. Res.* 2005, **38**, 784; mechanistic considerations: W. J. van Rensburg, C. Grové, J. P. Steynberg, K. B. Stark, J. J. Huyser and P. J. Steynberg, *Organometallics* 2004, **23**, 1207.

Section 5.6: Asymmetric cyclopropanation: T. Aratani, *Pure Appl. Chem.* 1985, **57**, 1839; H. Lebel, J. Marcoux, C. Molinaro, A. B. Charette, *Chem. Rev.* 2003, **103**, 977.

Asymmetric catalysis: August issue of *Chem. Rev.* 2003 and *Proceedings Nat. Acad. Sciences,* Asymmetric Catalysis Special Feature part I, 2004, **101**, 5347 and part II, *ibidem*, p. 5723; B. M. Trost and M. L. Crawley, *Chem Rev.* 2003, **103**, 2921.

Annex 2 Catalysis of coupling reactions with aryl halides: A. Zapf and M. Beller, *Chem. Comm.* 2005, 431; idem, *Topics in Catalysis* 2002, **19**, 101; C. E. Tucker and J. G. de Vries, *Topics in Catalysis* 2002, **19**, 111; R. A. de Vries, P. C. Vosejpka and M. L. Ash, in *Chemical Industries 75, Catalysis of Organic Reactions*, ed. F. Herkes, M. Dekker, New York, 1998, p. 467; catalysts for aryl coupling: A. M. Rouhi, *Chem. Eng. News* 2004, **82**, (36), 49; highly efficient catalysts: V. Farina, *Adv. Synth. Catal.* 2004, **346**, 1553; coupling with aryl chlorides: A.F. Linke and G. C. Fu, *Angew. Chem. Int. Ed.* 2002, **41**, 4176; C–N and C–O coupling reactions: B. Schlummer and U. Scholz, *Adv. Synth. Catal.* 2004, **346**, 1599.

Annex 3.2 Phase transfer: B. Cornils, *J. Mol. Catal. A: Chem.* 1999, **143**, 1; C. J. Li, *Acc. Chem. Res.* 2002, **35**, 533; micellar catalysis: M. F. Rouasse, I. B. Blagoeva, R. Ciri, L. Garcia-Rio, J. R. Leis, A. Marques, J. Mejuto and E. Monnier, *Pure Appl. Chem.* 1997, **69**, 1923; fluorous phase: I. T. Horvath, *Acc. Chem. Res.* 1998, **31**, 641; ionic liquids: *J. Organomet. Chem.*, special issue, 2005, **690** (15); supercritical fluids: P. G. Jessop, T. Ikariya and R. Noyori, *Chem. Rev.* 1999, **99**, 475; heterogeneous phase: *Handbook of Heterogeneous Catalysis*, ed. G. Erk, H. Knoetzinger and J. Weitkamp, vol. 5, Wiley-VCH, Weinheim, 1997. *Nanostructured Catalysts*, eds. S. L. Scott, C. M. Crudden and C. W. Jones, Springer, Berlin, 2003.

Annex 3.3 C–H activation: F. Kakiuchi and W. Chatani, *Adv. Synth. Catal.* 2003, **345**, 1077; A. E. Shilov and G. B. Shul'pin, *Chem. Rev.* 1997, **97**, 2879; G. Dyker, *Angew. Chem. Int. Ed. Engl.* 1999, **38**, 1698.

Annex 3.4: Multistep reactions: M. Catellani, G. P. Chiusoli and M. Costa, *Pure Appl. Chem.* 1990, **62**, 623; R. C. Larock, *Pure Appl. Chem.* 1990, **62**, 653; L. Tietze, *Chem. Rev.* 1996, **96**, 115; G. Dyker, *Chem. Ber/Recueil* 1997, **130**, 156; K. C. Nicolaou, T. Montagnon and S. A. Snyder, *Chem. Comm.*, 2003, 557; J. Tsuji in *Handbook of Organopalladium Chemistry for Organic Synthesis*, vol. 2, ed. E.-I. Negishi, Wiley, New York, 2002, 1669; R. Grigg, and V. Sridharan, *J. Organomet. Chem.* 1999, **576**, 65; cooperative catalysis: J. M. Lee, Y. Na, H. Han and S. Chang, *Chem. Soc. Rev.* 2004, 302.

General texts dealing with applications of homogeneous catalysis to C–C bond formation: B. Cornils and W. Herrmann, *Applied Homogeneous Catalysis with Organometallic Compounds*, 2nd ed., Vol 1, 2, and 3, VCH, Weinheim, 2004. S. Bhaduri and D. Mukesh, *Homogeneous Catalysis: Mechanisms and Applications*, Wiley, New York, 2000. M. Beller and C. Bolm, *Transition Metals for Organic Synthesis*, 2nd ed., vol. 1 and 2, Wiley-VCH, Weinheim, 2004. P. W. N. M. van Leeuwen, *Homogeneous Catalysis*, Kluwer, London, 2004. R. D. Larsen, *Organometallics in Process Chemistry*, Springer, Heidelberg, 2004.

CHAPTER 6
Metathesis of Olefins

CATHERINE L. DWYER

Sasol Technology Research & Development, Sasolburg 1947, South Africa

6.1 Introduction: History and Basic Chemistry of Metathesis

Alkene metathesis is a means of forming new carbon-carbon double bonds, and is a valuable integrating technology for any company producing alkenes or using them in further conversions. This truly novel reaction was discovered in the 1950's, and although its actual nature went unrecognized for over a decade, it has found application in diverse areas. The originality and the importance of the metathesis reaction was highlighted by the award of the Nobel Prize for 2005 to Chauvin, Grubbs and Schrock, who each made seminal contributions to the understanding and the development of the subject.

Depending on the nature of the substrate, several specific types of metathesis reactions are possible: *Cross Metathesis* (CM) for bulk chemicals, *Ring Opening Metathesis Polymerization* (ROMP) for speciality plastics, and *Ring Closing Metathesis* (RCM) for fine chemicals applications. They have very different characteristics and criteria for success, which have complicated effective process development. Despite this, a number of metathesis processes have been implemented industrially on large scale, and research into catalyst and process development is vigorous in both academia and chemical companies around the world. A focus on developing more selective and active catalysts coupled with new market and feedstock pressures is expected to lead to further applications of metathesis technology, as suggested by recent announcements of new processes and collaborations between industry and key research groups.

Metathesis is a versatile reaction applicable to almost any olefinic substrate: internal, terminal or cyclic alkenes, as well as dienes or polyenes. (Alkyne metathesis is a growing area, but will not be dealt with here.) The reaction is also known as *olefin disproportionation* or *olefin transmutation*, and involves the exchange of fragments between two double bonds. Cross metathesis (CM, Figure 1) is defined as the reaction of two discrete alkene molecules to form two new alkenes. Where the two starting alkene molecules are the same it is called *self-metathesis*. *Ethenolysis* is a specific type of cross metathesis where ethylene

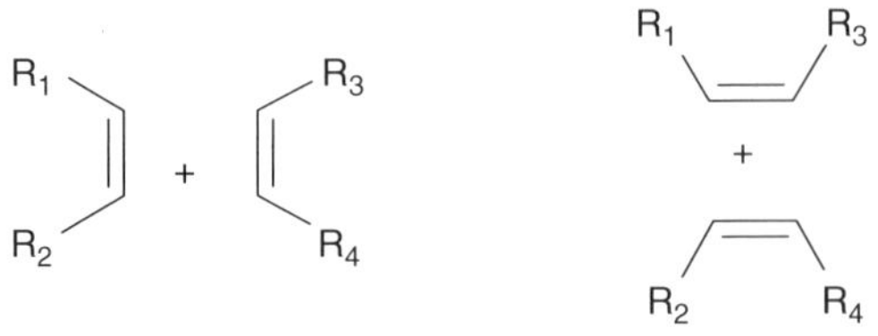

Figure 1 *Cross metathesis of alkenes.*

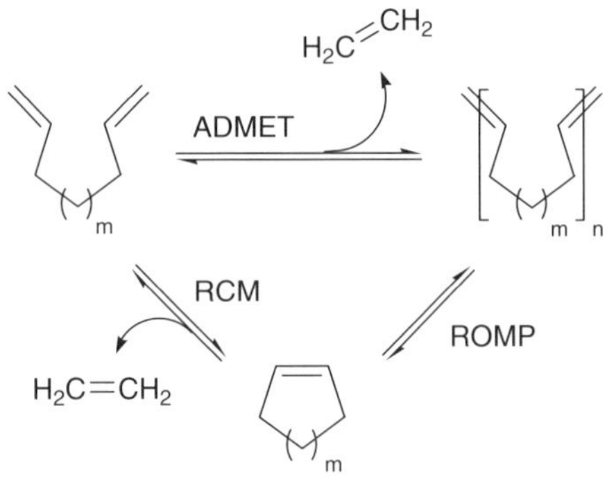

Figure 2 *Metathesis reactions of cycloalkenes and dienes.*

is employed as one of the alkene starting materials, and always affords alpha olefin products. Cross metathesis reactions are of use in the manufacture of commodity chemicals or in further conversion of oleochemicals.

Unsaturated polymers can be produced by means of ring-opening metathesis polymerization (ROMP) of cyclic alkenes. These unique polymers can also be produced via intermolecular *Acyclic Diene Metathesis* (ADMET). Dienes can also react intramolecularly via *Ring Closing Metathesis* (RCM) to afford cyclic products. RCM is often applied to synthesis of compounds for fine chemical and pharmaceutical application. Generic examples of these reactions are shown in Figure 2.

Metathesis has a relatively short but interesting history. In the 1950's huge research efforts in both industry and academia were directed towards investigating the reactions of olefins with transition metals. Much of this was spurred on by the research carried out by Ziegler and Natta and in industry into organometallic complexes, as well as into the polymerization of ethylene and other monomers. During independent studies at various companies including DuPont, Standard Oil and Phillips Petroleum, some odd results were obtained which could not be explained by then current knowledge. This included the formation of unsaturated polymers from cyclic olefins and the observation of disproportionation products of alkenes during reactions targeting other products. It was only in 1967 that the generic phrase *olefin metathesis* was coined by Calderon of the Goodyear Tire and Rubber Company, thereby linking the previously unrelated studies of alkene exchange or disproportionation processes and ring opening polymerization chemistry. The concept of a reversible reaction where double bonds could be broken and recombined in a different manner, led to an entirely new branch of catalytic chemistry, spurring significant mechanistic studies and new applications. Initial mechanistic proposals suggested that reaction proceeded in a pairwise manner, via coordination and reaction of two olefins at a metal centre. However, it is now accepted that the reactions are mediated by means of metal carbene complexes, and proceed via a metallacyclobutane intermediate. This 'non-pairwise' mechanism, initially proposed by Herisson and Chauvin in 1971, suggests that metathesis occurs via sequential reaction of olefins with a metal carbene and satisfactorily explains the formation of mixed products. Over the years a variety of elegant mechanistic studies have lent further support to this theory.

6.2 The Carbene-Metallacyclobutane Mechanism of Metathesis

The mechanism first proposed by Chauvin is illustrated in Figure 3. Formation of a metal alkylidene is therefore a critical step when employing catalysts formed *in situ* from supported metal oxides or homogeneous metal halides. After initial coordination of the olefin to the metal centre, rearrangement of the double bonds occurs to form a metallacyclobutane intermediate. This in turn undergoes rearrangement to eliminate an olefin (formed from half of the original olefin and the carbene fragment) and regenerate a new propagating metal alkylidene. It can be seen that if $R_1 \neq R_2, R_3$, two new propagating metal carbenes may form, and two products are possible depending on the orientation of the incoming olefin.

Some of the earliest catalysts for this transformation were heterogeneous metal oxides (typically of tungsten, molybdenum or rhenium) on a support such as silica or alumina. These are still the catalysts employed for all current large scale industrial processes. Ill-defined homogeneous systems comprising alcohol solutions of a metal halide in conjunction with a promoter were also employed extensively in the early years, and are currently still used for the

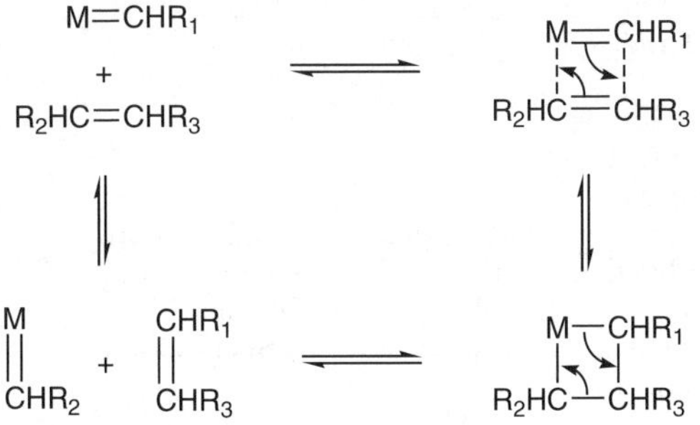

Figure 3 *The basic mechanism of metathesis via metal carbenes.*

production of speciality polymers. The synthesis and use of discrete metal alkylidene complexes by Grubbs, Schrock and others during the 1980's and 1990's was a breakthrough development for homogeneous metathesis technology. Such catalysts offer greater flexibility and improved performance via rational catalyst design; more particularly by varying the ligands around the metal centre to manipulate activity and selectivity. However these largely developmental catalysts are also less robust, and efforts to commercialize them are currently limited to high value products such as speciality chemicals and polymers and pharmaceuticals. A summary of the key catalyst types and their characteristics is given in Table 1. It is evident that the different catalysts operate at very different conditions and that performance also varies dramatically. As will be shown in the following sections, because metathesis reactions and substrates are so varied no single catalyst type is superior in all respects and the catalyst choice for a given process is not a trivial one.

Table 1   *Summary of the most common metathesis catalysts and their chief features*

	HETEROGENEOUS			HOMOGENEOUS
Type	WO_3/SiO_2	$MoO_3/CoO/$ Al_2O_3	Re_2O_7/Al_2O_3	Ru, W, Mo, Re halides and alkylidenes
Operating temperature	300–550°C	50–300°C	−20–100°C	−40–100°C
Operating pressure	1–30 bar	1–30 bar	1–60 bar; higher pressures are required to keep reagents in liquid phase	1–30 bar
Promoters	none	none	$R_mSnX_n/$ $R_mAlX_n/none$	$R_mSnX_n/R_mAlX_n/$ none
Activity	Good activity towards cyclic and acyclic olefins	Highly active for linear olefins	Highly active for simple olefins; active for functionalized olefins when activated with R_4Sn	High activity for ROMP and RCM. Good reactivity for linear olefins, but slower when branched
Selectivity	Poor: Double bond and skeletal isomerization, and cracking may occur	Good	Excellent	Generally excellent; isomerization may occur with some catalysts
Stability	Good: Coking is problematic but the catalyst can be readily regenerated	Catalysts do not readily tolerate functional groups or vinylidenes	Catalysts are more tolerant of certain functionalities	Some catalysts can tolerate functionality; alkylidenes undergo thermal and chemical decomposition

6.3   Industrial Applications of Metathesis

After the discovery of olefin metathesis in the 1950's, the reaction was taken up enthusiastically by a large number of petrochemical companies who could see opportunities to manipulate their olefin streams arising from primary processes such as naphtha cracking and Fischer-Tropsch conversion of syngas. Although numerous processes have been developed over the last 50 years, surprisingly few large scale ones have been commercialized to date. However improvements in the technology and the identification and implementation of processes targeting smaller volume, higher value products have ensured the continuation of active metathesis research. In these days of tightening economies, environmental concerns and rising feedstock costs, it is a technology which may hold the key to elegant, atom-efficient and targeted processes which give versatility to chemical companies seeking to leverage feedstocks and add value to their product slate. Although metathesis polymerization processes are unlikely ever to rival the massive polyethylene and polypropylene industries, smaller scale but higher value ROMP products find application in niche markets such as the motor and sporting goods industries due to their unique properties. In recent years the development of highly selective homogeneous catalysts has prompted extensive studies towards fine chemical and pharmaceutical manufacture. Although numerous papers discuss the use of metathesis for preparation of diverse and complex organic compounds, in many cases it appears that these processes are not yet in commercial production, and it is likely that further development will be required before metathesis becomes a standard process in this area (see below). The range of metathesis processes which are currently available and in use in the commodity chemicals and polymer industries is summarized in Table 2.

6.3.1   Production of Light Olefins

Surprisingly enough given the simplicity of the reaction scheme, cross metathesis of olefins (Figure 1) is one of the more difficult types of metathesis. This is in part because of the lack of an obvious driving force (such as release of ring strain) and the associated equilibrium limitations of such a process, and also because it is far more difficult to control the selectivity of a reaction between two different (but still similar) acyclic olefins. The first metathesis process to find industrial application was the Phillips Triolefin Process, with a plant which operated from 1966 to 1972 near Montreal in Canada. This technology was developed in an effort to increase ethylene and butene production from 'low value' cracker-derived propylene, to meet the growing market demand for polyethylene and polybutadiene. Production of 15 kt/a ethylene and 30 kt/a butene from 50 kt/a propylene (Figure 4, *iv*) was achieved over a WO_3/SiO_2 catalyst at temperatures of around 400°C. The catalyst was doped with sodium to decrease surface acidity and prevent double bond shift reactions, which might lead to lower metathesis selectivities. As it was long before the term 'metathesis' had been coined and when the technology was far from

Table 2　Past and present industrial metathesis processes

Process	Operated by	Feeds	Products	Catalyst; Conditions
Phillips Triolefin Process	Phillips; ran from 1966-1972	propylene	15 kt/a ethylene+30 kt/a 2-butene	WO_3/SiO_2; 400°C
Reverse Triolefin Process	Lyondell Petrochemical Co.; operated since 1985	ethylene+2-butene	136 kt/a propylene	WO_3/SiO_2; 400°C
ABB Lummus OCT®	BASF Fine Petrochemicals; in progress	ethylene+2-butene	400 kt/a propylene	WO_3/SiO_2; 350°C; 20 bar
	Mitsui Chemicals; in progress	ethylene+2-butene	140 kt/a propylene	
	Korea Petrochemical Co.; in progress	ethylene+2-butene	110 kt/a propylene	
Axens Meta-4®	Not yet commercialized	ethylene+2-butene	propylene	Re_2O_7/Al_2O_3; 35°C; 60 bar; $MoO_3/CoO/Al_2O_3$; 100-125°C; 10 bar
SHOP	Shell; 3 plants; 1190 kt/a	C_4 and $>C_{18}$ alkenes	C_{11}-C_{14} alkenes, remainder recycled	$WO_3/MgO/SiO_2$; 370°C; 30 bar
Phillips Neohexene Process	Phillips; operated since 1980	diisobutene+ethylene	1.4 kt/a neohexene; isobutene recycled	WCl_6; 25-100°C
Cyclooctene polymerization	Degussa-Hüls; operated since 1980	cyclooctene	12 kt/a polyoctenamer (Vestenamer®)	WCl_6; 25-100°C
Norbornene polymerization	CdF Chimie; Elf Atochem; operated since 1976	norbornene	polynorbornene (Norsorex®)	$RuCl_3/HCl/BuOH$; 25-100°C
	Nippon Zeon; 2 steps; operated since 1991	norbornene	hydrogenated norbornene (Zeonex®)	
Telene® Process	BF Goodrich	dicyclopentadiene	polydicyclopentadiene (Telene®)	$[R_3NH]_{2x-6x}Mo_xO_y/Et_2AlCl/ROH/SiCl_4$; 25-100°C
Metton® Process	Nippon Zeon	dicyclopentadiene	polydicyclopentadiene (Pentem®)	
	Hercules Inc.	dicyclopentadiene	>13.6 kt/a polydicyclopentadiene (Metton®)	$WCl_6/WOCl_4/Et_2AlCl/ROH$; 25-100°C
Other DCPD processes	Materia, Cymetech, Hitachi Chemical Co.	dicyclopentadiene	polydicyclopentadiene	Ruthenium based catalysts; 25-100°C
Production of pheromones	Materia	5-decene and 1,10-diacetoxy-5-decene	peach twig borer pheromone (5-decenyl acetate)	Ruthenium based catalysts; 25-100°C
Production of fragrances	Symrise	cyclooctene	cyclohexadecenone (Globanone) via cyclohexadecadiene	Ruthenium based catalysts; 25-100°C

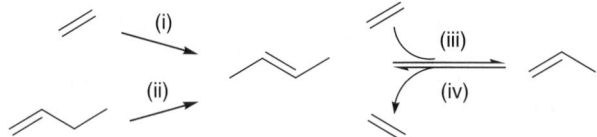

Figure 4 *Routes to C_2–C_4 olefins via metathesis: (i) Ni catalyzed dimerization; (ii) isomerization; (iii) ethenolysis of 2-butene; (iv) self metathesis of propylene. (i + iii) Lyondell Petrochemical Co.; (ii + iii) ABB Lummus (OCT®), Axens (Meta-4®); (iv) Phillips.*

understood, this demonstrates that while understanding is important, it need not stand in the way of successful industrial implementation.

From the early 1970's onwards, increased demand for propylene (as a feedstock for polypropylene, Oxo alcohols, acrylonitrile, cumene, propylene oxide and other value added chemicals) led to the closing of this plant, due to higher value outlets for the propylene feedstock. This trend has continued, and it is unlikely that such a process will be restarted. However, given the reversibility of the metathesis reaction, industry watchers predicted that the process might one day find application in reverse: namely to make propylene from ethylene and butene. In 1985, Lyondell Petrochemical Co. started up a plant in Texas based on this 'reverse' Phillips technology. This plant made economic sense as it was back-integrated into an ethane cracker, which produced the required ethylene feed. The overall process (Figure 4, i+*iii*) includes an initial ethylene dimerization step using a homogeneous nickel catalyst. The 2-butene product is then co-fed with additional ethylene to a metathesis reactor where propylene is produced over a WO_3/SiO_2 catalyst. ABB Lummus offer the reverse Triolefin process for license as Olefin Conversion Technology (OCT®; Figure 4, *ii+iii*; see also Box 1). Here a cracker-derived 1-butene feed undergoes preparation to remove impurities, followed by isomerization to 2-butene. Although pressure is not required for the reaction, running at higher pressures allows for a simpler, cheaper process by alleviating the need for decompression upstream (the cracker-derived ethylene is supplied at elevated pressures) and costly multi-stage compressors downstream (the downstream processes require propylene at elevated pressures). The high operating temperatures required with the tungsten catalyst lead to slow deactivation of the catalyst by coking. The process therefore also incorporates periodic catalyst regeneration. In recent years propylene demand has continued to rise, significantly outpacing that of ethylene. However the largest source of propylene is from steam crackers and fluid catalytic cracking (FCC) units which cannot meet the increased demand as they produce a fixed ratio of propylene to ethylene. On-purpose propylene technologies are therefore required to meet the future shortfall. Indications are that much of the further expansion of on-purpose propylene may come from metathesis. New plants based on OCT® have been announced by several companies, including BASF FINA Petrochemicals (300kt/a), Mitsui Chemicals (140kt/a), Korea Petrochemical Industry (110kt/a).

Box 1 The Meta-4 process

The Meta-4® process was developed by the Institut Français du Petrole (IFP) and the Chinese Petroleum Corporation, and is currently offered for license by Axens, an IFP subsidiary company. This process differs from OCT® technology in the catalyst used, namely Re_2O_7/Al_2O_3, although the overall process is the same. Despite the high oxidation state, rhenium is a "softer" (= larger and more polarizable) metal, and therefore such catalysts operate at significantly milder temperatures than their tungsten and molybdenum analogues (35°C). Reaction between soft metals and soft bases (ligands such as alkenes) occurs readily as the HOMO of the base is relatively close in energy to the LUMO of the metal, and there is a substantial lowering of energy for the resultant fully occupied bonding orbital. However, higher pressures (60 bar) are required to ensure the reactants are in the liquid phase. The rhenium catalyst is very sensitive to feed impurities, and the butene feedstock requires extensive pre-treatment to remove isobutene, butadiene and trace oxygenates. The feed preparation is costly, but is essential as oxygenates and butadiene would lead to unacceptable levels of catalyst poisoning, while isobutene could react with feed and products thereby lowering process yields. Although such catalysts can be regenerated in air at high temperatures, the relatively rapid deactivation coupled with high catalyst costs, have not made this technology competitive with OCT®, and hence it has not yet been commercialized.

6.3.2 The Shell Higher Olefins Process (SHOP)

By far the largest application of metathesis is in the multistep Shell Higher Olefins Process (SHOP). Ethylene is the primary feedstock, undergoing oligomerization over a homogeneous nickel catalyst to form a range of alpha olefins (C_4-C_{40}) with a Schultz-Flory distribution (see Sections 6.5.5 and 4.7.2). The C_6-C_{18} olefins are separated out and used as co-monomers or feedstocks for lubricants, plasticizers, and detergents. Isomerization of the light ($<C_6$) and heavy ($>C_{18}$) alpha olefins to internal olefins occurs over a solid potassium or magnesium oxide catalyst. These internal olefins are then redistributed by cross metathesis over a molybdate-alumina catalyst to afford a range of olefins of intermediate lengths. The desired C_{11}-C_{14} internal olefins are separated from the mixture, and are either hydroformylated to afford detergent alcohols (Section 6.4.6), or alkylated to afford linear alkyl benzenes (LAB's). The remaining olefins are recycled to the isomerization reactor. To date, four SHOP units have been built around the world, with a total capacity of > 1 Mt/a. Although mechanistic aspects of the homogeneous nickel oligomerization step have been well documented by the group of Keim, surprisingly little has been published regarding the intricacies of the metathesis step. Various generic studies have shown that it is actually Mo^{4+} and/or Mo^{5+} species which act as active centres for metathesis, and not the Mo^{6+} species of the original supported MoO_3 catalyst. These active species are formed through *in situ* reduction of the catalyst by the olefin feed, where the formation of a π-allyl complex is a key step in the initiation phase. Pre-reduction of the catalyst using an activator such as $SnMe_4$ can assist to ensure optimal performance of the catalyst from the start of the reaction, although this adds to process costs. Slow and steady deactivation appears to be typical of these molybdate catalysts, and is believed to occur via elimination of an alkene from the metallacyclobutane intermediate to form an inactive Mo species.

6.4 Homogeneous Ruthenium Alkylidene Complexes

Homogeneous ruthenium alkylidene catalysts have attracted enormous interest in recent years due to their generally greater robustness to process conditions combined with a lower oxophilicity and hence a greater tolerance towards functionalized substrates and feed impurities. The first such metathesis active complex (Figure 5(a), R = Ph, R' = CH=CPh$_2$) was discovered by Grubbs and Nguyen already in 1992, although it was only active for the ROMP of highly strained cycloalkenes. In contrast to the known Schrock systems, it was subsequently found that catalyst performance could be greatly improved by use of more bulky, electron donating phosphines such as tricyclohexyl-phosphine, and these first generation catalysts were exploited in numerous

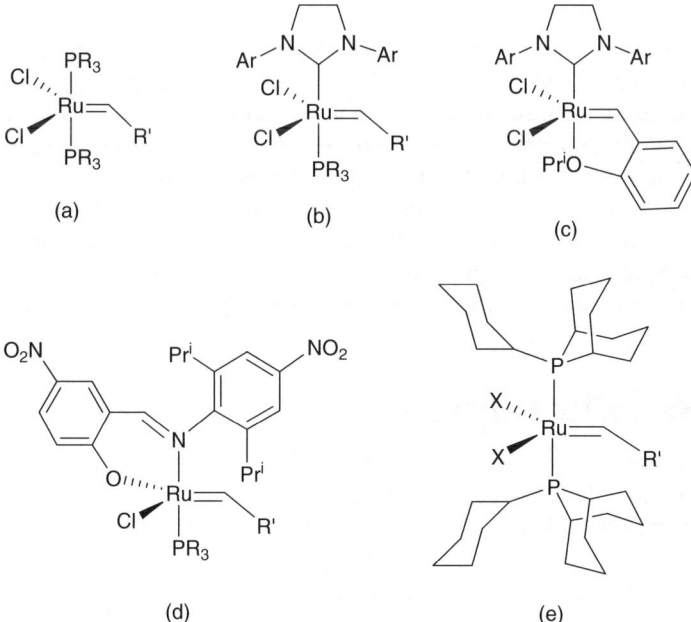

Figure 5 *Alkylidene metathesis catalysts. (a–e; R = Ph, Cy, Cyp; R' = Ph, CH=CMe$_3$; Ar = mesityl.)*

applications. In 1998, the near-simultaneous discovery of the N-heterocyclic carbene ligands by Herrmann and Grubbs afforded entry to a range of highly active second generation catalysts which could achieve greater turnovers (Figure 5(b)). The highly basic carbene ligand promotes lability of the phosphine, resulting in faster reaction of the olefin and leading to significant improvements in reaction rate. Variation of the NHC-aryl substituents can be used to manipulate selectivity and stability. However the catalysts suffer from some lack of selectivity as they degrade slowly to highly active isomerization catalysts which can react with the olefin products. Other developments include the Grubbs-Hoveyda phosphine-free catalyst (Figure 5(c)) which also affords excellent reactivity, and those employing Schiff bases as ligands (Figure 5(d)).

Recently Sasol Technology refocussed attention on the largely abandoned first generation bis(phosphine) catalysts, and achieved success in developing analogues which retain the good selectivities typical of these catalysts, while achieving activities in line with those of the second generation catalysts, even at ppm levels of ruthenium. The ligand which is key to this performance is the bulky and conformationally restricted cyclohexyl-phobane, which is closely related to the eicosyl-phobane employed by Shell in their cobalt-based hydroformylation process (the *phobanes* are bicyclic phosphine ligands, *e.g.*, 9-substituted-9-phosphabicyclononanes, see also Section 6.4.6.3). The resultant Grubbs catalyst (Figure 5(e) X = Cl, Br, I; R = H, Ph) shows distinctly improved stability, which may be linked to its unique structural behaviour in solution. Restricted rotation of the bulky phosphine leads to the formation of distinct rotational isomers, which may stabilize the alkylidene or prevent backbiting of the ligand at the ruthenium centre.

While these catalysts show the right behaviour in terms of activity and selectivity, there is still a long way to go in terms of obtaining catalysts which are long-lived and can be recycled. The key to success lies in a careful trade-off between activity and stability which can be achieved by careful manipulation of ligand sterics and lability. While a labile ligand is essential to allow rapid reaction of olefin at the reactive centre, that reactivity also translates to faster decomposition and a greater likelihood of unwanted side reactions. Encouragingly, developments in laboratories around the world suggest such catalysts are getting closer to achieving commercial viability in larger scale processes.

6.5 Speciality Polymers

ROMP processes are unique in that they offer unsaturated polymer products with properties between those of saturated polyethylene and the highly unsaturated polybutadienes. These *polyalkenamers* have been the subject of intense study by companies dealing in speciality polymers. In the 1970's the ROMP product of cyclopentene attracted attention as a replacement for natural rubber, due to its good strength and ageing properties. Although the elastomer was never commercialized, as its overall characteristics did not meet requirements, the work stimulated research into ROMP of other cyclic alkene monomers, with several commercial successes. *Vestenamer* is the ROMP product of cyclooctene, and is used as a blending material to impart greater hardness, elasticity and durability to other rubbers. The commercial product has a high trans content (80%). An alternative Vestenamer with only 62% trans content and a lower crystallinity has been developed, which is suitable for lower temperature applications where excessive hardening is undesirable. *Norsorex* is the trade name for the 90% trans polymer of norbornene, produced currently by Elf Atochem. It is sold as a moulding powder and the final vulcanized product is used in various vehicle fittings such as bumpers, arm-rests and engine mountings. Polymers of dicyclopentadiene are suitable for injection moulding applications, and have been produced by several companies (see Table 2). In 2001 Easton Sports initiated the commercialization of new archery and baseball products based on Materia's Poly-DCPD technology.

Although ROMP is generally extremely rapid and high yielding due to the release of ring strain, the chemistry of ROMP is not trivial. Catalysts have extremely high productivity, as the rate of chain propagation far exceeds that of any decomposition reactions which might occur. However due to the presence of repeating olefinic units in the polymer product, it is also important to ensure that propagation is more rapid than: *i*) intermolecular chain transfer (leading to break-up of existing polymer chains, Equation 1), and *ii*) intramolecular chain transfer (leading to macrocyclic products, Equation 2).

$$\qquad\qquad\qquad\qquad\qquad\qquad\qquad\qquad (1)$$

$$\qquad\qquad\qquad\qquad\qquad\qquad\qquad\qquad (2)$$

The intrinsic reactivity of strained cycloalkenes such as norbornene and cyclobutene ensures that they react as desired, and simple homogeneous metal halide catalysts are often effective for this transformation. However for less strained cyclic substrates, manipulation of catalyst activity/selectivity by means of modifying ligands is required. This is where the well-defined alkylidene catalysts pioneered by Grubbs and Schrock have come to the fore. An interesting example illustrating the range of catalyst reactivity is provided by the

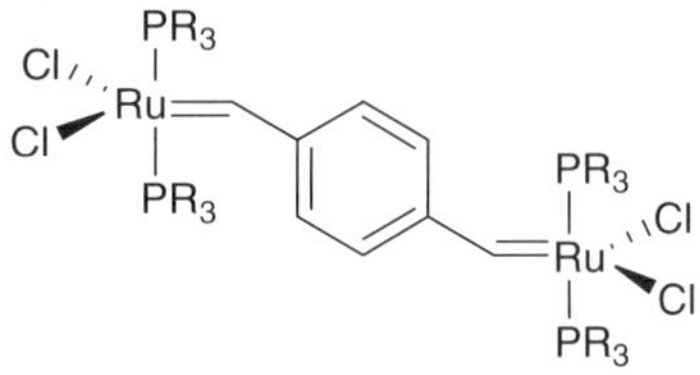

Figure 6 *Bimetallic complexes that exhibit ROMP activity: a: R = Ph; b: R = Cy; c: R = Cyp [M. Weck, P. Schwab, R. H. Grubbs, Macromolecules, 1996, 19, 1789].*

ROMP behaviour of bimetallic complexes Figures 6a–c. Complex 6a readily catalyzes the ROMP of norbornene, but is not active enough for ROMP of functionalized norbornene derivatives. In contrast, complexes 6b and 6c are active for the functionalized substrates, but suffer from the side reactions shown in Equations 1 and 2 when norbornene is the substrate.

A host of further issues complicate catalyst performance for ROMP reactions. Intrinsic polymer characteristics are not just dependent on the nature of the monomer and/or comonomer, but are also highly dependent on the *cis, trans* sequence of double bonds along the polymer chain, as well as on the tacticity of the polymer if a chiral or prochiral monomer is used, since the latter reflects the stereochemical sequence by which the chiral centres are linked. [See Chapter 7 and J. G. Hamilton in *"Handbook of Metathesis, Volume 3"*, R. H. Grubbs ed., Wiley-VCH, Weinheim, 2003].

6.6 Fine Chemicals and Pharmaceuticals

RCM and CM provide convenient access to a range of intricate products of relevance to the pharmaceutical, agrochemical and fragrance industries. While RCM can effectively create the functionalized carbo- and hetero-cyclic structures common in many such products, cross metathesis is a simple way of introducing often difficult combinations of functional groups.

Companies including Shell and Phillips have developed and operated several metathesis processes based on heterogeneous catalysts for the production of higher value fine chemical products. The Phillips Neohexene Process involves the ethenolysis of diisobutene (as a mixture of 2,4,4-trimethyl-1-pentene and -2-pentene) with ethylene. A $WO_3/MgO/SiO_2$ catalyst is employed, where MgO effects the *in situ* isomerization of 2,4,4-trimethyl-1-pentene to 2,4,4-trimethyl-2-pentene. The product, neohexene (3,3-dimethyl-1-butene), is used as an intermediate for production of a synthetic musk perfume as well as an antifungal agent. Isobutene is generated as a byproduct and can be recycled to the isobutene dimerization reactor which supplies the feed. Although in this case two catalysts are combined in one reactor, the combination of isomerization and metathesis is a powerful one which can be further exploited using 'non-selective' metathesis catalysts which afford isomerization as a side reaction. Several other ethenolysis processes have been employed in fine chemicals manufacture. *Further Exploitation of Advanced Shell Technology* (FEAST) is the rather grandiose name given to a Shell project aimed at producing small volumes of speciality olefins, particularly α,ω-dienes from ethenolysis of cyclic alkenes over a Re_2O_7 catalyst. Phillips have operated similar processes using a WO_3/SiO_2 catalyst. Such products find application as speciality co-monomers, cross-linking agents or precursors to a range of fine chemicals and pharmaceuticals. In the fragrance industry, Symrise is producing cyclohexadecadiene (a precursor to cyclohexadecenone, marketed as the fragrance Globanone) from cyclooctene over a Re_2O_7 catalyst.

Of late the use of homogeneous catalysts in fine chemicals manufacture has grown dramatically, due to the better selectivities which can be attained, and the robustness, particularly of the ruthenium-based catalysts, towards the range of functional groups commonly found in such molecules. In the agrochemicals industry, a range of pheromones is accessible via cross metathesis, and ruthenium-based Grubbs catalysts have proven particularly effective for these transformations. These can be used as biopesticides as they disrupt the breeding cycles of various crop pests in an environmentally benign manner. 5-Decenyl acetate, the major component of the peach twig borer pheromone, can be produced effectively from 5-decene and 1,10-diacetoxy-5-decene (Figure 7). In order to maintain high selectivities, the reaction is carried out at low temperatures (5°C), and the unreacted feed materials can be separated out by distillation and recycled, leading to high process efficiencies and minimal waste. Further interesting applications include production of a mosquito pheromone via a key metathesis step, which can be employed as non-toxic alternative to prevent the spread of mosquito-borne diseases.

Pharmaceutical applications of metathesis are widespread in the patent literature. While CM and ROM are applicable in some syntheses, it is RCM which predominates, as there are few reactions which are as effective for synthesis of highly functionalized medium ring compounds. Such macrolides are key components of a vast range of medicinally active compounds which can combat bacterial and viral infections as well as cancers, bone and neurological

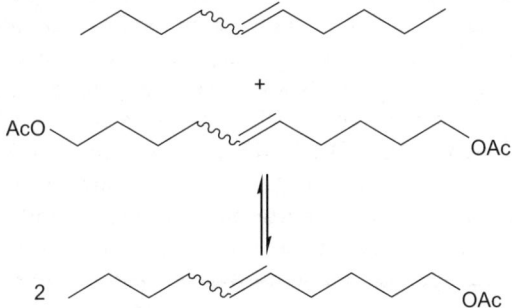

Figure 7 *Synthesis of 5-decenyl acetate, a key component of the peach twig borer pheromone.*

diseases. Epothilones are examples of a family of naturally occurring macro-cyclic anti-cancer agents which can be synthesized using RCM as a key step.

As pharmaceuticals are generally very high value products the catalyst cost is less critical for an economical process (see Chapter 1). However pharmaceutical applications have much higher selectivity requirements, as the testing of materials for human consumption is a rigorous one. For pharmaceuticals and fragrances, separation of homogeneous catalysts and ligands from the product is critical to meet standards set by control bodies such as the United States FDA. Apart from its toxicity residual metal may also lead to gradual degradation of the product. Published methods for removal of such residues generally involve the use of a complexing or oxidising agent, such as the water soluble tris(hydroxymethyl)phosphine (Grubbs), DMSO or tri-phenylphosphine oxide (Georg) and lead tetraacetate (Paquette). However as such methods require addition of a further reagent, further sources of product contamination are possible. A more effective manner of preventing ruthenium contamination is by 'heterogenizing' the highly selective homogeneous catalysts, and this has become a topic of intense research. A problem with such systems is the labile nature of organometallic complexes. This makes permanent anchoring of the metal via a ligand extremely difficult and leaching of the catalyst always remains a possibility.

A typical ruthenium alkylidene catalyst (Figure 5) can be anchored through the neutral ligand (phosphine or carbene), the anionic ligand (where X is usually a phenolate), or even the alkylidene (where R = phenyl), and Barrett and Hoveyda have reported strategies. However such work is in its infancy and further development is required before these techniques become commercially viable.

6.7   Recent Progress

The recent increase in industrial cross metathesis research has been driven largely by the immediate developments in propylene production (OCT® and others), as well as by a longer term drive to look at sustainable feedstocks. While OCT® technology is achieving success, it requires the use of ethylene, which is also a valuable intermediate commanding relatively high prices. Process economics are therefore very dependent on the relative swings in ethylene and propylene pricing. Sasol Technology have patented a process which uses only butene as feedstock [Patent WO 00/14038]. 1-butene is isomerized to 2-butene over a WO_3/SiO_2 catalyst, and the two undergo cross metathesis to produce propylene and 2-pentene in one reactor. 2-Pentene can react further with 1-butene to produce more propylene and 3-hexene, and so the process continues making use of the various equilibria to achieve a C_4 conversion of around 65% and a propylene yield of 30% per pass. ABB Lummus have also recently patented an autometathesis process based on one pot isomerization-metathesis of a C_4 stream without added ethylene [US Patent 6 777 582]. Their catalyst system is also WO_3/SiO_2, but MgO is added to facilitate isomerization. This process can be integrated with their back-isomerization technology which can take the 3-hexene by-product to 1-hexene, a valuable co-monomer. While such technologies benefit from lower feedstock costs, the lower product yields per pass and large recycles require increased capital costs from larger reactors.

The sequence of isomerization and metathesis is found in SHOP, neohexene and OCT®. However isomerization is sometimes a very unwanted side reaction, presumably mediated by metal hydrides arising from catalyst degradation. Most C_4-C_{14} olefins are produced via ethylene oligomerization processes which produce even-numbered olefins. The odd-numbered C_5, C_7, and C_9 olefins are generally not desirable as co-monomers, and such feedstocks are available for upgrading to higher value products. Sasol Technology have successfully piloted a process to take low value Fischer-Tropsch derived 1-heptene streams to higher value detergent range olefins. Isomerization of the terminal double bond would lead to a range of metathesis products shorter than the desired C_{12} product; thus a heterogeneous WO_3/SiO_2 catalyst operated at high temperatures and modified to reduce isomerization, proved to be best suited to the C_7 feedstock.

Unsaturated fatty acids, esters and alcohols derived from renewable feedstocks, are prime targets for ethenolysis reactions to produce high value shorter chain products, for example the ethenolysis of methyl oleate. The reaction is not trivial, as functionalized olefins are notoriously poor in cross metathesis reactions, but using heterogeneous rhenium oxide catalysts some success has been achieved. Materia and Cargill have received matched funding from the US Department of Energy to develop a platform of new industrial chemicals from soy and other oilseeds, further highlighting the drive towards sustainable feedstocks.

Discussion Points DP1*: What are the main concerns with homogeneous metathesis catalysts that currently prevent their application in large scale processes? What are the current issues regarding the heterogeneous catalysts? How would you tackle these issues?*

6.8   Future Outlook

Metathesis is a relatively new technology and is only now starting to reach its full potential. Homogeneous catalyst development is intense; however, improvement of the heterogeneous processes continues. The 'green' nature of metathesis is a further reason for the ongoing interest in this technology. ROMP and cross metathesis processes are extremely atom efficient and due to the reversible nature of the reaction, byproducts of the latter processes can be recycled to extinction, giving theoretically quantitative yields (cf. SHOP process).

Looking ahead, the future of large scale cross metathesis will be dictated by feedstock availability and market pulls. Additional sources of Fischer-Tropsch-derived olefins may be available from future gas-to-liquids (GTL) and coal-to-liquids (CTL) processes. As with linear alpha olefin (LAO) oligomerization processes, all of these technologies will produce a relatively fixed distribution of olefins, which may require metathesis in order to shift the distribution closer to market requirements. Metathesis of functionalized olefins will grow in importance as catalysts are developed that can tolerate different functional groups. Application in the oleochemicals arena linked to a drive towards renewable feedstocks, is a specific large scale opportunity.

In the area of ROMP, development of new polymers and identification of further specialized uses for existing polymers in the greatly expanding leisure markets are to be expected. Further relevant trends include the development of 'self-healing' polymers, telechelic polymers and development of polymer depolymerization processes. Much in the same vein, RCM, CM and ring opening cross metathesis will continue to feature as key synthetic steps for manufacture of fine chemicals and pharmaceuticals However the challenge therefore still remains for industry and academia to develop cheaper, more efficient catalysts by looking at new combinations of metals, ligands and promoters, as well as new technologies such as immobilization, biphasic catalysis and materials synthesis.

References

History of metathesis: A. M. Rouhi, *Chem.& Eng. News*, December 23, 2002, 34. J.-L. Hérisson, Y. Chauvin, *Makromol. Chem.*, 1971, 141, 161. R. H. Grubbs, *Tetrahedron*, 2004, 60, 7117. A. Furstner, *Angew. Chem., Int. Ed.*, 2000, 39, 3012. K. J. Ivin, J. C. Mol, *Olefin Metathesis and Metathesis Polymerization*, Academic Press, London, 1997. R. H. Grubbs (Ed.), *Handbook of Metathesis*, Volumes 1-3, Wiley-VCH, Weinheim, 2003.

Reviews of industrial processes: J. C. Mol, *J. Mol. Catal. A*, 2004, **213**, 39. R. Streck, *Chemtech*, August 1989, 498. A. M. Rouhi, *Chem. Eng. News*, December 23, 2002, 29. F. Lefebvre, J.-M. Basset, *NATO ASI Ser C*, 1998, **506**, 341. R. L. Pederson, I. M. Fellows, T. A. Ung, H. Ishihara, S. P. Hajela, *Adv. Synth. Catal.*, 2002, **344**, 728. Oleochemicals: K. A. Burdett, L. D. Harris, P. Margl, B. R. Maughon, T. Mokhtar-Zadeh, P. C. Saucier, E. P. Wasserman, *Organometallics*, 2003, **23**, 2027.

Catalyst development and catalysis: C. Pariya, K. N. Jayaprakash, A. Sarkar, *Coord. Chem. Rev.*, 1998, **168**, 1. E. L. Dias, S. T. Nguyen, R.H. Grubbs, *J. Am. Chem. Soc.*, 1997, **119**, 3887. R.H. Grubbs, *Acc. Chem. Res.*, 2001, **34**, 18. J. A. Love, M. S. Sanford, M. W. Day, R. H. Grubbs, *J. Am. Chem. Soc.*, 2003, **125**, 10103. J. S. Kingsbury, J. P. A. Harrity, P. J. Bonitatebus, A. H. Hoveyda, *J. Am. Chem. Soc.*, 1999, **121**, 791. L. Jafarpour, S. P. Nolan, *J. Organomet. Chem.*, 2001, **617–618**, 17. A. Furstner, M. Picquet, C. Bruneau, P.H. Dixneuf, *Chem. Commun.*, **1998**, 1315.

Cross metathesis: A. K. Chatterjee, T-L. Choi, D. P. Sanders, R. H. Grubbs, *J. Am. Chem. Soc.*, 2003, **125**, 11360. G.S. Forman; A.E. McConnell; M.J. Hanton; A.M.Z. Slawin; R.P. Tooze; W. Janse van Rensburg; W.H. Meyer; C.L.Dwyer; D.W. Serfontein; M.M. Kirk, *Organometallics*, 2004, **23**, 4824.

Biphasic or recyclable systems: D. M. Lynn, B. Mohr, R. H. Grubbs, L. M. Henling, M. W. Day, *J. Am. Chem. Soc.*, 2000, **122**, 6601 (Water). A. H. Hoveyda, D. G. Gillingham, J. J. Van Veldhuizen, O. Kataoka, S. B. Garber, J. S. Kingsbury, J. P. A. Harrity, *Org. Biomol. Chem.*, 2004, **2**, 8 (Recycling). A. Furstner, L. Ackermann, K. Beck, H. Hori, D. Koch, K. Langemann, M. Liebl, C. Six, W. Leitner, *J. Am. Chem. Soc.*, 2001, **123**, 9000 (s/c CO_2).

M. Mayr, M. R. Buchmeiser, K. Wurst, *Adv. Synth. Catal.*, 2002, **344**, 712. (Silica supported 2nd generation Grubbs catalyst). S. E. Gibson, V. M. Swamy, *Adv. Synth. Catal.*, 2002, **344**, 619. (Microencapsulated 2nd generation Grubbs catalyst). A. G. M. Barrett, D. C. Braddock, S. M. Cramp, P. A. Procopiou, *Tetrahedron Lett.*, 1999, **40**, 8657.

Removal of Ru residues: H. D. Maynard, R. H. Grubbs, *Tetrahedron Lett.*, 1999, **40**, 4137. Y. M. Ahn, K. L. Yang, G. I. Georg, *Org. Lett.*, 2001, **3**, 1411. L. A. Paquette, J. D. Schloss, I. Efremov, F. Fabris, F. Gallou, J. Mendez-Andino, J. Yang, *Org. Lett.*, 2000, **2**, 1259.

CHAPTER 7
Polymerization Reactions

GERHARD FINK[a] AND HANS-HERBERT
BRINTZINGER[b]

[a] *Max-Planck-Institut für Kohlenforschung, D-45470, Mülheim a. d. Ruhr*
[b] *Universität Konstanz, D-78457, Konstanz*

Dedicated to the memory of Professor Paolo Corradini

7.1   An Introductory Overview

The tremendous growth of polymer production since the middle of the twentieth century has been intimately connected with the development of new types of catalysts. A notable example in this regard is the field of polyolefin materials.

While catalysts are also used in the production of other types of polymers, the properties of most of these materials are not particularly dependent on the type of catalyst employed. Many *polycondensation* reactions, e. g. the formation of polyesters, polyamides or urea-formaldehyde resins, are speeded up by addition of some Brønsted or Lewis acids. Since relevant properties of these polymer products, such as their average chain lengths, are controlled by *equilibrium parameters*, primarily by the reaction temperatures and molar ratios of the monomers employed, and since their linkage patterns are dictated by the functional groups involved, addition of a catalyst has little leverage on the properties of the resulting polymer materials.

For typical *polyolefin* materials, on the other hand, the most relevant properties depend – in addition to the types and molar ratios of the monomers used – quite critically on the catalyst used for their production. This is due to the large numbers of different structural elements which can be formed – with practically equal free energies – by *polymerization* reactions even of simple olefins such as ethylene and/or propylene (Figure 1). The proportions with which each of these concatenation patterns occurs in a particular polymer product are thus controlled by the *relative rates* of their formation, i.e. by the selectivity with which these patterns are produced in the course of the polymerization process employed, rather than by any equilibrium parameters.

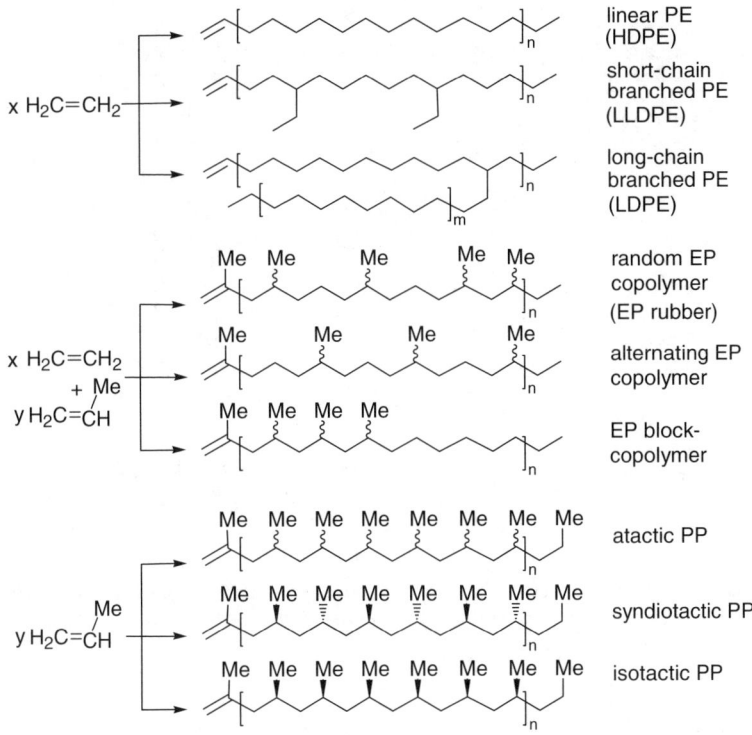

Figure 1   *Polymer enchainment patterns occurring in polyethylene (PE), ethylene-propylene copolymer (EP), and polypropylene (PP) chains (HDPE = crystalline high-density polyethylene, LLDPE = linear low-density polyethylene, LDPE = low-density polyethylene, EP rubber = elastomeric ethylene-propylene copolymer).*

   Among the processes used for the formation of polyolefins, the longest-known but least selective one is *free radical polymerization*. A free radical species X· produced e.g. by thermolysis of benzoyl peroxide or by photolysis of azabisisobutyronitrile (AIBN) – can react with the double bond of a vinyl derivative $H_2C\!=\!CHR$· to form a new radical of the type XCH_2-CHR which can then add another $H_2C\!=\!CHR$ unit; repetition of this process leads to polyolefin formation (Figure 2, top). This process works best for vinyl derivatives with unsaturated side groups, which provide *resonance stabilization* for an adjacent radical centre, e.g. with vinyl and acrylic esters, vinyl cyanides and vinyl chloride and with styrene and 1,3-dienes. It is extensively used in the emulsion polymerization of vinylic and acrylic derivatives and in the light-induced formation of photoresists for the nanofabrication of semiconductor chips and integrated electronic circuits.

   Formally similar reaction sequences occur in *anionic polymerization*. Here, a $H_2C\!=\!CHR$ double bond reacts with a strongly nucleophilic anion X⁻ to form a new carbon-centred anion XCH_2-CHR⁻. Continuation of this process leads to the formation of polymer chains, especially again for those vinyl derivatives

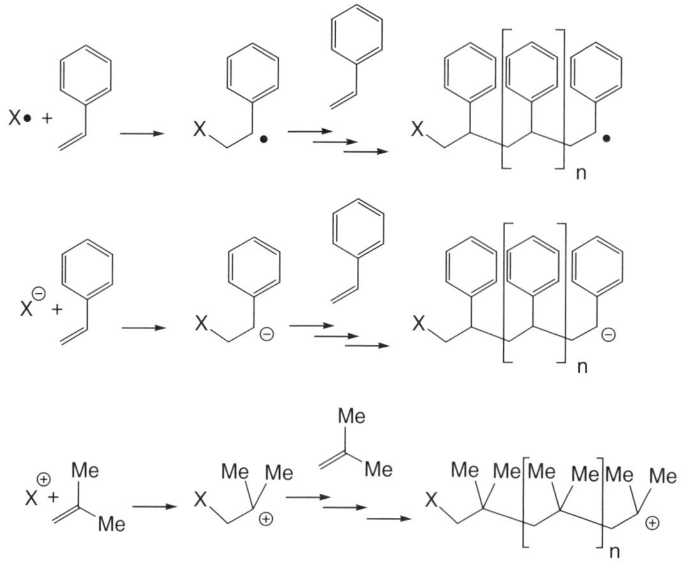

Figure 2 *Free radical polymerization (top) and anionic polymerization (middle) of styrene and cationic polymerization of iso-butene (bottom). For the nature of initiators X see text.*

where an unsaturated substituent R stabilizes the adjacent anionic charge (Figure 2, middle). Typical initiators for anionic polymerizations e.g. of styrene or of 1,3-dienes, are potassium amide or n-butyl lithium. In contrast to free radical polymerizations, where active species disappear rather rapidly (e.g. by radical combination), resonance-stabilized anions often are quite long-lived in aprotic media. As a consequence, *living polymerization* is frequently observed in these reaction systems: The anionic chain ends will continue to grow upon addition of new monomer until they are quenched, e.g. by addition of a protic reagent.

Atom-transfer and *group-transfer* polymerizations are variants of these processes, in which radical or anionic chain ends – instead of occurring freely as such – are temporarily released by breaking a suitably labile, but otherwise protective bond (e.g. to a Cu atom or to a Me_3Si group) which then gets reattached to the newly-formed chain end.

In *cationic polymerizations*, initiation occurs by attachment of a proton or some other Lewis-acidic cation X^+ to the $H_2C{=}CR_2$ double bond of a vinyl monomer to form a new carbon-centred cation of the type $XH_2C\text{-}CR_2{}^+$, which then grows into a polymer chain by subsequent $H_2C{=}CR_2$ additions (Figure 2, bottom). This type of polymerization works well – and is used in practice – only for olefins such as isobutene, where 1,1-disubstitution stabilizes the formation of a cationic centre. Since side reactions, such as release of a proton from the cationic chain end, occur rather easily, cationic polymerization usually gives shorter chains than anionic polymerization.

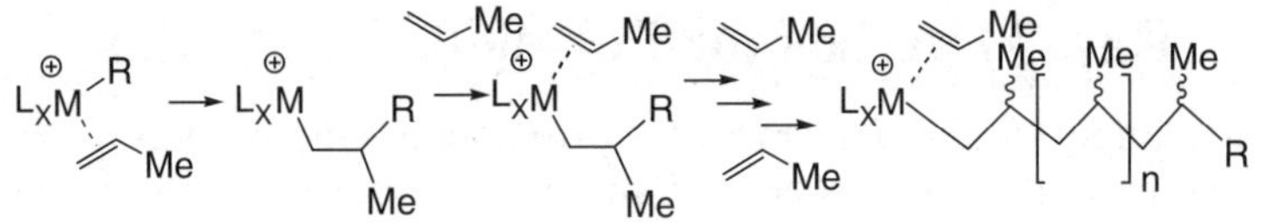

Figure 3 *Propylene polymerization by successive olefin insertions into metal-alkyl bonds.*

In all the processes described so far, the initiator is – in principle – not attached to the carbon centre by which a polymer chain continues to grow and is thus not in a position to control the type or the stereochemical orientation of an entering olefin substrate. These processes are, in addition, not well-suited for the polymerization of ethylene, propylene and other simple α-olefins, as these do not efficiently stabilize radical, anionic or cationic centres. These industrially abundant and least costly monomers are best polymerized by catalysts which operate by way of *insertion polymerization*. Here a positively charged metal catalyst centre is connected to a chain end which bears a partial negative charge, while at the same time binding – and thus suitably polarizing – a monomer molecule (Figure 3). Due to its position adjacent both to the growing chain end and to the entering monomer, these insertion-type polymerization catalysts are – in principle – best suited to control the selective formation of one or other of the polymer enchainment patterns outlined in Figure 1. Catalysts of this type, which are available today for practically all of the *3d* transition metals and some of their *4d* homologues, will be the main topic of this chapter.

Titanium-based *solid-state catalysts* for the industrial production of poly-olefin materials were discovered in the early 1950's and have been continually improved since then (see Section 7.3). Due to the high degree to which they have been perfected for the production of large-volume polyolefin commodi-ties, they continue to dominate the processes presently used for polyolefin production. Despite (or because of) this product-oriented perfection, only limited degrees of variability with regard to some relevant polymer properties appear to be inherent in these solid-state catalysts.

Access to polyolefins with a wider choice of properties has more recently been provided by various homogeneously soluble organometallic catalysts. Some of these catalysts, in particular those based on sandwich and half-sandwich com-plexes of zirconium and titanium and on nitrogen-containing complexes of group 4 and of some of the group 8–10 metals (see Section 7.4), are thus likely to be increasingly used for the production of polyolefins for special-purpose applica-tions, which require properties not easily accessible otherwise.

7.2 Industrial Aspects of Polyolefin Production

During the last five decades, industrial production of polyolefin materials has experienced strong increases in production volumes as well as changes in production procedures. Here we give an overview of the situation in 2005 (Boxes 1 and 2).

Box 1 Raw Materials for Polyolefin Production

The most important monomers for the production of polyolefins, in terms of industrial capacity, are ethylene, propylene and butene, followed by isobutene and 4-methyl-1-pentene. Higher α-olefins, such as 1-hexene, and cyclic monomers, such as norbornene, are used together with the monomers mentioned above, to produce copolymer materials. Another monomer with wide application in the polymer industry is styrene. The main sources presently used and conceivably usable for olefin monomer production are: petroleum (see also Chapters 1 and 3), natural gas (largely methane plus some ethane, *etc.*), coal (a composite of polymerized and cross-linked hydrocarbons containing many impurities), biomass (organic wastes from plants or animals), and vegetable oils (see Chapter 3).

Box 2 Polyolefin Production Volumes and Major Producers

About 50% of the present world-wide plastics production (>200 Mt/a) is based on polyethylene and polypropylene. When polystyrene is included, this percentage rises to 60%. With regard to their total production volumes, polyolefin materials thus are among the top 10 of all products generated in chemical industry. Major producers of polyethylene and polypropylene are shown, together with their production capacities, in Figure 4.

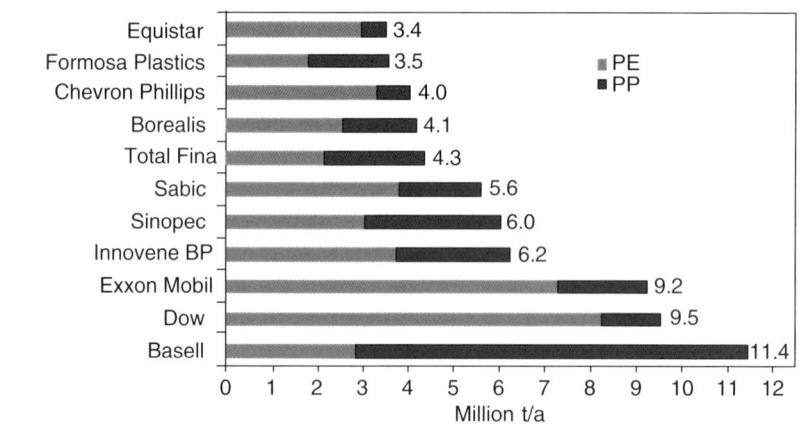

Figure 4   *Major producers of polyethylene (PE) and polypropylene (PP) in 2004/ 2005, with best-guess production capacities (including joint ventures).*

Almost all current large-scale polymerization plants for catalytic polyethylene and polypropylene production are suited only for the use of solid-state catalysts, which are introduced into the reactor in the form of grains or pellets by a suitable feed system and activated *in situ* by addition of an aluminium alkyl activator (see Section 7.3). Reaction conditions, such as pressure, temperature and monomer composition, are varied to afford polymers with different properties and morphologies.

Discussion Point DP1: *At present the production of polyolefin materials is based almost exclusively on petroleum. However further increases in crude-oil prices might make other potential sources competitive. Identify three alternative olefin sources, formulate the essential chemical reactions necessary for each production process and try to assess advantages, disadvantages and relative likelihoods of industrial implementation for such processes.*

7.2.1 Polyethylene Production

The catalytic polymerization of ethylene is usually conducted by one of the following methods:

Phillips Particle-forming process (Figure 5): In a double-loop reactor, constructed from wide-bore jacketed pipe, the catalyst and growing polymer particles are suspended in a slurry and kept in rapid circulation to avoid polymer deposits on the reactor walls. Due to its high surface-to-volume ratio, this reactor facilitates heat removal and allows short residence times. Typical reaction conditions are 100°C and 30–40 bar. Isobutane, a poor solvent for polyethylene, is used as a diluent and as a vehicle to introduce the catalyst into the reactor. The solid polymer is collected from a sedimentation leg and passed to a flash tank where the monomer and isobutane diluent are separated by evaporation and subsequently recondensed and recycled, while the polymer powder is fed into an extruder and formed into pellets.

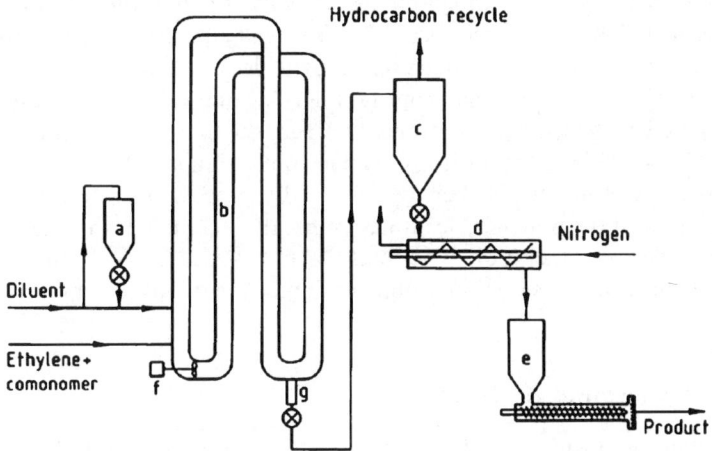

Figure 5   *Phillips Particle-forming process: a) catalyst hopper; b) double loop reactor; c) flash tank; d) purge drier; e) powder-fed extruder; f) impeller; g) sedimentation leg.*

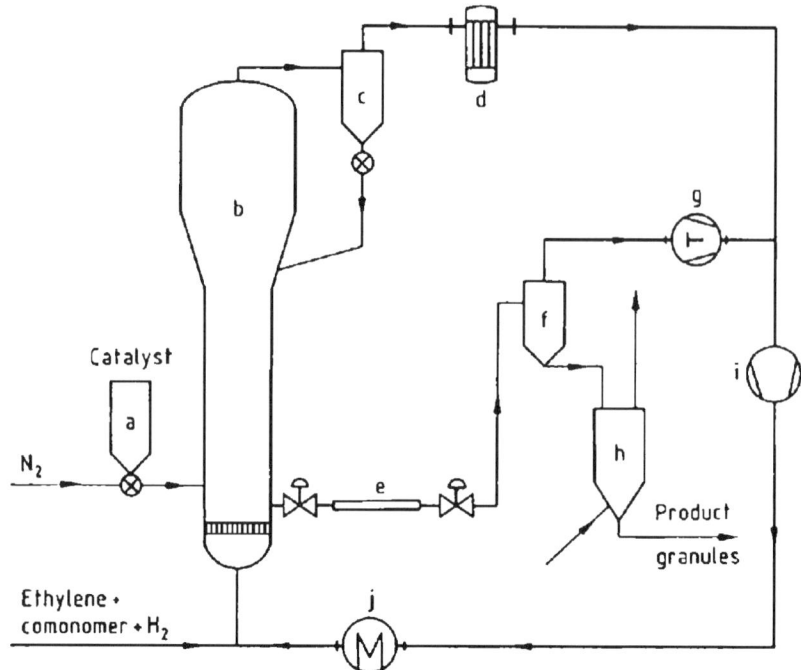

Figure 6   *Union-Carbide process. a) catalyst hopper; b) fluidized-bed reactor; c) cyclone;*
d) filter; e) polymer take-off system; f) product recovery cyclone; g) monomer
recovery compressor; h) purge hopper; i) recycling compressor; j) recycle gas
cooler.

Union Carbide fluidized-bed process (Figure 6): The reactor, about 30 m high,
has a characteristic shape with a lower cylindrical reaction section, and an
upper expanded section in which the gas velocity is reduced to allow entrained
particles to fall back into the bed. The feed gas enters the reactor from the
bottom through a distributor plate which provides an even upward flow of gas
and prevents polymer powder from falling. Depending on the product being
made, reaction temperatures range between 80 and 100°C and pressures
between 7 and 20 bar. Most often ethylene conversion is only *ca.* 2% per pass;
the unreacted monomer is then recycled. The process operates close to the
melting point of the polymer; accurate temperature control is thus necessary to
avoid particle agglomeration. The final reaction mixture is fed into a powder
cyclone from which residual monomers are recovered and recompressed.

7.2.2   Polypropylene Production

This is often conducted to make impact-resistant polyolefin blends. For this purpose, isotactic polypropylene, which is tough but somewhat brittle, is produced
on the catalyst pellet in a first reaction step, using highly active stereospecific

catalysts. Within this polymer matrix further polymer, usually a softer copolymer, is then deposited in a second reaction chamber. Four large-scale polymerization processes are used.

Spheripol Process: In a first reaction step homopolymerization is conducted in liquid propylene at 70°C in two loop reactors (Figure 5) connected in series. Since the first few seconds of polymerization with a highly active catalyst are decisive for the polymer morphology, catalyst particles are in some cases prepolymerized under milder conditions and charged into the loop reactor together with the cocatalyst to improve the particle forming process. By connecting two loop reactors in series, a more uniform residence time distribution of the catalyst grains is obtained. The polymer particles are collected in a cyclone and then either deactivated by treatment with steam, to obtain the final granular product, or else fed into a second polymerization, fluidized-bed reactor (Figure 6). Most often, ethylene and propylene are copolymerized onto the previously formed polypropylene matrix at 15–35 bar and 80°C in this second reactor. The resulting impact-resistant copolymer is fed into a cyclone and then worked up by deactivating catalyst residues with steam.

Novolen Process (Figure 7): With this technology, which comprises two vertical stirred gas-phase reactors in series, homopolymers as well as impact and random copolymers are produced. The first reactor operates at 80°C and 20–35 bar monomer pressure and is used exclusively for the homopolymerization of propylene. Propylene is injected as a liquid and cools the exothermic polymerization by its

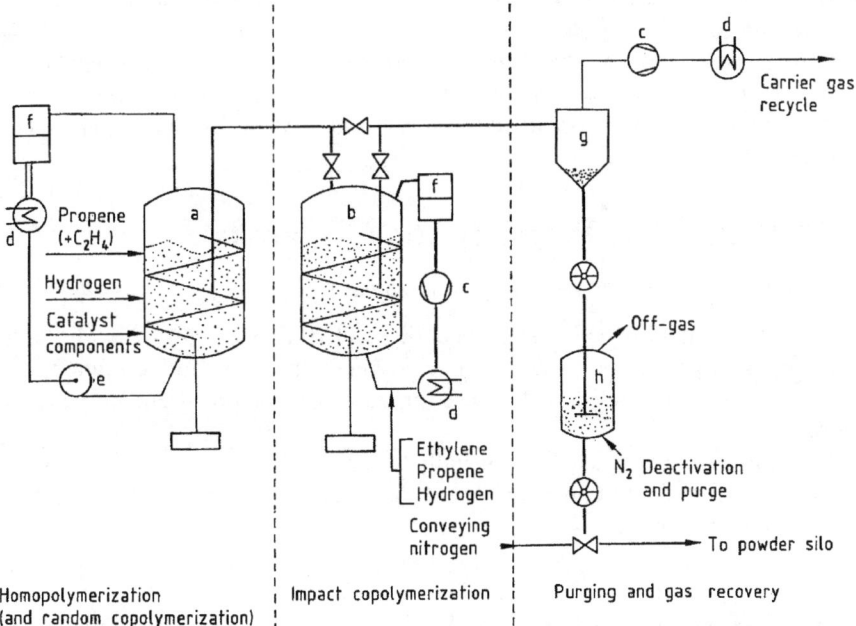

Figure 7   *Novolen process: a) first reactor, b) second reactor for impact copolymers, c) compressor, d) condensation, e) pump, f) filter, g) cyclone, h) deactivation and purge.*

evaporation in the reactor. The product mixture is continuously transferred from the reactor into a cyclone and then either deactivated or further polymerized in a second gas-phase reactor to produce impact-resistant copolymers under milder reaction conditions (10–25 bar and 60°C).

Amoco-Chisso Process: This process resembles the Novolen process except that the two reactors connected in series are stirred horizontally with blades and operated at 90°C and 20 bar. Polymers produced with the Amoco-Chisso technology show high uniformity, as particle residence time distribution in the reactor is particularly narrow.

Fluid-Bed Process: Union Carbide and Shell technologies are combined in this process: Two fluidized-bed reactors are used in series to produce impact-resistant polypropylene. As with other processes, reaction conditions are 50–100°C and 10–40 bar, with particle residence times between 1 and 4 hours.

Due to the use of advanced, highly active and selective solid-state catalysts (sections 7.3), the processes described above produce polymers from which neither stereoirregular polymer components nor catalyst residues need be removed. This has resulted in substantial reductions in the costs of investments, energy and maintenance, compared to slurry processes with first-generation catalysts. Ongoing developments are aimed at increased process flexibility and at process adaptation to the use of supported metallocene catalysts (Section 7.4).

Discussion Point DP2: *Olefin polymerization is an exothermic process. Estimate the heat of reaction released per day by a polymerization reactor with a typical production capacity of about 1000 t of PE or PP per day. Identify the means which can be used to remove this amount of heat from the reactor. What are their relative merits (and limits) in terms of energy use or recycling? Some (e.g. cationic) polymerization reactions proceed rapidly and give products with excellent properties when conducted at temperatures below 0°C. What makes such processes uneconomical?*

7.3   Solid-State Polymerization Catalysts

This category comprises two types of catalysts:

–Titanium-containing catalysts, generally called *Ziegler-Natta* catalysts, in honour of Karl Ziegler, who discovered them in 1953, and Giulio Natta, who initiated and developed their use for stereospecific propylene polymerization, and

–chromium-containing catalysts, usually called *Phillips* catalysts, with reference to the U.S. petroleum company where they were discovered in 1951 by Paul Hogan and Robert Banks.

7.3.1   Ziegler-Natta Catalysts

In their presently used form, Ziegler-Natta catalysts are typically prepared by adsorbing $TiCl_4$ onto small grains of a *$MgCl_2$ support*, with diameters of ca 50 μm, which are obtained either by prolonged grinding of $MgCl_2$ in a steel-ball mill or by its precipitation from a soluble precursor in the presence of some Lewis base, a so-called "*inner donor*", such as ethyl benzoate or a 1,3-diether. Subsequently a second, so-called "*outer donor*", e.g. a silyl ether or a phthalic acid diester, is often added. Upon being injected into the polymerization reactor these catalysts, which contain about 5–20% (w/w) of $TiCl_4$, are activated by reaction with aluminium alkyls, mainly *triethylaluminium*, to very high levels of *productivity*. Each gram of catalyst will typically produce 50–100 kg of polyethylene or polypropylene in the course of a few hours, its normal residence time in the reactor. The growth of the polymer also takes place inside the catalyst grains; thus each grain is disintegrated into several hundred minuscule particles with diameters of only 1–2 nm, which remain imbedded in a polymer pellet with a diameter of about 1 mm, i.e. with a volume several thousand times greater than the catalyst grain from which it had grown (Figure 8). Because of their small mass and size, the catalyst fragments can remain in the final polymer product without interfering with its chemical, mechanical or optical properties.

Box 3 Typical Polyolefin End Uses

Polyolefin materials have entered into so many varied applications in everyday life that a complete overview is not possible here. Table 1 summarizes the sectors which consume the greatest proportions of ethylene- and propylene-based polyolefin materials.

Table 1 *Uses of ethylene- and propylene-based polyolefin materials (estimated percentage of total plastic application in Europe in 2003, from www.plasticseurope.org; for list of abbreviations see Glossary)*

Packaging	37%	Films for cooking (HDPE), food packaging (LDPE, PE-co-norbornene, PMP), paper laminating (PP, PIB), milk packaging (LDPE, PP), chewing gums (PIB) and cookery (PMP), containers (HDPE) and caps (LDPE)
Building and construction	18%	Pipes (HDPE, LLDPE, PP, isotactic-PB, PMP, ABS), carpets (PP), storage tanks (HDPE, PB), asbestos replacement (PP), hot-melt adhesive (atactic PB), insulating foils (PP, PIB)
Transportation	6%	Transport tanks and containers (HDPE), lubricant (PIB), seals (EPD)
Electronic and electrical devices	9%	Electrical home devices (PP, PS, ABS), cables (PMP), technical parts (PP), lamps (PS)
Agriculture	2%	Crop protection (PE), twines, stripes and strings (PE, PP)
Medical	1%	Blister packaging (PE-co-norbornene), medical devices (PMP)
Sports		Sporting goods (ABS, PS)

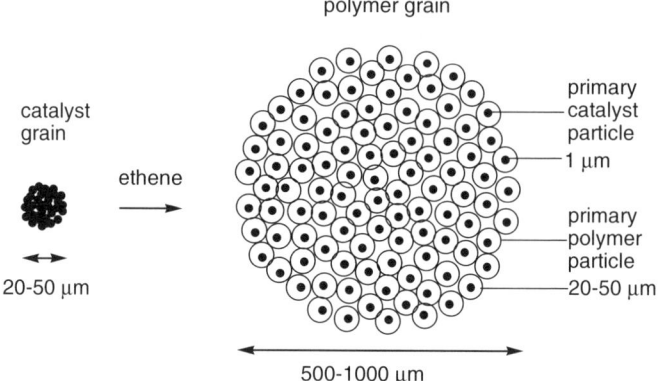

Figure 8 *Simplified model for the fragmentation of MgCl₂-supported Ziegler-Natta catalyst grains to primary catalyst particles and shape conservation of growing polymer grains.*

Box 4 Polyolefin Waste and Recycling

As plastic wastes are proliferating at increasing rates, infrastructures for waste recycling are developing in almost all areas of major polymer consumption. Policies in this regard are based on four main steps: *i*) prevention/reduction of plastic waste, *ii*) material recycling (mechanical recycling, feedstock recycling), *iii*) energy recovery and *iv*) dumping.

Feedstock recycling (chemical recycling) is the chemical reconversion of polymers to raw materials, i.e. to monomeric olefins. Suitable pyrolysis and gasification processes are being tested but have not yet become commercially competitive with the use of petrochemicals from crude oil. On the other hand, energy recovery, the combustion of polyolefin wastes in incineration plants to produce energy in the form of heat and electricity, appears to be the most efficient and unproblematic polyolefin waste utilization, since the chemical composition of these materials (which are free of chlorine, sulfur and nitrogen) is practically indistinguishable from that of heavy oils, which are burned in incineration plants, together with normal waste, to ensure temperatures sufficiently high for the complete destruction of toxic effluents.

MgCl₂ is ideal as a support for polymerization catalysts since it consists of loosely aggregated, crystalline sub-particles, which are extensively fragmented already by initial polymer formation; hence high polymerization rates are reached immediately. Disadvantageous however, is the high fragility of the initial MgCl₂ support, which can lead to formation of polymer fines (dust) due to the turbulence of gas-phase polymerizations. Good morphology control of the polymer grains is achieved however when a pre-polymerization step is applied under mild conditions. Since the catalyst support fragments are now kept embedded in the polymer matrix, polymer grains form good replicas of the original catalyst grains in the subsequent main polymerization.

Even the most advanced Ziegler-Natta catalysts contain distinctly non-uniform catalytic sites. Although most of the less selective (and hence undesirable) sites present in catalysts made by older recipes appear to be eliminated or blocked by the "inner" and "outer" donors now used to condition these catalysts, the polymers they produce still show large variations in molar mass, stereoregularity and comonomer distributions, which indicate that they originate from distinct catalytic sites.

Despite their complexity, Ziegler-Natta catalyst systems have been intensively studied by experimental and theoretical methods and the following model of these catalysts appears to be commonly accepted [G. Monaco, M. Toto, G. Guerra, P. Corradini, L. Cavallo, *Macromolecules* 2000, **33**, 8953; M. Seth, T. Ziegler, *Macromolecules* 2003, **36**, 6613]. After reduction by the alkylaluminium activator, the active sites of the catalyst contain *Ti(III) ions*, which are deposited on disordered surface layers of the minute $MgCl_2$ crystallites. Essential in this regard appears to be the structural similarity of $MgCl_2$ and $TiCl_3$ crystals, both of which form layer structures with similar lattice constants. The presence of "inner" and "outer" donors appear to favour, at the expense of alternative species, Ti(III) centres placed at a 110 edge of the $MgCl_2$ layer lattice, where each Ti(III) ion is surrounded by five Cl^- ions and of related Ti(III)/Ti(IV) pairs placed at a 100 edge. To comply with the stoichiometry $TiCl_3 = TiCl_{4/2}Cl$, four Cl^- ions are in bridging positions and thus not easily exchanged, while the fifth is a terminal Ti-Cl bond and is thus readily exchanged for an alkyl group upon reaction with the alkylaluminium activator (Figure 9).

The vacant sixth coordination site of these Ti centres can take up an olefin molecule to form the reaction complex required for the initiation and subsequent growth of polyolefin chains. Due to their octahedral dichelate-type structure, these Ti(III) centres are *chiral* and thus able to steer each incoming molecule into a preferred enantiofacial orientation. The *stereospecificity* with which subsequent propylene units insert into the growing polymer chain is most likely based on a mechanism analogous to that determined for soluble polymerization catalysts (Section 7.4.3).

Present-day Ziegler-Natta catalysts are supremely suitable for the production of linear polyethylene and of highly isotactic polypropylene. They are also used to produce the softer ethylene-propylene *copolymers*, used for packaging and related purposes. Due to the presence of distinct catalyst sites in typical Ziegler-Natta catalysts, these copolymers suffer from non-uniformity however, and copolymers which contain increased amounts of higher α-olefins, desirable for certain applications, cannot easily be made with these catalysts.

7.3.2 Phillips Catalysts

Chromium-containing Phillips catalysts are prepared by adsorption of a chromium compound, mostly *chromium trioxide*, onto an amorphous silica support and a subsequent reduction by exposure to ethylene. The resulting catalysts are

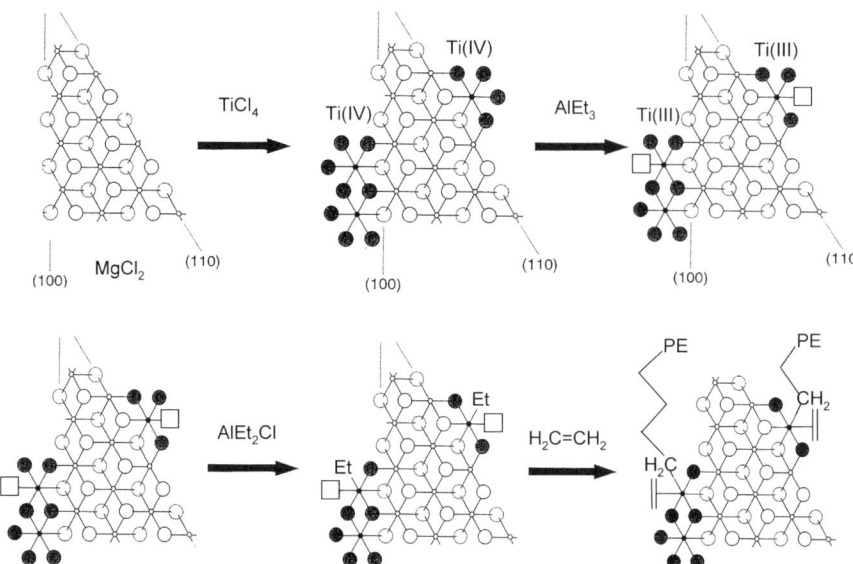

Figure 9   *Simplified model for the formation of active centres of Ziegler-Natta catalysts by adsorption of TiCl₄ on solid MgCl₂, reduction of Ti(IV) to Ti(III) chloride-alkyl exchange with aluminium alkyls and polyethylene (PE) formation by coordination and insertion of ethylene molecules.*

highly active and do not require further activation. These catalysts, while not able to generate polypropylene or higher polyolefins, are used extensively for the production of the strictly linear, so-called *high-density polyethylene* (HDPE) with particularly high molar mass and, hence, optimal mechanical and thermal strength. Their active sites are generally considered to consist of *Cr(II) centres*, but the ligand environment of these centres and the pathways by which they induce the insertion of ethylene molecules into growing polymer chains are still very incompletely understood [E. Groppo, C. Lamberti, S. Bordiga, G. Spoto, A. Zecchina, *Chem. Rev.* 2005, **105**, 115].

7.4   Soluble Olefin Polymerization Catalysts

Polymerization catalysis with soluble complexes of group IV transition metals, in particular with hydrocarbon-soluble titanocene complexes, was discovered in the 1950's, shortly after the appearance of Ziegler's and Natta's reports on solid-state catalysts, and rather thoroughly studied from then on. Alkylaluminium compounds, such as AlEt₂Cl, are required to activate also these soluble catalysts. In distinction to their solid-state counterparts, however, early soluble catalysts were able to polymerize only ethylene, and not any of its higher homologues. After their activation by methylalumoxanes had been discovered (Section 7.4.1), soluble catalysts became as efficient as solid-state catalysts – in

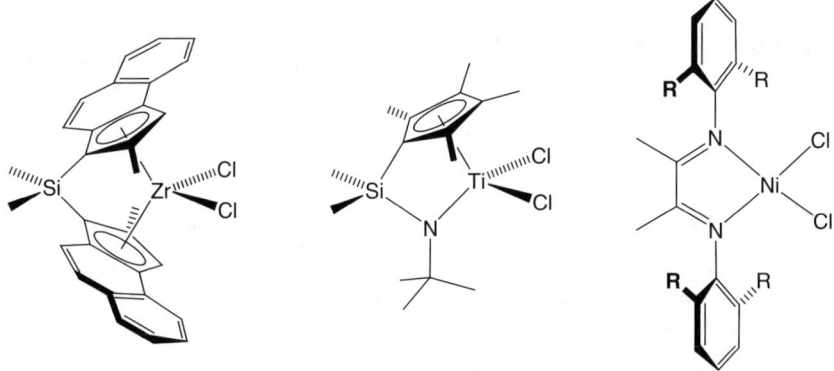

Figure 10 *Precursors for soluble polymerization catalysts: ansa-zirconocene complex (left), constrained-geometry titanium complex (middle) and nickel diimine complex (right).*

some aspects even more so – for the polymerization of all types of olefins. Today, soluble olefin polymerization catalysts are available from practically all $3d$-transition metals as well as from some of their $4d$ homologues.

The presently most advanced, with regard to industrial applications, are variously substituted and/or ring-bridged zirconocene catalysts, and related, so-called "constrained geometry" catalysts (Figure 10). Most of our understanding with regard to the basic functioning and limitations of polymerization catalysis in general comes from studies on soluble catalysts of this type and, in particular, on catalysts derived from a series of nickel and palladium complexes (see Figure 10). These catalysts are thus included in the following discussion, although their industrial applications are still lagging behind those of the group 4 catalysts.

7.4.1 Activation Reactions

For practical reasons, catalysts for olefin polymerizations are usually prepared *in situ* from a stable *pre-catalyst* by exposure to some suitable *activator* (or *co-catalyst*). The resulting *catalyst system* then contains highly reactive catalyst species which are capable of inducing the growth of polymer chains in the presence of suitable monomers. The elementary reactions involved in activation processes of this kind are particularly well-defined for zirconocene-based catalysts.

Air-stable zirconocene dichlorides of the type $Cp^x_2ZrCl_2$ (Cp^x_2 being a particular pair of substituted and/or bridged cyclopentadienyl ligands) are converted to active catalysts, for example by reaction with an aluminium trialkyl compound and dimethylanilinium tetrakis(perfluorophenyl) borate, $PhNMe_2H^+$ $(C_6F_5)_4B^-$, one of a number of so-called *cationization reagents*. Spectroscopic data indicate that the main species in the resulting catalyst system is an ion pair which contains an alkyl *zirconocenium cation* $Cp^x_2Zr\text{-}R^+$ in direct coordinating contact with its counter-anion $(C_6F_5)_4B^-$ (Figure 11). $(C_6F_5)_4B^-$ is a *weakly coordinating anion*, due to a high degree of delocalization of its anionic charge and to the inert C-F

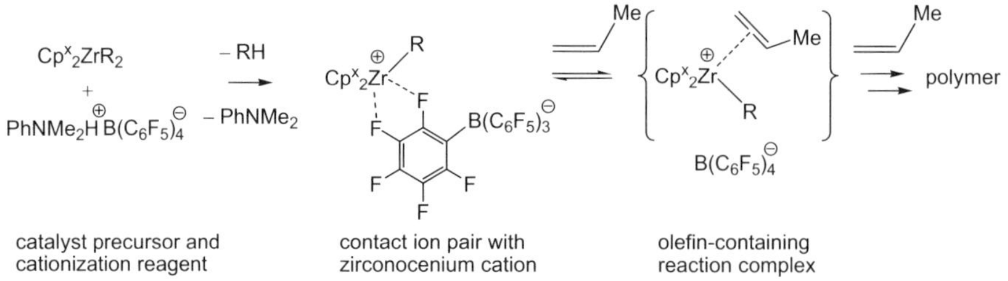

Figure 11   *Formation of a cationic olefin-containing reaction complex (right) by reaction*
of a zirconocene dialkyl precursor complex with a cationization reagent (left),
via a highly reactive contact ion pair containing an alkyl zirconocenium cation
(middle).

lining of its surface. Such an anion can be expected to be rather easily displaced from the Zr centre by an olefin monomer to form an *outer-sphere ion pair*, which contains the cationic *reaction complex* Cp^x_2Zr-R(olefin)$^+$. This is thought to be the site of polymer growth by olefin insertions first into the Zr-R bond and then into successively formed Zr-polymer bonds.

Cations of the same type are also produced, with release of Ph_3C-R, by reaction of $Cp^x_2ZrCl_2$/AlR_3 mixtures (or of the dimethyl complex $Cp^x_2ZrMe_2$) with Ph_3C^+ $(C_6F_5)_4B^-$, trityl perfluorotetraphenyl borate. In addition to its action as an *alkyl donor* to the Zr centre the trialkylaluminium AlR_3 functions as a *scavenger* by freeing the reaction mixture from protonic or oxidizing impurities. Higher concentrations of AlR_3, in particular of $AlMe_3$, tend to interact with cations such as Cp^x_2Zr-Me$^+$ with formation of binuclear adducts, Cp^x_2Zr-$(\mu$-Me$)_2AlMe_2^+$, which stabilize these cations against destructive side reactions [M. Bochmann, *J. Organomet. Chem.* 2004, **689**, 3982]. However they also diminish their reactivity *vis-a-vis* olefin monomers, since the olefin now has to displace the trialkylaluminium from the Zr center to form the reaction complex Cp^x_2Zr-R(olefin)$^+$. Triisobutylaluminium, $Al(CH_2CH(CH_3)_2)_3$, is less prone to form adducts with a zirconocene alkyl cation – due to the diminished tendency of its bulky alkyl groups to act as bridges between Zr and Al centers – and is thus often used as an activator/scavenger reagent.

Active catalyst systems are also formed when a *Lewis acid*, LA, reacts with a stable dialkyl zirconocene complex $Cp^x_2ZrR_2$ – provided that the anion R-LA$^-$, formed by alkyl abstraction by the Lewis-acid LA, is only weakly coordinated. An example is the Lewis acid perfluorotriphenylboron, $(C_6F_5)_3B$. It reacts with dimethyl zirconocenes to form an ion pair Cp^x_2Zr-Me$^+$... MeB$(C_6F_5)_3^-$. These catalyst systems usually require higher temperatures to achieve their full activities (and often give shorter polymer chains) than those described above, most likely due to the more "sticky" (*i.e.* more strongly coordinating) nature of MeB$(C_6F_5)_3^-$ as compared to B$(C_6F_5)_4^-$ anions.

Catalyst sytems with activities less sensitive to impurities are obtained when a zirconocene dichloride complex is reacted with *methylalumoxane*, the Lewis-acidic product of partial hydrolysis of trimethylaluminium, often abbreviated as *MAO*

[W. Kaminsky, *J. Chem. Soc., Dalton Trans.*, **1998**, 1413]. Hydrocarbon solutions of MAO contain various species of general composition $(AlMe)_m(AlMe_2)_nO_{m+n/2}$ with $m \approx n \approx 10$–100, with six- and four-membered $(-AlO-)_x$ rings and with Al centres of coordination number four as well as three. Some of these Al centres appear to be highly Lewis-acidic. A *MeMAO$^-$* anion is thus generated, together with the cation Cp^x_2Zr–Me^+, by methide abstraction from $Cp^x_2ZrMe_2$. MeMAO$^-$ appears to be displaced from Cp^x_2Zr–Me^+ by olefin monomer about as easily as $B(C_6F_5)_4^-$ is from Cp^x_2Zr-$Me^+\ldots B(C_6F_5)_4^-$. The most highly Lewis-acidic species appear to be present in normal MAO mixtures to only a small extent. Rather high [Al]:[Zr] ratios of ≥ 1000 are thus required to induce maximal activity in MAO-activated catalyst systems.

Since substantial amounts of "free" trimethylaluminium (*i.e.* Al_2Me_6) are present in normal MAO preparations, cationic *trimethylaluminium adducts* Cp^x_2Zr-$(\mu$-Me$)_2AlMe_2^+$, in outer-sphere association with their MeMAO$^-$ counteranions, are in general the dominant species in MAO-activated catalyst systems [D. E. Babushkin, N. V. Semikolenova, V. A. Zakharov, E. P.Talsi, *Macromol. Chem. Phys.* 2000, **201**, 558]. While being quite stable against destructive side reactions MAO-activated catalyst systems are thus often less active than those obtained with the cationization reagents described above.

In a few cases, active catalysts can be obtained without any activator, when release of a neutral stabilizing ligand entity from a precursor complex generates a neutral, coordinatively unsaturated and hence highly reactive metal centre. Examples of such *single-component catalysts* are a number of Ni(II) catalysts which are formed by release of a phosphine or pyridine ligand from neutral nickel alkyl or aryl precursors containing imino-phenolate or other monoan-ionic N,O or P,O chelate ligands (Figure 12). Some representatives of this class of catalysts, while not as productive as those described above, are inert enough to be usable as catalysts for *emulsion polymerization* of ethylene in water-hydrocarbon mixtures [E. F. Connor, T. R. Younkin, J. I. Henderson, A. W. Waltmann, R. H. Grubbs, *Chem. Comm.* **2003**, 2272; L. Kolb, V. Monteil, R. Thomann, S. Mecking, *Angew. Chem. Int. Ed. Engl.* 2005, **117**, 433].

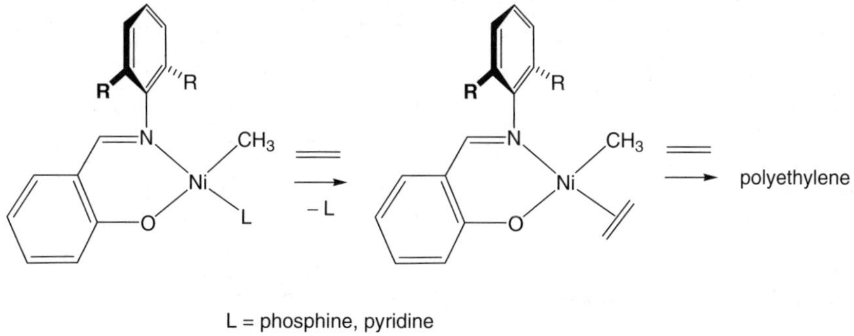

L = phosphine, pyridine

Figure 12 *Activation of phenoxy-imine nickel complexes by loss of a phosphine or pyridine ligand.*

Discussion Point DP3: *Organic derivatives of the group 13 elements aluminium and boron are needed as essential components for almost all of the insertion-catalyzed olefin polymerizations. List four such compounds of interest and describe for each of them the structural and reactivity properties relevant to its action as activator/cocatalyst. Outline some of the features of polymerization catalysts that do not require any Al- or B-containing cocatalysts.*

7.4.2 Polyolefin Chain Growth

The processes by which growth of a polymer chain is engendered in a homogeneously soluble catalyst system are best documented for some of the cationic alkyl Ni(II) and Pd(II) complexes with neutral diimine ligands. Low-temperature NMR studies show that the *resting state* of the catalyst (*i.e.* the form in which most of the catalyst accumulates in the presence of ethylene) is an outer-sphere ion pair in which a weakly coordinating $B(3,5-(CF_3)_2C_6H_3)_4^-$ anion has been displaced from the cationic metal centre by an ethylene molecule. The *rate-limiting step* for polymer production is the *migratory insertion* of ethylene into the adjacent metal-alkyl or metal-polymer bond. This step occurs more easily in Ni(II) catalysts than in their Pd(II) analogues. The electron-deficient primary insertion product is probably stabilized by a *β-agostic bond;* in the presence of ethylene however this bond is immediately broken under formation of a new ethylene complex (Figure 13).

Propylene and higher olefins, on the other hand, do not bind to the metal centre strongly enough to induce complete formation of the olefin-containing reaction complex from the β-agostic insertion product. The latter appears to be more stable here and will thus form the catalyst resting state. Olefins other than ethylene must form the required olefin-containing reaction complex in a *pre-equilibrium* reaction step. The β-agostic resting state appears to be destabilized by the presence of bulky ortho-substituents at both imino-aryl groups; olefin uptake can then occur more easily, as indicated by increased activities of the correspondingly substituted catalysts for the polymerization of propylene.

A crucial property of a polyolefin is its *regioregularity*. It is governed by the catalyst used for its production, i.e. by the *regioselectivity* with which it controls the direction of olefin insertion into a metal-alkyl bond. For example, propylene can insert either with its CH_2 or its CHMe end toward the metal. The first

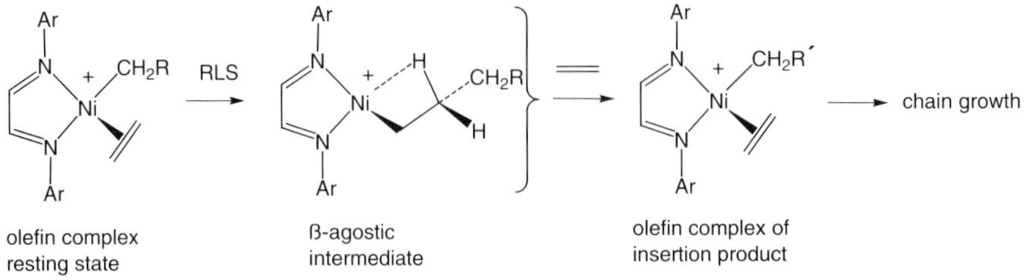

olefin complex   β-agostic   olefin complex of
resting state   intermediate   insertion product

Figure 13   *Chain growth by ethylene insertion in diimine nickel catalysts (RLS = rate-limiting step).*

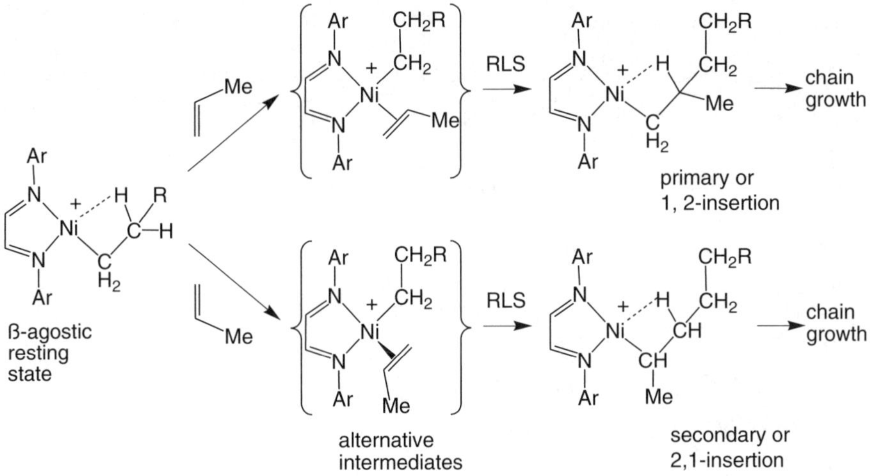

Figure 14 *Alternative propylene insertion modes in diimine-nickel catalysts (RLS = rate-limiting step).*

mode is called *primary* or *1,2-insertion*, whereas attachment of the CHMe end at the metal centre leads to *secondary* or *2,1-insertion* (Figure 14). As with other types of catalytic reactions (e.g. olefin hydrogenation or hydroformylation) this choice is mainly decided by steric factors. Narrow coordination sites will generally favour formation of the slimmer primary insertion product, while the less-hindered olefin end will otherwise be the preferred site of attack of a migrating alkyl group. Both reaction modes occur to comparable extents for most Ni(II) and Pd(II) polymerization catalysts.

A remarkable feature of Ni(II)- and Pd(II)-based catalysts is their amazing propensity for *chain migration*, i.e. for the stepwise movement of the metal centre along the newly formed polymer chain. By repeated β-H transfer with elimination of a polyolefin and reinsertion of the latter with reversed orientation (Figure 15), the metal is able to migrate to each of the enchained C atoms, passing in some cases even tertiary C atoms. Each of the ensuing metal-attachment sites can, in principle, become the starting point of another chain segment by renewed olefin insertions. A number of otherwise unexpected structural elements are thus found in polymers produced with catalysts of this type.

Due to this chain-migration process ethylene is polymerized to macromolecules containing multiple branches – rather than to the linearly enchained polymer obtained with classical solid-state catalysts. In propylene polymerization with these catalysts 1,2-insertions give the normal methyl-substituted polymer chains, but after each 2,1-insertion the metal centre is blocked by the bulky secondary alkyl unit and can apparently not insert a further propylene. Instead the metal must then first migrate to the terminal, primary C atom before chain growth can continue by further propylene insertions. By this process, also called *1,ω-enchainment* or *polymer straightening*, some of the methyl or (in the case of higher olefins) alkyl substituents are incorporated into the chain.

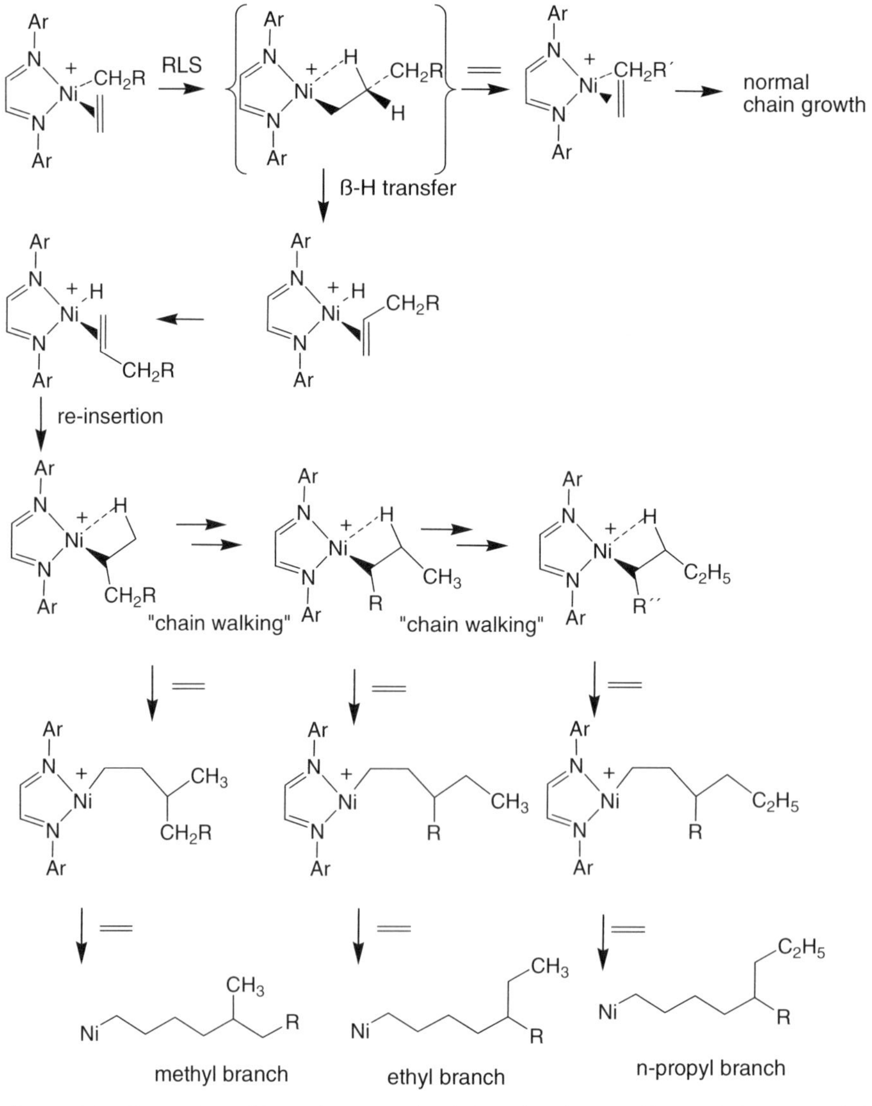

Figure 15 *Formation of branches in polyethylene by chain walking of diimine-nickel catalyst.*

Some Ni(II)-catalysts polymerize higher olefins by *2,ω-enchainment*. Here even the primary insertion product appears too bulky for further chain growth due to its adjacent alkyl branch; the metal thus has to migrate to the unencumbered end of the alkyl side chain before another insertion can occur [V. Möhring, G. Fink *Angew. Chem. Intern. Ed. Engl.* 1985, **24**, 1001].

In zirconocene-catalyzed olefin polymerizations similar processes are involved. Here polymerization rates depend at least linearly on olefin concentrations; an olefin-containing reaction complex thus cannot be the catalyst resting state. Instead the fourth coordination site of the metal centre might be occupied by an agostic bond to the metal-bound polymer chain, as in Ni(II)- and Pd(II)-catalyzed α-olefin polymerizations, by direct contact to a weakly coordinating Me-MAO⁻ or perfluoroarylborate anion, or by the formation of a trialkylaluminium adduct. Low-temperature NMR studies on relatively slow catalyst systems point to contact-ion pairs as the dominant resting-state species [C. R. Landis, K. A. Rosaeen, D. R. Sillars, *J. Am. Chem. Soc.*, 2003, **125**, 1710]. In faster catalyst systems at above-ambient temperatures where an insertion occurs every few milliseconds, the nature of the predominant species, from which the reaction complex is formed by uptake of an olefin, is still under discussion [I. E. Nifantiev, L. Y. Ustynyuk, D. N. Laikov *Organometallics* 2001, **20**, 5375; Z. Xu, K. Vanka, T. Ziegler, *Organometallics* 2004, **23**, 104; F. Song, S. J. Lancaster, R. D. Cannon, M. Schormann, S. M. Humphrey, C. Zuccaccia, A. Macchioni, M. Bochmann, *Organometallics* 2005, **24**, 1315].

Transition states for α-olefin insertions have been shown from measurements of *kinetic isotope effects* to involve an agostic interaction of an α-H atom of the growing chain with the metal centre of the metallocene catalyst [W. E. Piers, J. E. Bercaw, *J. Am. Chem. Soc.* 1990, **112**, 9406; H. Krauledat, H.-H. Brintzinger, *Angew. Chem. Int. Ed. Engl.* 1990, **29**, 1412]. This interaction which becomes a γ-agostic bond in the primary insertion product, appears to relieve some of the increasing electron deficiency as the complex proceeds along the insertion reaction coordinate.

Due to their narrow aperture, zirconocene-based catalysts insert olefins almost exclusively in the 1,2- or primary direction. Small proportions of 2,1-inserted propylene units and, for some catalysts, 1,3-inserted units derived from them by chain straightening, are a cause of melting-point lowering in some metallocene-produced polypropylenes.

Group 4 metal catalysts of the constrained-geometry type generally follow similar patterns in their polyolefin formation reactions. These catalysts, which can be operated at rather high temperatures, are able – probably due to their more open coordination sites – to incorporate into a polymer chain also the unsaturated ends of polyolefins, i.e. of *macromonomers*, which are hardly touched by metallocene-based or by Ni(II)- or Pd(II)-based catalysts (Figure 16). Consequently, constrained-geometry catalysts form a peculiar type of long-chain branched polyethylene (Figure 1), which possesses remarkable elasticity and toughness [J. C. Stevens in *Catalyst Design for Tailor-Made Polyolefins*, ed. K. Soga, and M. Terrano, Kodansha Elsevier, Tokyo 1994, p. 277; K. W. Swogger, *ibid.*, p. 285).

Discussion Point DP4: *Figures 13 and 14 represent reaction schemes for the polymerization of ethylene and of propylene by diimine-nickel catalysts. From these schemes predict how the rates of polyethylene and of polypropylene formation should depend on the concentrations of the respective monomers. What influence should the kind of anion present be expected to have on the rates of polymer formation in each of these cases? How would these answers differ from those to the same questions with regard to zirconocene-based polymerization catalysts (Figure 11)?*

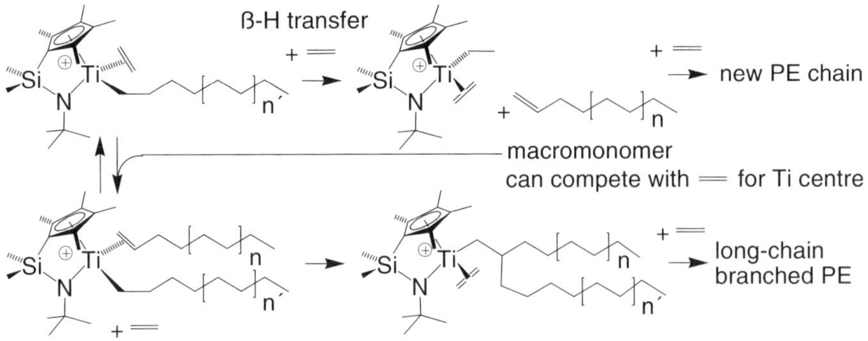

Figure 16 *Formation of long-chain branched polyethylene by a constrained-geometry catalyst through macromonomer insertion.*

Discussion Point DP5: *The previously unexpected observation of side-chain branches in diimine-nickel catalyzed polyethylene formation is explained by the reaction scheme represented in Figure 15. Propose related "chain migration" schemes which explain i) the "chain straightening", i.e. the incorporation of propylene methyl substituents into the backbone of polypropylene chains produced by these catalysts, ii) the 2,ω-concatenation of higher α-olefins by some Ni-based catalysts, and iii) the introduction of stereoerrors in isotactic polypropylene by chain-migration of chiral ansa-zirconocene catalysts.*

7.4.3 Stereochemistry of α-Olefin Enchainment

The thermal and mechanical properties of polypropylene and other polymers produced from α-olefins depend – just as much as on their regioregularity – on the *stereoregularity* of consecutive olefin insertions, i.e. on the relative configuration of the tertiary C atoms which occupy every other position in the polymer backbone. Some of the diastereomeric structures are represented as Fischer-type projections in Figure 2.

If consecutive olefin insertions all occur with the same stereochemical orientation, i.e. on the same *olefin enantioface*, all alkyl substituents at the polymer backbone will have the same orientation. In such an *isotactic* polymer chain, all the tertiary C atoms inside its backbone are of the same configuration. If all consecutive olefin insertions occur with opposite enantiofacial orientation, the resulting *syndiotactic* polymer contains in its backbone tertiary C atoms of strictly alternating configuration. An *atactic* polymer will finally result when olefin insertions occur randomly, without enantiofacial preference.

Isotactic polypropylene is a rather stiff and tough solid material with a melting point of 164°C. Closely packed, CH_3-studded helices (Figure 17), rigidly interwoven in crystalline domains (Figure 18), account for the mechanical and thermal resistance of isotactic polymers. Syndiotactic polypropylene has a related crystalline structure, but atactic polymers are amorphous and form oily or waxy materials depending on chain lengths.

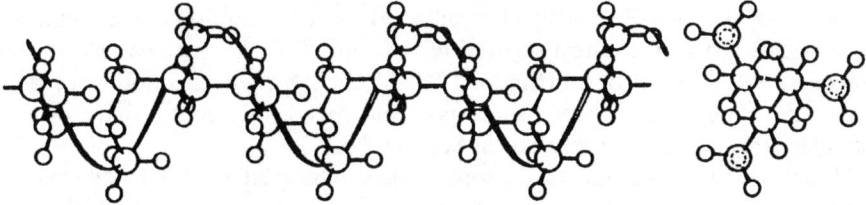

Figure 17   *Helical structure of isotactic polypropylene.*

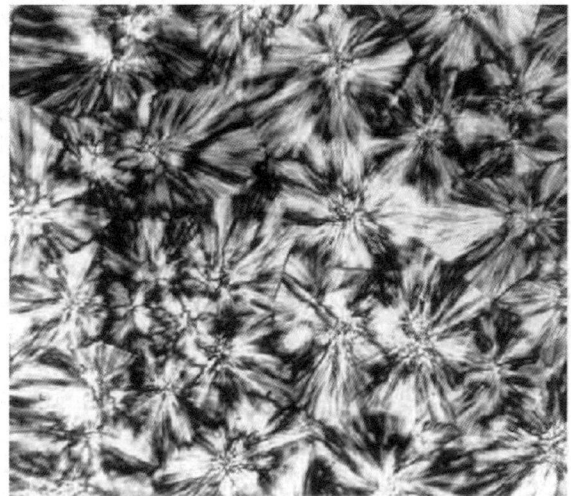

Figure 18   *Spherulite crystals of isotactic polypropylene.*

Highly isotactic polymers, with stereoregularities similar to those obtained with the solid-state catalysts discussed in Section 7.3, are produced by zirconocene catalysts which have a bridged and appropriately substituted ligand framework of C_2 symmetry, such as that shown in Figure 16. The Me_2Si-bridge of these *ansa-zirconocene* complexes renders the ligand framework particularly stereorigid, while the spatially demanding extensions of each C_5 ring provide for efficient chirality at each of the coordination sites.

Studies on catalysts carrying ^{13}C-labeled alkyl groups at their metal centre have shown that a chain with at least two C atoms is required for an efficient control of the enantiofacial orientation of an inserting olefin [A. Zambelli, C. Pellechia, *Makromol. Chem., Macromol. Symp.* 1993, **66**, 1]. This observation, together with molecular-mechanics model calculations, has led to the concept that catalytic-site control is *chain-segment mediated* [G. Guerra, P. Corradini, L. Cavallo, M. Vacatello, *Makromol. Chem., Macromol. Symp.* 1995, **89**, 77]. The metal-bound $C(\alpha)$-$C(\beta)$ chain segment will orient itself into the most open sector of the catalyst's ligand framework so as to minimize steric repulsions. An incoming olefin then must adopt that enantiofacial orientation which places its alkyl substituent *trans* to the C(α)-C(β) segment along the incipient C(2)...C(α) bond (Figure 19). All available evidence indicates that analogous mechanisms are also responsible for the stereoselectivity of the Ziegler-Natta catalysts described in Section 7.3.1.

Further studies on kinetic isotope effects, observed with 1-D labeled α-olefins, have added to this picture the notion of α-*agostic assistance*: Of the two C(α)-H bonds only one can interact with the electron-deficient metal centre so as to place the C(α)-C(β) chain segment in an open ligand sector. The resulting four-membered cyclic transition state acquires added conformational rigidity by this agostic stabilization, which probably contributes to the high stereoselectivities of these catalyst systems [M. K. Leclerc, H.-H. Brintzinger, *J. Am. Chem. Soc.* 1996, **118**, 9024].

Similar considerations hold also for syndio-specific polymerization catalysts, for which the C_S-symmetric zirconocene complex shown in Figure 19 is a prototype. Here the two coordination sites have opposite chirality. The preferred orientation of the C(α)-C(β) segment of the polymer chain and hence the preferred enantiofacial orientation of the inserting olefin will thus alternate with each consecutive insertion, by which the Zr-CH$_2$(polymer) bond moves from one coordination site to the other.

In distinction to C_2-symmetric, iso-specific catalysts, however, stereoerrors can arise here also when the growing chain moves from its original coordination site to the other, without the intervention of an olefin insertion. These *skipped insertions* become frequent at low olefin concentrations for syndiospecific catalysts shown in Figure 19, since then site-exchange of the polymer chain without insertion becomes competitive with further chain growth.

In other cases, for example in syndiotactic propylene polymerization with certain unbridged O,N-chelated Ti complexes, chain-end chirality can control the stereochemistry of 2,1-olefin insertions by adjusting the chirality of a ligand framework so as to minimize mutual repulsions [P. Corradini, G. Guerra, L. Cavallo, *Acc. Chem. Res.* 2004, **37**, 231].

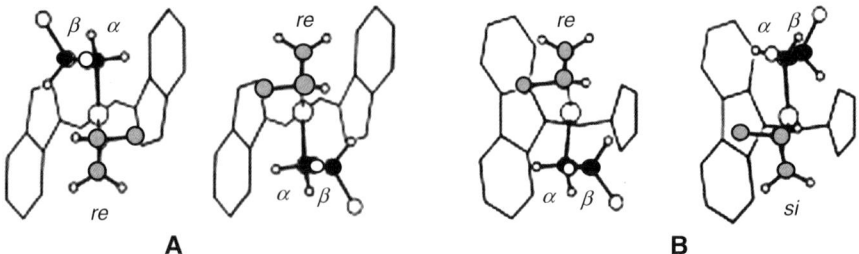

Figure 19   *Stereoselective insertions of propylene (grey) under catalytic-site control, mediated by the α,β segment of the growing polymer chain (black), for isospecific polymerization by a C_2-symmetric catalyst (A, left) and for syndiospecific polymerization by a C_S-symmetric catalyst (B, right).*

7.4.4 Chain-Growth Termination and Re-initiation

Simple zirconocene catalysts, such as the system $(C_5H_5)_2ZrCl_2/MAO$, produce relatively short-chain polypropylenes with several hundred to a few thousand monomer units. More practically useful polymers, with chain lengths of some ten to hundred thousand monomer units, i.e. with molar masses of about half a million to several millions, are available using more elaborated zirconocene catalysts such as that represented in Figure 10. For polymers of the type considered here, average chain lengths, as measured by the mean degree of polymerization, P_N, are approximately equal to the ratio of insertion and termination rates, $P_N = v_i/v_t$. In addition to high rates of chain growth, reduced rates of chain-growth termination by release of the polymer from the catalyst centre, are thus essential for the production of long-chain polymers.

Release of the unsaturated chain end of a polyolefin can occur by *β-H transfer to the metal* or *to a monomer* molecule (see Appendix 1 for backgound material). A metal-alkyl species, i.e. the starting unit for a new polymer chain, arises from the metal-hydride species formed in the first case by insertion of an olefin, or it can be formed directly by *β-H transfer to a monomer* (Figure 20). While the results are thus identical, the two reaction paths differ in their respective kinetics: In the first case, the rate-limiting β-H transfer is independent of the olefin concentration, while the rate of β-H transfer to a monomer requires the formation of an olefin-containing reaction complex and will thus increase linearly with olefin concentration.

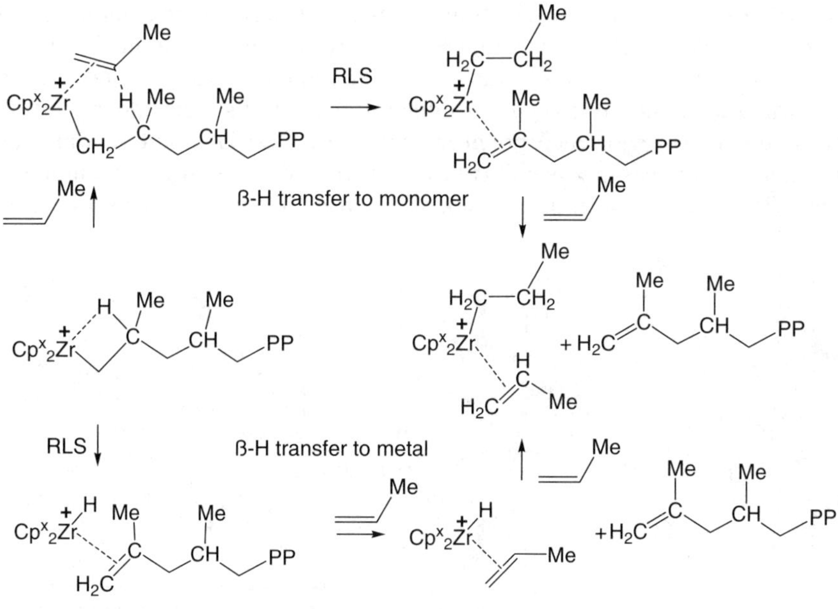

Figure 20 *Chain growth termination and start of a new chain by β-H transfer to a coordinated monomer (top) and to the metal centre (bottom), followed by olefin insertion into the metal-hydride bond (RLS = rate-limiting step).*

Since its dependence on olefin concentration is the same as that observed for chain-growth by olefin insertion, β-H transfer to a monomer has the consequence that average chain lengths P_N are not affected by changes in olefin concentration, since the latter cancels between the numerator and denominator of the expression $P_N = v_i/v_t = k_i \cdot [M]/k_t \cdot [M] = k_i/k_t$. For β-H transfer to the metal, however, chain lengths increase with olefin concentration, since the latter affects only the numerator of the expression $P_N = v_i/v_t = k_i \cdot [M]/k_t$.

Of the two mechanisms, the quasi-degenerate β-H transfer to an olefin requires lower activation energies and is thus likely to occur at higher rates. But this reaction path requires a more highly organized and spatially more demanding arrangement of its reaction participants. It is thus rather efficiently suppressed in catalysts as that shown in Figure 10 which carry space-filling methyl or other alkyl substituents in their lateral positions. Accordingly these catalysts are apt to produce particularly long-chained polypropylene and other polyolefins, especially at high olefin concentration, e.g. in liquid propylene.

Space-filling substituents induce increased chain lengths also with Ni-based polymerization catalysts. Here a spatially demanding transition state, required for β-H transfer to a monomer molecule and subsequent release of the unsaturated polymer chain end, appears to be suppressed by ligand substituents which block the coordination positions above and below the tetragonal coordination plane (Figure 21).

Other processes also contribute to chain growth termination under special conditions. In particularly crowded catalysts, *β-methyl transfer* to the metal centre can occur instead of β-H transfer. When other reaction paths are blocked, *σ-bond metathesis*, i.e. transfer of an H atom from a monomer to the metal-bound alkyl C atom can release a polymer with a saturated chain end with formation of a new unsaturated metal-bound chain start. Saturated chain ends will also result when H_2 gas is added to a catalyst system thus leading to the production of shortened polymer chains. Such an H_2 addition will often also cause an increase in overall catalyst activity, since H_2 will predominantly react with species – such as occasional 2,1-inserted units – which are rather

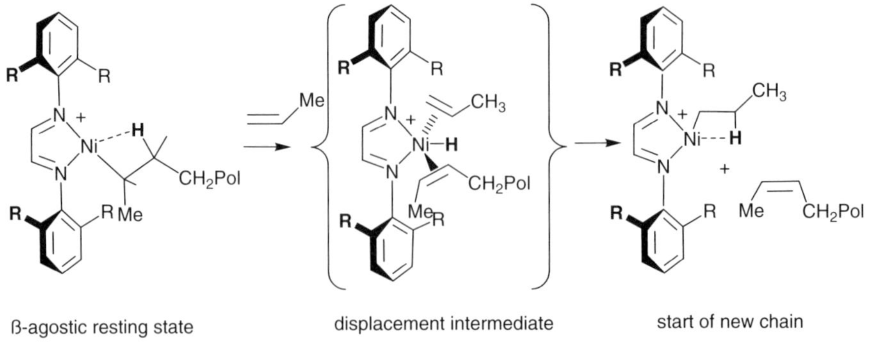

β-agostic resting state   displacement intermediate   start of new chain

Figure 21   *Chain growth termination by β-H transfer and displacement of unsaturated chain end from nickel centre by a monomer, hindered by bulky substituents R.*

unreactive to olefin insertion and would thus tend to slow further polymer production [V. Busico, R. Cipullo, S. Ronca, *Macromolecules* 2002, **35**, 1537].

In many instances, particularly with MAO-activated zirconocene catalysts carrying bulky ring ligands, a major pathway is *polymer transfer to Al centres*; in exchange a methyl group is transferred to the catalyst centres. New chains started there thus contain an extra CH_3 group, while the polymer product accumulates as Al-, mostly Me_2Al-capped chains. Upon hydrolysis of these polymerization systems, completely saturated polymer chains are obtained. In principle, the Al-bound polymer chains represent a stable but rather reactive kind of organometallic reagent and might be used for interesting purposes, such as the introduction of polar chain ends, or for transfer to a different catalyst with the aim of obtaining polymers with distinctly structured chain segments [C. Przybyla, G. Fink, *Acta Polym.* 1999, **50**, 77; S. Lieber, H.-H. Brintzinger, *Macromolecules* 2000, **33**, 9192; D. J. Arriola, E. M. Carnahan, P. D. Hustad, R. L. Kuhlman, T. T. Wenzel, *Science* 2006, **312**, 714].

Discussion Point DP6: *Alternative reaction sequences for the release of unsaturated chain ends from a catalyst centre are represented in Figure 20. Determine for each of these sequences how the termination rate v_t depends on the monomer concentration. Write reaction equations for several alternative chain-release and re-start reactions mentioned in Section 7.4.4. Can you think of polymerization systems for which increased ethylene concentrations might lead to polymers with reduced molar mass?*

For typical homogeneous polymerization catalysts the rates of chain growth by olefin insertion and of chain termination, by some combination of the processes discussed above, is likely to be rather independent of the length of the metal-bound polymer chain, except for the very first insertion steps. When a catalyst system meets this condition, its polymer product will have a relatively narrow, so-called Schulz-Flory molar mass distribution and its *polydispersity index* PDI, defined as the ratio of the weight average and the number average of its molar mass, M_W/M_N, will have a value close to 2. PDI values, experimentally determined e.g. by size-sensitive gel exclusion chromatography, are thus often used to test whether a given catalyst system is to be considered a *single-site catalyst* with uniform catalyst centres. In contrast to these – mostly soluble – polymerization catalysts the "classical" solid-state catalysts usually have rather broad molar mass distributions with PDI values of 5–15, due to the non-uniformity of their catalyst centres.

Polymers with even narrower mass distributions, e.g. with PDI values close to 1, arise in *living polymerization* systems, in which no chain termination processes can occur at all, such that all chains remain bound to the metal centre from which they have started to grow at the same time. Living polymerizations, which offer useful opportunities, *e.g.* with regard to the production of block copolymers by exchange of one monomer for another, occur in *anionic polymerizations* of styrenes or butadienes such as are induced by simple lithium alkyls. For α-olefin polymerization catalysts of the type discussed above, living polymerizations are rare. These more elaborate catalysts can thus release a newly formed polymer chain within a time interval of typically less than one second, such that each catalyst centre can on average produce many thousand polymer chains during its lifetime.

7.5 Supported Metallocene Catalysts

Metallocenes immobilized on solid support materials have been successfully introduced in industry as polymerization catalysts for the production of new application-oriented polymer materials. Industrial polymerizations, which are carried out either as a slurry process in liquid propylene or as a gas-phase process (Section 7.2.3), require that catalysts are in the form of solid grains or pellets; soluble metallocene catalysts thus have to be supported on a solid carrier.

These catalyst carriers have to be mechanically stable to avoid the formation of polymer fines (dust); on the other hand, they have to be sufficiently fragile to permit fragmentation into *primary particles* of sub-micrometer size by the hydraulic forces exerted by the growing polymer. At present amorphous, porous SiO_2 gels appear to be best suited as support for MAO-activated metallocene catalysts, since they possess high surface area and porosity, good mechanical properties, as well as stability and inertness under reaction and processing conditions.

For use as catalyst supports, silica gels are calcined at temperatures of 400–500°C so as to remove surface-bound water as well as most geminal and vicinal Si-OH groups. The remaining, mostly isolated Si-OH groups, are then deactivated by reaction with MAO and/or an aluminium alkyl. Upon contact with such a pretreated support the zirconocene catalyst precursor becomes immobilized, i.e. firmly attached to the support grains. Spectral studies show that ion pairs of the type discussed in Section 7.4.1 are then present. Properly chosen contact conditions leave the zirconocene deposited rather evenly throughout the interior of the porous SiO_2 grains.

As a result of this immobilization the $[Al]_{MAO}/[Zr]$ ratio required for full activation of the catalyst can be decreased by about two orders of magnitude compared to homogeneous systems. In certain cases an $[Al]_{MAO}/[Zr]$ ratio of 40 is sufficient to obtain reasonable activities. Such a high polymerization activity requires however that all catalyst components are uniformly distributed throughout the volume of the support particles. The shell-like deposit of the MAO cocatalyst on the particle surface and its absence in the interior of a particle (Figure 22; Al mapping) has the consequence that polymerization takes place only in this outer shell, such that an unfragmented SiO_2 core remains in the product and hinders later polymer processing.

Initiation of polymerization and individual phases of polymer growth on SiO_2-supported catalyst particles can be followed by a combination of kinetic and microscopic methods. A few minutes after exposure to propylene the polymerization rate reaches an initial maximum, which is followed by a period of low activity (Figure 23). In a third phase the polymerization rate rises again and in a final, fourth phase, a broad maximum of activity is reached. This

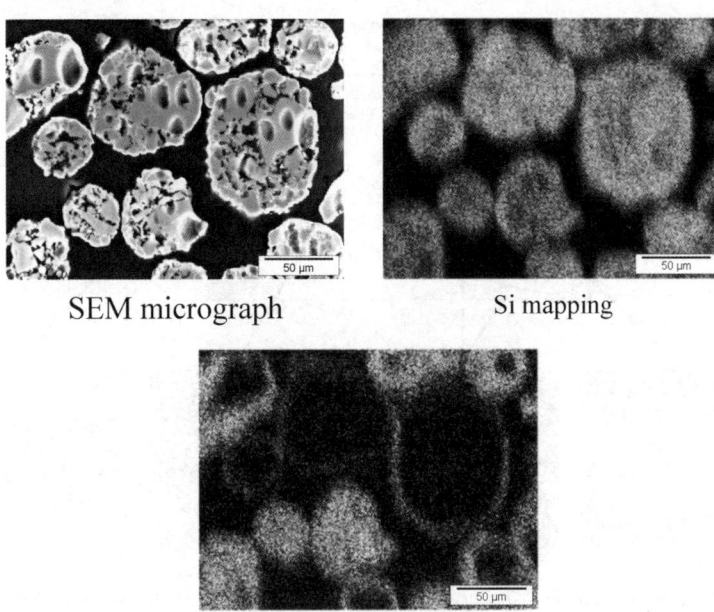

SEM micrograph   Si mapping

Al mapping

Figure 22 *SiO$_2$-supported, MAO-activated zirconocene catalyst grains, Scanning Electron Microscopy (SEM) micrograph and element mapping by Energy-Dispersive X-ray Microanalysis.*

activity profile differs from that of the MgCl$_2$-supported solid-state catalysts described in Section 7.3, where high activities are immediately reached and continuously maintained.

Electron microscopy shows that polymerization starts at active centres on the surface of the particle. During this initial stage, a thin polymer cover is formed on and just below the outer surface of the silica support. This thin cover consists of highly crystalline polypropylene, which acts as a diffusion barrier for the monomer. Diffusion of propylene through this layer thus becomes rate-limiting for polymer formation; consequently the high initial polymerization activity decreases sharply after a few minutes and a period of relatively low activity is reached.

During this period, the length of which depends on temperature, monomer concentration and particle diameter, polymer is growing towards the centre of the carrier, thus breaking down increasing portions of the latter (Figure 23, bottom). As more and more active centres in the interior of the particle become exposed to monomer, the polymerization activity increases again and the silica particle is progressively fragmented by the hydraulic forces of the growing polymer. This process continues towards the centre of the particle, which is finally broken down into fragments with diameters of about 50 nm or less, evenly distributed in the polymer matrix. Since most active sites are now accessible to monomer, polymerization activity reaches its maximum and a further,

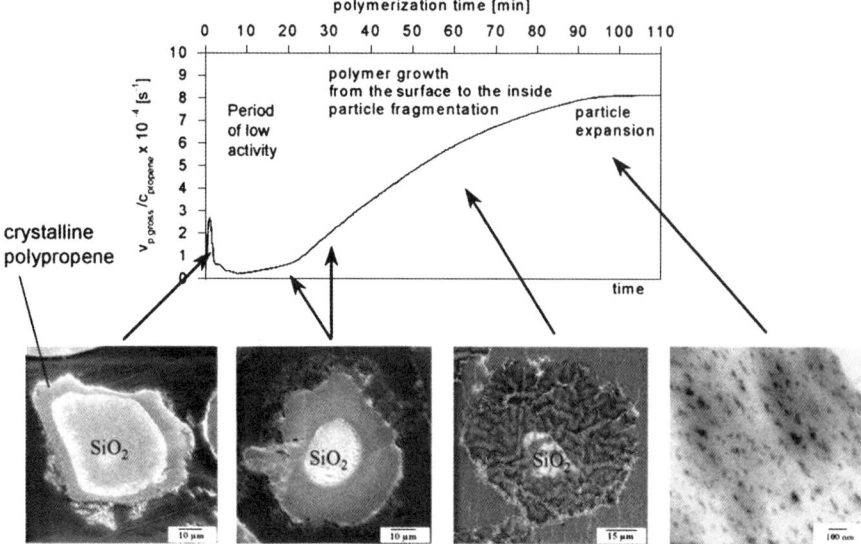

Figure 23   *Polymerization rate profile (top) and electron-micrographs of SiO₂-supported
zirconocene catalyst grains (bottom) at successive stages of particle growth.
(Reflection scanning-electron micrographs of embedded-block cross sections;
SiO₂ light, polymer dark grey; last picture: high-resolution transmission electron
micrograph of a microtomed thin Section; SiO₂ dark, polymer light grey).*

continuous particle expansion takes place, similar to that observed immediately
after addition of monomer to Ziegler-Natta catalysts on more easily fragmented
MgCl₂ supports.

Based on these kinetic and microscopic observations, olefin polymerization
by supported catalysts can be described by a "shell by shell" fragmentation,
which progresses concentrically from the outside to the centre of the support
particles, each of which can thus be considered as a discrete *microreactor*. A
comprehensive mathematical model for this complex polymerization process,
which includes rate constants for all relevant activation, propagation, transfer
and termination steps, serves as the basis for an adequate control of large-scale
industrial polymerizations with SiO₂-supported metallocene catalysts [A. Alex-
iadis, C. Andes, D. Ferrari, F. Korber, K. Hauschild, M. Bochmann, G. Fink,
Macromol. Mater. Eng. 2004, **289**, 457].

7.6   Copolymerization of Linear and Cyclic Olefins

Copolymers of ethylene with α-olefins, such as the short-chain branched LLDPE
(linear low-density polyethylene) "impact" materials or the EPD (ethylene-
propylene-diene copolymer) rubbers represent major percentages of the total
polyolefin production, due to their desirable mechanical properties. Solid-state
MgCl₂-supported Ziegler-Natta catalysts however, have unfavourable reactivity

ratios for α-olefin comonomers. As these catalysts strongly prefer ethylene, a large excess of the α-olefin – the longer the side chain the higher the excess – has to be fed to these systems to achieve sufficient comonomer incorporation.

Metallocene-based catalysts, on the other hand, have rather comparable reactivity ratios for ethylene and even for the higher α-olefins. Accordingly, high comonomer incorporation (uniform in each polymer chain and independent of its molar-mass) can be achieved with these single-site catalysts. This is of great advantage with regard to polymer processability and application.

Single-site metallocene catalysts are also highly reactive vis-à-vis *cycloolefins* such as cyclobutene, cyclopentene or norbornene. While homopolymers of these cycloolefins have melting temperatures ($>380°C$), much too high for technical processability, ethylene-cycloolefin copolymers (COC's) – e.g. ethylene-norbornene copolymers – are amorphous materials with glass transition temperatures, above which they become soft and processable [W. Kaminsky, *J. Polym. Sci. A, Polym. Chem.*, 2004, **42**, 3911].

Metallocene catalysts insert norbornene into metal-polymer bonds stereoselectively with *cis-2,3-exo* orientation and without any ring opening (Figure 24). Different micro-structures of the copolymers result from various possible concatenation patterns – e.g. from alternating or norbornene-norbornene diad and triad block structures – and from alternative relative configurations of neighboring norbornene units (Figure 25). Accordingly, the macroscopic properties of a copolymer product depend on the selectivity of the particular metallocene catalyst used for its generation [D. Ruchatz, G. Fink, *Macromolecules* 1998, **31**, 4669, 4674, 4682, 4684; M. Arndt, I. Beulich, *Macromol. Chem. Phys.* 1998, **199**, 1221].

Ethylene-norbornene copolymers, which have thermoplastic properties when heated above their glass transition temperatures of *ca.* 200–250°C, have been commercialized by Ticona GmbH under the trade name TOPAS (*T*hermoplastic *O*lefin *P*olymer of *A*morphous *S*tructure). Their properties – exceptional transparency, low double refraction, high stiffnes and hardness, low permeability for moisture and excellent biocompatibility – make these ethylene-norbornene copolymers particularly valuable as engineering polymers, for optical applications and as materials for food and medical packaging.

Discussion Point DP7: *While ethylene and norbornene give essentially alternating, amorphous copolymers, attempts to copolymerize ethylene and cyclohexene give only crystalline polyethylene. Which factors might contribute to these observations? Unsaturated norbornenyl chain ends cannot arise by β-H transfer*

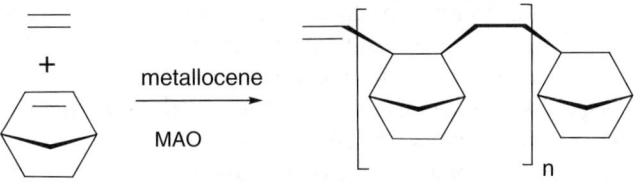

cis-2,3-*exo*-insertion

Figure 24 *Stereochemistry of ethylene-norbornene enchainment.*

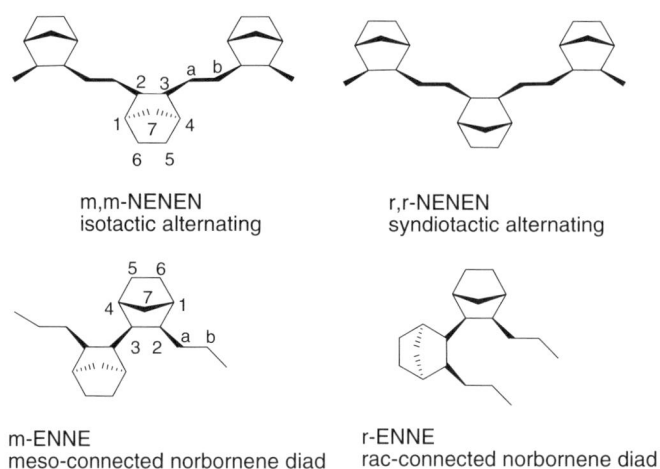

m,m-NENEN
isotactic alternating

r,r-NENEN
syndiotactic alternating

m-ENNE
meso-connected norbornene diad

r-ENNE
rac-connected norbornene diad

Figure 25 *Alternative concatenation microstructures in ethylene-norbornene copolymers.*

from a metal-bound norbornyl unit. Try to verify this postulate by considering the stereochemistry of norbornene insertion and the consequences of Bredt's rule. Delineate how unsaturated norbornenyl chain starts arise by σ-bond metathesis.

7.7 Copolymerisation of Olefins with Polar Monomers and with CO

Despite many attempts, copolymers of olefins with polar monomers, such as vinyl or acrylic esters, which would combine the low price, toughness and easy processability of polyolefins with desirable surface properties of polar polymers – e.g. with the possibility to glue these polymers to other materials – have so far not been obtained using any of the catalysts based on group 4 transition metals, since these oxophilic catalysts are deactivated by oxygen-containing substrates. Soluble catalysts based on one of the less oxophilic group 10 transition metals nickel or palladium, however incorporate finite fractions of polar monomers, albeit at the expense of diminished activity [A. Sen, M. Kang, in *Late Transition Metal Polymerization Catalysis*, eds. B. Rieger, L. Saunders Baugh, S. Kacker, S. Striegler, Wiley-VCH, Weinheim, 2003, p. 307].

Another interesting example of the tolerance of group 10 transition metal catalysts toward oxygen functionality is the copolymerization of olefins with carbon monoxide. Exposure of an acidified methanol solution of a Pd(II) diphosphine chelate to an olefin-CO mixture leads to the formation of polymers, which contain these monomers concatenated in strictly alternating fashion, with carboxylate ester groups occupying the chain ends [E. Drent, P. H. M. Budzelaar, *Chem. Rev.* 1996, **96**, 663; G. Consiglio, in *Late Transition Metal Polymerization Catalysis*, eds. B. Rieger, L. Saunders Baugh, S. Kacker, S. Striegler, Wiley-VCH, Weinheim, 2003, p. 279.

This reaction may proceed by a sequence of steps such as are represented in Figure 26: CO inserts into a Pd-OCH$_3$ unit, present in equilibrium with the starting compounds, to form a cationic carboxylate complex, which takes up an olefin to form, by insertion of the latter into the Pd-carboxylate bond, a rather stable five-ring chelate, containing a Pd-bound C=O group. This stable chelate can be broken up only by CO, which then inserts with formation of a ketyl species. In turn the latter reacts selectively with olefin to form a new, stable five-ring chelate, rather than with CO to some unstable species containing adjacent C=O groups. Repetition of this sequence leads to polymer growth until it is terminated, e.g. by attack of CH$_3$OH at the cationic ketyl centre. Interesting background to this and the relationship to the Pd(II) catalyzed carbonylation of ethylene to methyl propionate is in Chapter 4, Section 4.3.1.

The resulting polyketone, which has remarkable mechanical and thermal resistance, has been commercialized as a valuable engineering plastic, but general applications have so far been hindered by its sensitivity to photo-degradation.

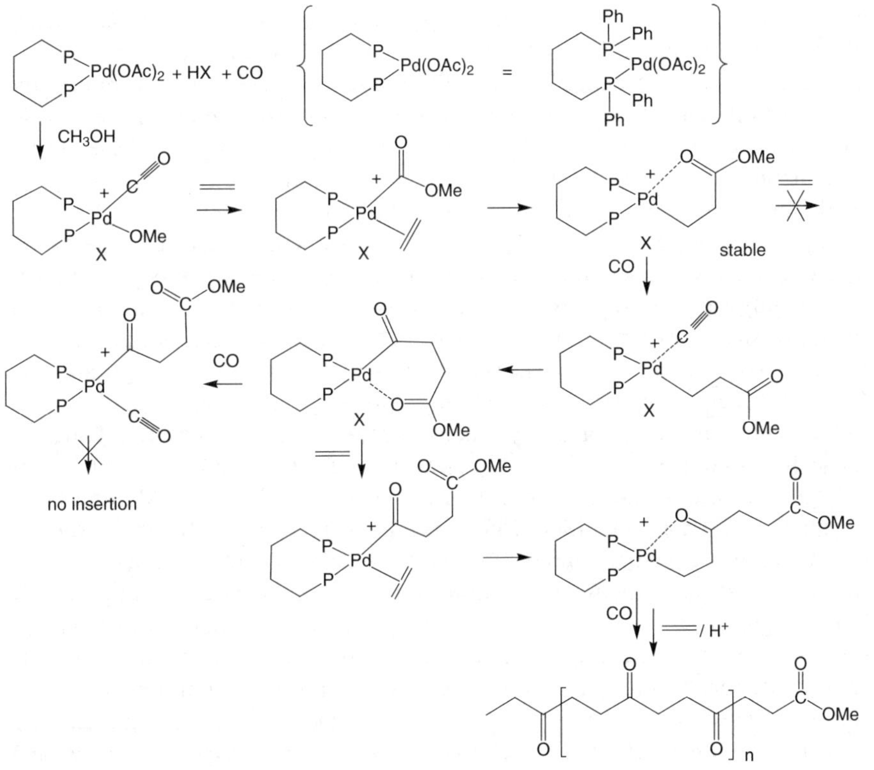

Figure 26 *Proposed steps in Pd(II)-catalyzed polyketone formation from ethylene and CO in MeOH.*

Annex 1 Polymer Stereochemistry Studied by ^{13}C NMR Spectroscopy

The *microstructures* of polyolefin materials are most effectively studied by ^{13}C NMR spectroscopy in solution. In particular C atoms at branching positions and at the ends of chains and branches are easily distinguished from those inside a chain. In this way, the numbers and mean lengths of branches, e.g. in polyethylene chains, can be determined rather reliably.

Another important practical application concerns the *microstructure* of polypropylene [V. Busico, R. Cipullo, *Progr. Polym. Sci.* 2001, **26**, 443]. The chemical shift of each CH$_3$ substituent in a polypropylene chain depends on the orientations of neighbouring substituents, i.e. on the relative configurations of the adjacent $-$C(CH$_3$)CH$_2-$ units. Pairs of adjacent CH$_3$ substituents are called *meso* or *m dyads* if their orientations are identical, while two CH$_3$ neighbors with opposite orientations form a *racemo* or *r dyad*. Together with its two neighbouring CH$_3$ groups, a given CH$_3$ group can form a mm *triad*, a rr triad or a mr = rm triad. Present standard spectrometer resolution distinguishes CH$_3$ resonances at the *pentad* level, where the position of each CH$_3$ signal is influenced by two adjacent units on each side. For atactic polymers with statistical distribution of relative configurations (such as are obtained by the simple system (C$_5$H$_5$)$_2$ZrCl$_2$/MAO) this resolution yields, instead of the three triad signals mm, rm and rr, nine pentad signals, the mm-centred set mmmm, mmmr and rmmr, the mr-centred set, mmrr, mmrm = rmrr and rmrm, and the rr-centred set rrrr, rrrm and mrrm (Figure 27). Completely isotactic polypropylene, on the other hand, would give rise only to the mmmm pentad signal, while a completely syndiotactic one would yield only the signal of the rrrr pentad.

In practice, however, ^{13}C NMR spectra of polymers produced with a particular catalyst always reveal finite proportions of various types of *stereoerrors*. Even highly isotactic polypropylene gives rise, in addition to its main mmmm signal, to small mmmr, mmrr and mrrm pentad signals. These signals indicate that a tertiary C atom will occasionally occur with the "wrong" configuration, its CH$_3$ substituent being oriented in a sense opposite to its neighbors.

The stereoselectivity of C_2-symmetric ansa-zirconocene catalysts, such as those discussed in Section 7.4 can be shown to have its origin in the chirality of the catalysts and not in the chirality of the last-inserted CH$_2$CHMe-unit (J. A. Ewen, *J. Am. Chem. Soc.* 1984, **106**, 6355). If such a *chain-end control* would be operative, an occasional stereoerror would be expected to perpetuate itself in the configuration of the following insertions, thus giving rise – in addition to a mmmr signal – to a stereoerror signal of the type mmrm. Instead, the actually observed signals of the double-r type, mmrr and mrrm, indicate that insertions return, after each stereoerror, to their previous enantiofacial preference, due to the *catalytic-site control* exerted by the chiral catalyst (Figure 28, p 252).

Discussion Point DP8: *Typical stereoerror patterns occurring in isotactic polypropylene are represented in Figures 27 and 28. Try to delineate the stereoerror patterns and the associated ^{13}C pentad signals which are to be expected for*

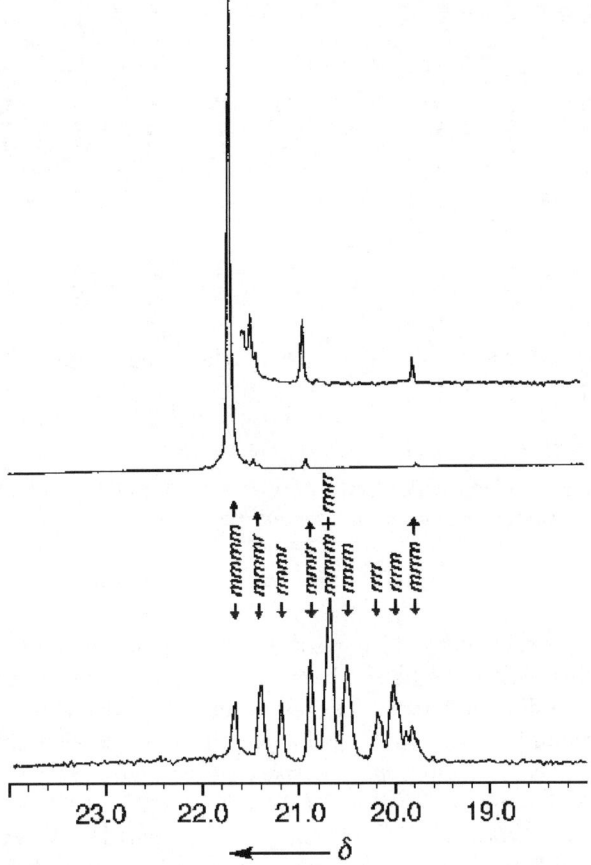

Figure 27 *Pentad ^{13}C NMR signals of essentially isotactic polypropylene (top) and of atactic polypropylene (bottom).*

syndiotactic polypropylene generated i) under chain-end control ii) under catalytic site control, and iii) under catalytic site control with occasional skipped insertions (i.e. change of coordination sites without insertion). Which effects might a change in monomer concentration have on the relative sizes of each of these error signals?

Annex 2 Stereospecific Polymerization of Conjugated Diolefins: Butadiene and Isoprene

For reasons of space diolefin polymerization has not been included in this Chapter. Some information and pertinent references are summarized here. 1,3-Dienes can be polymerized by lithium alkyls or by Ziegler-Natta type catalysts, containing titanium or cobalt, nickel, and neodymium. Industrially important products are 1,4-cis-polybutadiene (>2 Mt/a) and 1,4-cis-polyisoprene (>1 Mt/a). They are

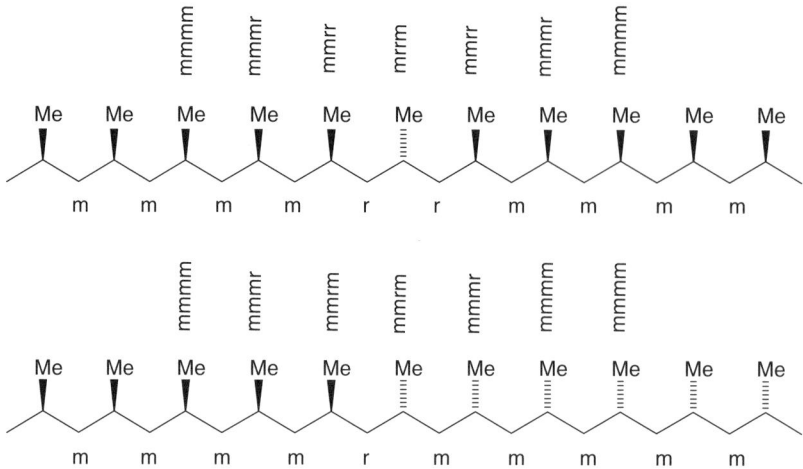

Figure 28 *Stereoerror pentads expected for essentially isotactic polypropylenes, gener-ated under catalytic-site control (top) and under chain-end control (bottom).*

used as elastomers; that derived from isoprene corresponds to natural rubber. The chemistry of their polymerizations is based on multiple insertions, as in the case of the simple olefins discussed above. The intermediates are allylmetal complexes, see Chapter 5, Section 5.4. Detailed discussions can be found in the following reviews: R. Taube and G. Sylvester in *Applied Homogeneous Catalysis*, 2nd ed, vol.1; ed. B. Cornils and W. Herrmann, Wiley-VCH, Weinheim 2002, p. 285; L. Porri and A. Giarrusso in *Comprehensive Polymer Science*, vol. 4, part II, ed. G.C. Eastmond, A. Ledwith, S. Russo, and B. Sigwalt, Pergamon, Oxford, 1989, p. 53; W. Kaminsky and B. Hinrichs, *Plastics Engineering*, 2005, **70** (*Handbook of Polymer Science*, 2nd ed. Marcel Dekker), 333.

Annex 3 Some Hints to Help Start the Discussions

DP 1 Using suitable terms look up methods for the production of olefins from alternative sources in the internet or in one of the encylopedias of technical/ industrial chemistry.

DP 2 For heats of polymerization assume *ca.* 100 kJ per mol of ethylene or propylene (disregard finer details). Consult e.g. Section 7.2 and references given at its end for heat removal aspects.

DP 3 Consider qualities such as "Lewis-acidic" and "weakly coordinating" with regard to the B- and Al-containing activators mentioned in Section 7.4.1. Consider also single-component catalysts.

DP 4 Find the species responsible for the respective rate-limiting steps: In which way might their concentrations depend on olefin concentrations and (if applicable) on the type of anion present?

DP 5 Let the metal freely migrate along the polymer chain attached to it and determine which site of renewed monomer insertion will lead to formation of the products specified in (i), (ii) and (iii).

DP 6 How are the species responsible for chain growth (Figure 14) and its termination (Figure 20) affected by monomer concentrations? Consider also Figure 16 for effects of monomer concentration.

DP 7 Consider structural and enthalpic differences between cyclohexene and norbornene and consult a text on organometallic reactions for causes and possible courses of "σ-bond metathesis".

DP 8 How would exchanging "m" and "r" in the last paragraph of Annex1 answer questions (i) and (ii)? For (iii), consider that "skipped insertions" are competing with normal olefin insertions.

Acknowledgements

Helpful suggestions for this Chapter from the editors and from Professor Ludwig Böhm, Professor Paolo Corradini and Dr. David Fischer are gratefully acknowledged. Figures 5-7 have been reproduced from an article by K. S. Whiteley and et al. in *Ullmann's Encyclopedia of Industrial Chemistry*, Figures 8 and 9 (with modifications) from an article by L. Böhm in *Angewandte Chemie*, Figure 23 from an article by G. Fink and collaborators in *Macromol. Chem. Phys.* 2003, **204**, 607, and Figures 27 and 28 from an article by H. H. Brintzinger et al. in *Angewandte Chemie* (all cited below). We thank the publishers for their consent to reproduce these figures, and BASELL GmbH for providing Figure 18 and data for Figure 4.

References

Section 7.1: Books on polymer syntheses and properties: J. M. G. Cowie, *Polymers: Chemistry and Physics of Modern Materials*, second edition, Chapman & Hall, London 1991; H.-G. Elias, *An Introduction to Plastics*, second edition, Wiley-VCH, Weinheim, 2003; J. R. Fried, *Polymer Science and Technology*, second edition, Prentice Hall. Englewood Cliffs, NJ, 2003.

Section 7.2: Reviews on industrial polyolefin production: K. S. Whiteley, G. T. Heggs, H. Koch, R. L. Mawer, W. Immel, Polyolefins, in *Ullmann's Encyclopedia of Industrial Chemistry*, sixth edition, VCH Weinheim, 2003, vol. 28, p. 393; G. Cecchin, G. Morini, F. Piemontesi, Ziegler-Natta Catalysts, in *Kirk-Othmer Encyclopedia of Chemical Technology*, John Wiley & Sons, Inc, 2003; L. L. Böhm, *Angew. Chem. Int. Ed. Engl.* 2003, **42**, 5010.

Section 7.4: Metallocene catalysts: H.-H. Brintzinger, D. Fischer, R. Mülhaupt, B. Rieger, R. M. Waymouth, *Angew. Chem. Int. Ed. Engl.* 1995, **34**, 1143; L. Resconi, L. Cavallo, A. Fait, F. Piemontesi, *Chem. Rev.* 2000, **100**, 1253.

Ni- and Pd-based polymerization catalysts: S. D. Ittel, L. K. Johnson, M. Brookhart, *Chem. Rev.* 2000, **100**, 1169; V. C. Gibson, S. K. Spitzmesser, *Chem. Rev.* 2003, **103**, 283.

Activation reactions of soluble polymerization catalysts: E. Y.-X. Chen, T. J. Marks, *Chem. Rev.* 2000, **100**, 1391.

Section 7.5: Supported metallocene catalysts: G. Fink, B. Steinmetz, J. Zechlin, C. Przybyla, B. Tesche, *Chem. Rev.* 2000, **100**, 1377.

Basic Organometallic Chemistry Related to Catalytic Cycles

PETER MAITLIS AND GIAN PAOLO CHIUSOLI

A1.1 The Key Steps

Our understanding of organic reactions catalyzed by soluble metal complexes ("homogeneous catalysis") is based on the properties and stoichiometric reactions of organometallic complexes, defined as molecules containing metal–carbon bonds. Significant aspects are summarized below, but for details the reader is recommended to one of the excellent texts cited at the end of this Appendix.

A1.1.1 Metal Complexes

A *metal complex* is composed of a central metal core surrounded by *ligands*. Ligands are the substituents (often acting as 2-electron donors), on the metal in a complex; many common ligands are neutral (H_2O, CO, olefins (CH_2=CHMe), organo-phosphines (PPh_3), and -phosphites ($P(OMe)_3$) or mono-anionic for example, halides (Cl^-, Br^-, I^-), hydride (H^-), hydroxo (OH^-). Many complex ligands are known and play vital roles in catalysis (for example in determining selectivity) and in bio-inorganic chemistry. The ligands often determine the properties and reactivities of the complex; changing them can lead to subtle changes in catalytic character. The *oxidation state* is defined as the charge associated with the metal when the ligands are removed in their closed shell configurations. A useful rule of thumb is that reactivity frequently decreases down a triad of metals ($3d > 4d > 5d$); thus while the more labile $3d$ and $4d$ complexes are used catalytically, the more inert $5d$ complexes are often used in model stoichiometric processes. However like most generalizations, there are exceptions and some important catalytic reactions of $5d$ metals are known.

A1.1.2   Ligands in Coordination Complexes

Ligands are atoms, ions, molecules, or groups of atoms that are attached to metals and that occupy defined positions in the coordination spheres of the metal ions. Typical arrangements are for four ligands to be arranged tetrahedrally (or square-planar), or six ligands octahedrally, about the metal, but many variations are known. Common ligands include inorganic species and organic molecules including amines (eg., $NH_2C_6H_{13}$ and N-heterocycles such as pyridine), phosphines (eg., PPh_3), sulfides (SMe_2), alcohols, olefins, dienes, and acetylenes. Many cyclic ligands are also employed including benzene (η^6-C_6H_6) and cyclopentadienyl (η^5-C_5H_5). For organic ligands such as hydrocarbons, the symbols η^2-, η^3- *etc* are used to denote that two, three, *etc* carbons are within bonding range of the metal; complexes containing η^2-, η^3- *etc* bonded organics are sometimes referred to as π-complexes. The situation defined by a single attachment, η^1- is often called a σ-bond. Many organic ligands that are used to control chemical reactions consist of organic molecules with coordinating head groups such as PR_2-, NR_2-, -CO_2H attached to non-coordinating and non-polar hydrocarbon tails. Solvents, especially polar ones such as water and alcohols, can also act as ligands.

Box 1 The 16-/18-Valence Electron (NVE) Formalism

It is found that among the more catalytically active later transition metals (Fe, Co, Ni; Ru, Rh, Pd; Ir, and Pt) many of the complexes involved, or postulated to be involved in the cycles, follow the 16- or 18-electron Number of Valence Electrons (NVE) formalism. This states that the total number of electrons associated with the metal valence shell is the number of d- electrons attributed to the metal in the relevant oxidation state plus the number of electrons donated by the ligands. For many catalytically active species of the later transition metals the NVE is 18 for five- or six-coordinate and 16 for four-coordinate square planar complexes, based on each simple ligand (e.g. CO, PR_3, H^-, Cl^-) donating 2 electrons. Thus $Ni(CO)_4$ is a 4-coordinate complex of Ni(0), d^{10}, with each CO "donating" 2 electrons, making a total of 18; Zeise's anion $Pt(C_2H_4)Cl_3^-$ is a complex of square planar Pt(II), d^8, each Cl donates 2 electrons as does the π-bonded (or η^2-) ethene; thus the electron count is $8 + 3 \times 2 + 2 = 16$; similarly $Co(CH_3)(CO)_4$ is an 18-e complex of Co(I), d^8. Organic ligands such as methyl, phenyl, acetyl, vinyl, *etc* are usually (η^1-) monodentate, uni-negative 2-electron donors (like chloride). However in the 18-e $Fe(NO)_2(CO)_2$ NO is formally regarded as a 3-electron donor to Fe(0), d^8. Furthermore, allyl, C_3H_4, is a uni-negative ligand which can be either a 2-electron or a 4-electron donor (bound η^1- and η^3- respectively) as illustrated by $Mn(\eta^1$-$C_3H_4)(CO)_5$ and $Mn(\eta^3$-$C_3H_4)(CO)_4$ which both obey the 18-electron NVE. Other electron counting conventions are sometimes used but this one is self-consistent and convenient for organometallic compounds.

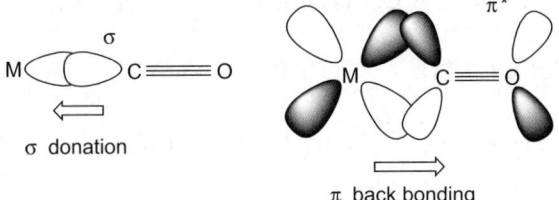

Figure 1 *Diagrammatic representation of the synergic bonding in a metal carbonyl: forward σ-donation from a filled CO orbital, balanced by back-donation from metal orbitals of appropriate symmetry into the CO π* orbitals.*

A1.1.3 Carbon Monoxide and Metal Carbonyls

Complexes that contain carbon monoxide, CO, coordinated to a metal are known as carbonyls and are relevant to the many known carbonylation reactions. In addition to the mono-nuclear carbonyls, $M(CO)_x$, a vast number of di- and poly-nuclear carbonyls containing metal-metal bonds are also known (e.g. $Fe_2(CO)_9$, $Ru_3(CO)_{12}$).

CO acts as both a σ-donor (via the lone pair of electrons on carbon) and a π-acceptor ligand in transition metal complexes. CO is usually depicted as having a triple bond (one σ- and two π-) between the C and the O as well as lone pairs on both the C and the O. The lone pair on C is used for donation into a suitable metal centred σ-orbital. However, the strongest M-CO bonds are formed (in simple terms) when some of the electron density donated by the carbon to the metal is directed back from a filled metal *d*-orbital of the correct symmetry into an antibonding π* of the CO. Thus the M-CO bond has two parts, the forward (C → M) donation, and the (M → C) back donation (Figure 1).

A sensitive tool to detect and measure the binding of CO to a metal is IR spectroscopy. CO normally shows a very strong band at 2143 cm^{-1}; when coordinated to a metal the CO bond strength, and hence the stretching frequency, ν(CO), decreases due to population of the CO π*-orbitals. Thus for example in the isostructural series, $V(CO)_6^-$ (1859 cm^{-1}), $V(CO)_6$ (1973 cm^{-1}); $Cr(CO)_6$ (2000 cm^{-1}); and $Mn(CO)_6^+$ (2095 cm^{-1}) there is a steady increase in ν(CO) going from V to Mn as the metal centre becomes less electron rich and less able to back-donate to the CO.

CO can bridge two or more *d*-block metals and some cases are known when the metal carbonyl oxygen also binds to strongly electropositive metals.

A1.1.4 Making Metal Carbonyls

Some of the carbonyls such as $Ni(CO)_4$ and $Fe(CO)_5$ can be made simply by reacting the metal and CO; however, the majority need an extra reducing agent, such as H_2, or an electropositive metal,

$$Ni + 4CO/1\ atm;\ 25°C \rightarrow Ni(CO)_4$$

$$\text{Fe} + 5\text{CO}/100 \text{ atm, } 150°\text{C} \rightarrow \text{Fe(CO)}_5$$

$$\text{CoCO}_3 + 2\text{H}_2 + 8\text{CO}/300 \text{ atm, } 130°\text{C} \rightarrow \text{Co}_2(\text{CO})_8$$

$$\text{CrCl}_3 + \text{Al} + 6\text{CO}/70 \text{ atm, } 140°\text{C}/\text{C}_6\text{H}_6/\text{AlCl}_3 \rightarrow \text{Cr(CO)}_6$$

$$\text{WCl}_6 + \text{Al}_2\text{Et}_6 + 6\text{CO}/\text{C}_6\text{H}_6/70 \text{ atm, } 50°\text{C} \rightarrow \text{W(CO)}_6$$

More specialized routes have been devised for other carbonyls, especially the polynuclear ones. Many use rather high temperatures and pressures; however recent researches indicate that much milder procedures can often be implemented by suitable choice of solvent and promoter.

A1.1.5   Metal Olefin Complexes

The reactions of olefins and related unsaturated compounds play key roles in many of the reactions described in this book. Such reactions proceed via metal-olefin complexes, the bonding in which is related to that in the carbonyls: here the forward donation is from the filled olefin C-C π-orbital, while the back donation is from metal d-orbitals of the correct symmetry into the empty π*-orbital of the olefin (Figure 2). The geometric constraints mean that ethylene for example lies perpendicular to the coordination plane of, and binds η^2- to, the metal. The classic example of this is found in Zeise's anion, $\text{Pt}(\text{C}_2\text{H}_4)\text{Cl}_3^-$.

The synergic bonding in the metal-olefin complexes also reduces the bond order of the coordinated C-C bond, but these changes are less easily detected than those in the metal carbonyls.

An important and dramatic effect of coordinating an olefin to a metal is that the olefin is now more positively charged and has become more susceptible to attack by nucleophiles. This is particularly so when the complex has an overall positive charge and the metal is in a higher formal oxidation state. By contrast, uncoordinated olefins are generally attacked by electrophiles. It should however be emphasized that much depends on the substitution of the olefin and of the metal and such simple generalizations are only partially valid.

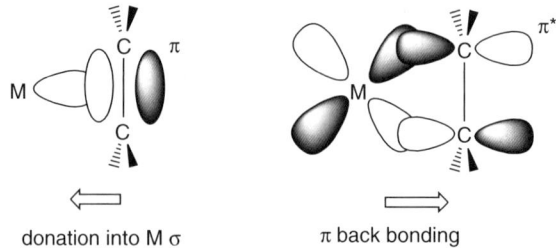

donation into M σ   π back bonding

Figure 2   *Diagrammatic representation of the synergic bonding in a metal olefin complex: forward σ-donation from a filled olefin π-orbital, balanced by back-donation from metal orbitals of appropriate symmetry into the olefin π* orbitals.*

A1.1.6 Making Metal Complexes of Mono-Olefins and other Unsaturated Organics

Metal-olefin π-complexes, the starting materials for many reactions are formed directly from a suitable metal salt and an olefin, diene, or polyolefin; since many are in low oxidation states a reducing agent is often added. In many cases metal-olefin π-complexes are formed *in situ* and are not isolated prior to further reaction; that is what happens in many catalytic reactions. Similar strategies are applied to reactions involving aromatic molecules.

$$PdCl_2 + C_2H_4 \rightarrow \{Pd(C_2H_4)Cl_2\}_2 \tag{1}$$

$$C_4H_6 + Fe(CO)_5 \rightarrow Fe(C_4H_6)(CO)_3 + 2\ CO \tag{2}$$

$$C_6H_6 + W(CO)_6 \rightarrow W(C_6H_6)(CO)_3 + 3\ CO \tag{3}$$

The general patterns of structures and bonding in metal poly-olefin complexes are similar to those in olefin complexes. Linear, branched and cyclic polyolefin complexes are known. The bonding is delocalized over the carbon atoms and the metal(s). These species also play important roles in organic syntheses.

A1.2 Catalytic Cycles

Catalytic reactions can be analyzed into a cyclic series of stoichiometric steps, for each of which there are many well-understood model systems. The most frequently encountered steps are: ligand substitution; oxidative addition; ligand migration (or migratory insertion); nucleophilic attack; reductive elimination; and β-and α-elimination. Catalytic cycles are defined by a sequence of several such reactions at the metal centre; the organometallic steps are often preceeded or followed by purely organic reactions.

Since such catalytic cycles involve transformations of organic species at a metal centre, a first step must be the formation of an organometallic. These can be formed by a direct reaction in which an organic reactant replaces a solvent or ligand molecule in the coordination sphere of a metal complex. For example, the ethylene-palladium(II) complex $\{Pd(\eta^2-C_2H_4)Cl_2\}_2$ can be formed by the direct reaction of $PdCl_2$ with ethylene (Equation 1), and the butadiene-iron complex is also formed directly by replacement of two CO's on Fe (Equation 2).

Compounds with metal – carbon η^1 or σ-bonds are of prime importance in catalysis. Reactions forming such bonds include oxidative addition to the metal, a ligand migration onto a coordinated CO, an olefin, or other π-complexed molecule, or a nucleophilic attack at such a liganded molecule. The resulting σ-bonded species can rearrange to give new ligands which, in turn, yield the organic products by displacement. The last steps often involve reductive elimi-nation, β-, or α-elimination. Many variants of these basic reaction types are also known.

Solvents can act as ligands to metals, to reagents and intermediates, as well as towards anions and cations involved in the reaction. Although solvents often only bond weakly their role can be critical: since they are present in large excess during a reaction, they will have large effects on the rates of many steps.

In addition to the ligands that actively participate in catalytic reactions, and the solvent, many metal complexes also contain so-called *spectator ligands* that play more subtle roles. These ligands do not directly participate in the catalytic cycle, but exert crucial influences on the rates and specificities of reactions by virtue of their size and shape (steric interactions) and their electronic (electron-withdrawing or -releasing) characters. Examples of such ligands are tertiary phosphines and phosphites, amines (including pyridines), and also some strongly π-bonded organics such as benzenes (η^6-C_6R_6) and cyclopentadienyls (η^5-C_5R_5, abbreviated Cp for R = H).

A1.2.1   Ligand Substitution

The material that is initially added to the reaction is often a pre-catalyst in a "resting state": it contains solvent or other ligands which must first be replaced before the catalytic cycle can begin. The first step is therefore the introduction of the reactant molecule to the metal centre, by a ligand substitution. To end the catalytic cycle the organic product must be liberated from the coordination sphere of the metal by another substitution reaction. The ease of these substitutions vary greatly, depending on the metal, its oxidation state, the substituting group, the group being substituted, and the conditions of the reaction. In mechanistic debates ligand substitutions of this type are sometimes ignored as the rates are presumed to be fast, but the preliminary (or the final) step can sometimes be the slowest step of the cycle and can determine the overall rate.

A1.2.2   Formation of Metal-Carbon σ-Bonds by Oxidative Addition

A very common method for forming metal carbon σ-bonds is by an oxidative addition reaction; this usually involves the addition of R-X (R = alkyl, aryl, *etc*; X = halide, *etc*) to a metal complex in a low oxidation state. In such a reaction, the oxidation state increases by 2; the coordination number of the metal also increases, usually also by 2. A good example is the addition of Me-I to $[Rh(CO)_2I_2]^-$ (4-coordinate square planar, Rh(I), d^8, NVE, 16, Box 1) to give $[Rh(Me)(CO)_2I_3]^-$ (6-coordinate Rh(III), d^6, NVE, 18) in the rate determining step of the Monsanto cycle for making acetic acid (Equation 4 and Chapter 4, Section 4.2.5),

$$[Rh(CO)_2I_2]^- + Me\text{-}I \rightarrow [Rh(Me)(CO)_2I_3]^- \qquad (4)$$

Many examples of d^8/d^6 oxidative addition reactions similar to the above are known; 5-coordinate species undergo the reaction but often with loss of a ligand, for example, Fe(0) d^8 to Fe(II) d^6, (Equation 5)

$$Fe(CO)_5 + I_2 \rightarrow Fe(I)_2(CO)_4 + CO \qquad (5)$$

Other common oxidative additions involve the transformation of d^{10} metal centres to d^8 (Equation 6),

$$Pt(PPh_3)_3 + MeI \rightarrow Pt(PPh_3)_2(Me)(I) + PPh_3 \quad (6)$$

(Pt(0), NVE 16/Pt(II), NVE 16). Oxidative additions generally occur most readily for low valent complexes, and for metals in the order $5d > 4d \simeq 3d$. In addition to those that formally cleave C–X, H–X, and X–X (X = halide) bonds, oxidative addition reactions are also known where the metal is inserted into C–O, C–H, and some strained or activated C–C bonds. Another reaction which is effectively an oxidative addition is the formation of metallacycles from a low valent metal and an olefin (Equation 7).

$$Fe(CO)_5 + CF_2 = CF_2 \longrightarrow$$

$$+ \quad CO \quad (7)$$

In this case the formation of two Fe-C σ-bonds has increased the formal oxidation state of the metal from Fe(0), d^8, to Fe(II), d^6.

The reverse of the oxidative addition reaction is reductive elimination (see below); indeed many oxidative additions are reversible.

A number of intimate mechanisms have been found for the oxidative addition reaction, including S_N2 nucleophilic attack, as in the addition of MeI to Rh(I) (Section 4.2.5).

A mechanistically similar reaction is MeI + Na[Mn(CO)$_5$] $\rightarrow$ Mn(CO)$_5$Me + NaI; the electron count also goes from d^8 to d^6, Mn(−I) to Mn(+I), but here the coordination number only increases by 1.

Other mechanisms for oxidative additions are also well-established for example, the concerted addition of H_2 to Ir(PPh$_3$)$_2$(CO)Cl to give Ir(PPh$_3$)$_2$(CO)(H)$_2$Cl (Ir(I) to Ir(III), also d^8/d^6), and a variety of radical and radical chain processes, for example, in the addition of MeI to Pt(PPh$_3$)$_3$.

A kinetic analysis can often be carried out on the reactions involved in a catalytic process to show which is the step that dictates the overall rate of the cycle; this is often known as the *rate determining step* (often abbreviated as rds) or *rate limiting step*. One way to accelerate the overall process is then to accelerate the rate determining step; to implement this requires considerable information about it. If for example, as has been found for the Monsanto process (Section 4.2.5), the oxidative addition is rate determining and the intimate mechanism of that step involves an S_N2 attack of the nucleophilic metal centre on the electrophile MeI, then that step, and hence the overall reaction, can be promoted by making the metal centre more nucleophilic by attaching strongly electron releasing ligands, such as trialkylphosphines, to the metal.

Complexes of metals in low oxidation states are generally electron rich and can act as bases, *e.g.* towards protons. If the complex bears a negative charge and contains electron−releasing ligands such as trialkylphosphines, the basicity is increased. Such complexes can then even react with hard-to-add Chloro- or oxy-compounds HX and RX (X = Cl, OAc, OCOCF$_3$, OSO$_2$Ph; R = alkyl, aryl) (see Chapters 3 and 4) and even HCN (Section 5.4.4).

A1.2.3   Cleavage of Metal-Carbon σ-Bonds by Reductive Elimination

Reductive eliminations are the reverse of oxidative additions; they also typically occur in 2-electron steps, *eg* ($d^6 \rightarrow d^8$, Rh(III) to Rh(I)), for example Equations 8–9,

$$[\text{Rh(MeCO)(CO)}_2\text{I}_3]^- \rightarrow [\text{Rh(CO)}_2\text{I}_2]^- + \text{MeCOI} \tag{8}$$

$$\text{Rh(PPh}_3)_3(\text{CH}_2\text{CH}_2\text{R})(\text{H})(\text{Cl}) \rightarrow \text{Rh(PPh}_3)_3(\text{Cl}) + \text{CH}_3\text{CH}_2\text{R} \tag{9}$$

or ($d^8 \rightarrow d^{10}$ Ni(II) to Ni(0) Equation 10).

Reductive elimination can again be promoted by ligands; there are examples where it is accelerated by ligand loss and examples where ligand attachment promotes.

A1.2.4   Formation of Metal-Carbon σ-Bonds by Nucleophilic Attack or by Migratory Insertion Reactions

A key aspect of any catalytic cycle is the transformation of an organic ligand. Two common and related such processes are the attack by an external nucleophile (for example on a coordinated olefin), and the migratory insertion (or ligand migration) reaction. The migratory insertion involves a σ-bonded ligand moving from the metal onto a ligand, and is sometimes regarded as an internal nucleophilic attack. In either case the usual product is a new σ-bonded organic ligand. Thus metal-olefin complexes are readily transformed into metal alkyls by attack on the coordinated olefin by a nucleophile (H$^-$, OMe$^-$, *etc*; Equation 11) or by migratory insertion of a metal-bonded group (H, alkyl) onto the coordinated olefin.

$$[\text{PdCp}(\eta^2\text{-C}_2\text{H}_4)(\text{PPh}_3)]^+ + \text{MeO}^- \rightarrow \text{PdCp(PPh}_3)(\eta^1\text{-CH}_2\text{CH}_2\text{OMe})$$

$$\text{Mn}(\eta^2\text{-C}_2\text{H}_4)(\text{CO})_5^+ + \text{BH}_4^- \rightarrow \text{Mn}(\eta^1\text{-Et)(CO)}_5 \tag{11}$$

Rather similar reactions can occur onto coordinated CO in metal carbonyl complexes, where acyl- or formyl-metal complexes are formed by nucleophilic attack, eg.,

$$\text{Fe(CO)}_5 + (\text{MeO})_3\text{BH}^- \rightarrow [\text{Fe(CHO)(CO)}_4]^- + (\text{MeO})_3\text{B}$$

or by a migratory insertion, Equation 12,

$$Ir(Me)(CO)_3I_2 + CO \rightarrow Ir(MeCO)(CO)_3I_2 \qquad (12)$$

Migration of alkyl onto coordinated CO is often promoted by dissociation of a ligand, especially an anionic ligand. Calderazzo and Noack in 1967 showed that the migratory insertion that occurs in the reactions of ligands with Mn(Me) $(CO)_5$ involved an intramolecular migration of methyl onto a *cis*-coordinated carbonyl. Later studies have shown this to be a general mechanism for migratory insertions.

Reactions involving nucleophilic attack on coordinated ligands are promoted by very polar solvents that stabilize anions. Metal carbonyls are also attacked by strong nucleophiles (RLi) to give the Fischer carbene complexes (*qv* below, Section A1.4),

A1.2.5 Cleavage (Transformation) of Metal-Carbon σ-Bonds by β-Elimination

The reverse of the 1,3-migration of H from metal to coordinated olefin to form metal alkyls is the β-elimination reaction, where the hydrogen on the β-carbon migrates to the metal,

$$Pt(PEt_3)_2(C_2H_5)(Cl) \rightarrow Pt(PEt_3)_2(H)(Cl) + CH_2{=}CH_2$$

$$Cu(Bu_3P)CH_2CD_2C_2H_5 \rightarrow Cu(Bu_3P)D + CH_2{=}CDC_2H_5$$

Organo-transition metal complexes themselves are more easily studied when they are stabilized by making their reactions less facile. Various ways to do this are possible. The β-elimination is one of the most common decomposition paths especially for complexes bearing longer chain alkyl ligands. Thus organic ligands without β-H's (e.g., Me, CH_2Ph) are more robust thermally and decompose at higher temperatures. However methyl, benzyl and similar complexes can show other decomposition routes, for example by α-elimination (below, Section A1.4), or by attack by external electrophilic or nucleophilic reagents. In those cases steric hindrance can stabilize and facilitate isolation of complexes as the bulky ligands prevent external attack by blocking sites at the metal.

A1.2.6 Analysis of a Model System: the Monsanto Carbonylation of Methanol to Acetic Acid Catalyzed by Rh/I⁻

The Monsanto carbonylation of methanol to acetic acid catalyzed by Rh/I⁻ is a well-understood example of an organometallic catalytic cycle and can act as a good model with well defined steps (shown schematically in Chapter 4, Section 4.2.4). The starting material is the square planar Rh(I) complex, $[Rh(CO)_2I_2]^-$ which is easily accessible by reaction of rhodium trichloride in solution with CO in the presence of iodide. This undergoes oxidative addition with MeI very readily to give the methyl-Rh(III) complex $[Rh(Me)(CO)_2I_3]^-$ as an unstable and short-lived intermediate (Equation 4), which is immediately transformed into the acetyl-Rh(III) complex, $[Rh(COMe)(CO)I_3]^-$, by migratory insertion. On carbonylation this gives the dicarbonyl-acetyl-rhodium(III) complex, $[Rh(COMe)(CO)_2I_3]^-$, (Equation 13) which undergoes reductive elimination of MeCOI to regenerate the square planar Rh(I) complex, $[Rh(CO)_2I_2]^-$ (Equation 4) which can start the whole cycle again,

$$[Rh(CO)_2I_2]^- + MeI \rightarrow [Rh(Me)(CO)_2I_3]^- \tag{4}$$

$$[Rh(Me)(CO)_2I_3]^- \rightarrow [Rh(COMe)(CO)I_3]^- \tag{13}$$

$$[Rh(COMe)(CO)I_3]^- + CO \rightarrow [Rh(COMe)(CO)_2I_3]^-$$
$$\rightarrow [Rh(CO)_2I_2]^- + MeCOI \tag{14}$$

$$MeCOI + H_2O \rightarrow MeCOOH + HI \tag{15}$$

$$HI + MeOH \rightarrow MeI + H_2O \tag{16}$$

In addition to the organometallic steps the methanol must be activated; this is done by reaction with HI to convert it into MeI (Equation 16). To end the reaction the acetyl iodide produced from the reductive elimination is hydrolyzed to acetic acid; this also regenerates the HI (Equation 15).

A1.3   Ligands for Asymmetric Catalysis

Since the biological activities of the two enantiomers of a single substance are often very different (one can be beneficial, the other harmful), single enantiomers of many pharmaceutical and agro-chemical compounds are now required. Making such compounds by catalytic enantioselective reactions has considerable advantages over traditional stoichiometric procedures. Synthesis of optically active organic compounds from non chiral starting materials is perhaps the most elegant application of homogeneous catalysis and represents the greatest possible challenge in selectivity required for a catalyst. With a few exceptions in heterogeneous hydrogenation, the field is dominated by homogeneous catalysts. Certain arrangements of ligands confer asymmetry on metal complexes; such complexes can promote reactions on prochiral molecules that lead to the preferential formation of one enantiomer over another. Thus much research is now devoted to the problem of designing ligands and catalysts for better enantioselective reactions. During reaction the ligands and the substrates in the complex are kept in place (by hydrogen bridges, and by ligand steric and electronic effects) to minimize loss of enantioselectivity. A given ligand set is often quite specific for a given enantioselective reaction on a given substrate, and there is no "universal" ligand. However bidentate ligands, especially with phosphorus-based donor groups, have often been successful.

The principle of enantioselective catalysis (shown in simple terms in Figure 3) uses a chiral complex catalyst which binds the substrate to give different diastereoisomeric intermediates from which different enantiomeric products are formed. The chiral information is generally placed on a ligand. The final

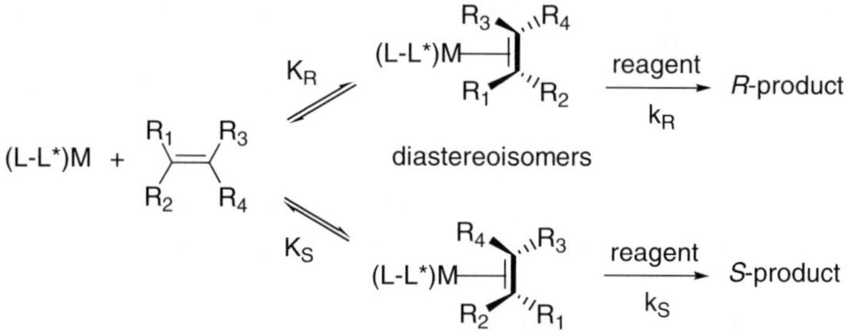

Figure 3 *Schematic representation of the metal catalyzed formation of R- and S enantiomers starting from a non-chiral olefin and using a metal complex containing a chiral bidentate ligand (L-L*); the rate constants k_R, k_S, refer to the reagent-promoted transformation of the diastereoisomer into the enantiomer.*

enantioselection is determined by the difference between both the equilibrium constants (K_R *vs.* K_S), leading to different diastereoisomeric concentration, and the kinetic constants (k_R *vs.* k_S) leading to the formation of the products. Enantioselectivity is generally expressed by the *enantiomeric excess* (or *ee*) defined as the % major enantiomer – the % minor enantiomer, with respect to their sum. Spectacular enantioselection has been obtained in a wide variety of organic transformations.

The asymmetric ligand will cause one of the two diastereoisomers to form preferentially, but only the rate with which they react with the desired reagent (for example H_2 for the hydrogenation or hydrogen peroxide for the oxidation of an olefinic substrate) will determine whether the *d* or the *l* product will predominate. To obtain a high enantiomeric excess the diastereoisomer that reacts more slowly must equilibrate sufficiently rapidly with the more reactive diastereoisomer through substrate dissociation and re-association. Achieving suitable conditions for asymmetric synthesis thus requires a subtle tuning of all the ligands involved, including the solvent.

Enantioselective catalysis exemplifies how a complex synthetic route to a chiral molecule can be simplified by catalytic methods, reducing the overall amount of waste and increasing yields. Only a few of the metal-catalyzed reactions have yet been applied industrially, mostly for the synthesis of pharmaceuticals, vitamins, agrochemicals, flavours and fragrances. They include, asymmetric oxidation, hydrogenation, isomerization, hydroformylation, and cyclopropanation (Chapters 2, 3, 4 and 5; Sections 2.7, 3.5, 3.6, 3.7, 4.6.7, 5.6).

Such processes have the following characteristics:

- multistep synthesis (5-10 steps or more for pharmaceuticals) and short product life (often less than 20 years);
- relatively small scale products (1-1000 t/a for pharmaceuticals, 500-10000 t/a for agro-chemicals) usually produced in multipurpose batch equipment;
- high purity requirements (>99% and <10 ppm metal residue);
- high added values to compensate for high process costs;

- short development time for the production process, since time to market affects the profitability of the products;
- synthetic route often designed around key enantioselective steps.

On this basis the choice of a specific catalytic step is usually determined by an economic analysis of the "catalytic" route *vs.* alternative routes (such as the separation of enantiomers by resolution, organic synthesis starting from chiral natural products used as building blocks (the *chiral pool* concept), or the use of enzymatic and microbial transformations) and time to market considerations (see also Chapter 1).

The following critical factors determine the viability of an enantioselective process:

- enantioselectivity (expressed as *ee*): $>99\%$ for pharmaceuticals and somewhat lower for agrochemicals if further enrichment is easy;
- the catalyst productivity as it determines catalyst costs: TON should be >1000 for small scale, high value products, and >50000 for larger scale, less expensive products;
- catalyst activity (as TOF for $>95\%$ conversion): >500 h^{-1} for small scale, >10000 h^{-1} for larger scale;
- availability and cost of ligands: in most cases they are chiral diphosphines which are expensive (US\$100-500 /g for lab quantities, US\$500-20000 /kg for large scale) and require special synthetic know-how to make. Early transition metals generally require cheaper ligands;
- the development time: this is crucial if the ligand has to be optimized and if little is known about the catalytic mechanism.

A1.4   Metal-Carbene, -Methylene, -Carbyne and -Methylidyne Complexes

While the metal σ-alkyls are examples of compounds of classical organometallic chemistry, it is now recognised that there is a whole range of non-classical species involving metal carbon double and triple bonds that participate in many reactions, especially olefin metathesis and polymerization reactions. Isolable complexes include those stabilized by steric interactions, Ta=CHCMe₃ (Me₃CCH₂)₃, (alkylidenes, or Schrock carbenes) and those stabilized electronically by hetero-atoms, Cr(CO)₅(CPh(OMe)) (Fischer carbenes). The Fischer carbenes react mainly with nucleophiles; by contrast the carbene carbon in the alkylidenes is nucleophilic and adds electrophiles. It is suggested that this may be due to strong M-C π-backbonding and the absence of electron-withdrawing substituents on the carbene.

In addition to complexes containing carbenes bonded to a single metal atom (M=CRR'), complexes are also known where the carbene ligand bridges two metal atoms (M–CRR'-M', μ-methylenes): similarly there is an extensive chemistry of metal-carbynes (M≡CR, with a formal triple bond between the metal and the ligand) and μ-methylidynes where the ligand is again bridging, as well as others such as vinylidenes (M=C=CRR'). All of these have been completely characterized as organometallic ligands and have been invoked as reaction intermediates, for example in metathesis and Fischer-Tropsch reactions.

A1.4.1 Synthesis, Structure and Bonding in Schrock Carbenes

Carbene complexes not containing hetero-atoms X and or Y but of the form, M=CHR, are known as alkylidene complexes and were first made by Schrock in 1975. A number of routes have been developed; one of the most interesting is by α-elimination, Equation 17,

$$Ta(Np)_3Cl_2 + 2LiNp \rightarrow Ta(Np)_3(=CHCMe_3)$$

$$[Ta(Cp)_2Me_2]^+ + NaOMe \rightarrow Ta(Cp)_2(CH_3)(=CH_2) + MeOH \quad (17)$$

$$(Me = CH_3; Np = CH_2CMe_3)$$

Another route is by incorporation of carbenoid moieties, Equation 18,

$$Os(PPh_3)_3Cl(NO) + CH_2N_2 \rightarrow Os(PPh_3)_2(Cl)(NO)(=CH_2) \quad (18)$$

In these *alkylidene* complexes the M=C bond is significantly shorter than normal M-C single bonds, and the bonding is illustrated by formula (I). The partial double bond character confers special properties and such carbenes are intermediates in olefin metathesis reactions

$$M=C \underset{R^2}{\overset{R^1}{<}} \qquad \overset{\delta-}{M}-\overset{\delta+}{C}\underset{Y}{\overset{X}{<}} \qquad \overset{\delta-}{M}-C\underset{Y}{\overset{X}{<}}$$

(I)   (II)   (III)

In addition to complexes of terminal alkylidenes, many complexes containing alkylidenes bridging two metal atoms have been made; these are commonly known as μ-methylene complexes. A wide range of complexes containing other metal-carbon bonds are known such as the *carbyne* complexes (for example, Cr≡CPh(CO)₅) which contain metal-carbon triple bonds. While such species may well be involved in some catalytic reactions, their importance in industrially catalyzed processes still has to be evaluated.

A1.4.2 Synthesis, Structure and Bonding in Fischer Carbenes

The Fischer carbene complexes contain hetero-atoms in X and/or Y, and were originally synthesised by E O Fischer and Massböl in 1964, by reaction of strong nucleophiles with metal carbonyls, followed by reaction of the resultant anion with an electrophile to give the neutral carbene (Equation 19),

$$W(CO)_6 + LiR \longrightarrow \underset{LiO}{\overset{R}{\underset{\diagdown}{C}}} - W(CO)_5 \xrightarrow{Me_3O^+} \underset{MeO}{\overset{R}{\underset{\diagdown}{C}}} - W(CO)_5$$

$$(19)$$

The Fischer carbene complexes can be described by resonance forms such as (II) and (III), and are characterized by quite long M—C and by rather short C—X bonds, which show some double bond character. The ionic forms give the complexes increased reactivity towards polar reagents, for example to attack by nucleophiles at the carbene carbon.

References

Organometallic Chemistry: Ch. Elschenbroich and A. Salzer, *"Organometal-lics"*, second edition, VCH, Weinheim, 1992; A. Yamamoto, *"Organotransition metal chemistry"* Wiley-Interscience, New York, 1986; J. P. Collman, L. S. Hegedus, J. R. Norton, and R. G. Finke, *"Principles and Applications of Organotransition metal chemistry"*, University Science Books, Mill Valley, CA, 1987. J. E. Huheey, E. A. Keiter and R. L. Keiter; *"Inorganic Chemistry: Principles of Structure and Reactivity"*, Harper Collins College Publisher, fourth edition, New York, 1993.

The Rh/I⁻ catalyzed methanol carbonylation, a model system for a catalytic cycle: P. M. Maitlis, A. Haynes, G. J. Sunley, and M. J. Howard, *J. Chem. Soc., Dalton Trans.*, 1996, 2187; see also Ir/I⁻-catalyzed methanol carbonylation: A. Haynes, P. M. Maitlis, G. E. Morris, G. J. Sunley, H. Adams, P. W. Badger, C. M. Bowers, D. B. Cook, P. I. P. Elliott, T. Ghaffar, H. Green, T. R. Griffin, M. Payne, J. M. Pearson, M. J. Taylor, P. W. Vickers and R. J. Watt, *J. Am. Chem. Soc.*, 2004, **125**, 2847.

Carbonyl syntheses are collected in: *Inorg. Synths.* **28**, 1990, ed Robert J. Angelici; *Inorg. Synths.* **34**, 2004, ed. John R. Shapley, Wiley-Interscience, Hoboken, NJ; further organometallic procedures are collected in other volumes of *Inorg. Synths.*

Asymmetric synthesis: I. Ojima, *Catalytic asymmetric synthesis*, Wiley, New York, 2000. Chirality in organometallic chemistry: R. D. Adams, ed., *J. Organomet. Chem.*, 2006, **691**, 10.

APPENDIX 2

Some Basic Aspects of Surface Science Related to Heterogeneously Catalyzed Reactions

MARIO G. CLERICI AND PETER MAITLIS

A2.1 Background

The chemical industry is heavily dependent on catalysis. This book concentrates on the transformations brought about by metals; thus enzymatic and simple acid/base catalysts are not covered. Homogeneous catalysis, in which the metal complex catalysts and promoters, the reagents, and the products, are all soluble in the reaction medium, is easier to understand and explain in chemical terms (see Appendix 1). However heterogeneously catalyzed reactions dominate many commercial processes. In such systems the reagents (as gases or liquids) are led over or through the catalyst, and it is relatively simple to separate out the product(s). Most of the industrial heterogeneous catalysts are based on solid metals, oxides or halides, though some special systems (such as ion-exchange and polymer supported metal complexes, supercritical fluids, ionic liquids, *etc.*) have shown promise under laboratory conditions.

The many very large-scale applications of heterogeneous catalysis include the hydrogenation of vegetable and animal oils and fats into edible foodstuffs (Section 3.3.1), and the catalytic cracking of large hydrocarbons in crude oil into smaller molecules (petrol or gasoline; diesel) that are effective in internal combustion engines (Section 3.2), as well as making ammonia (for fertilizers) from nitrogen and hydrogen, and catalytic converters to clean up car exhausts.

Commodity organic chemicals (Chapter 1) largely produced by heterogeneously catalyzed routes include: acrylic acid (Section 2.8), acrylonitrile (Section 2.10), adipic acid (Section 2.2.2), cumene (Section 5.2.3); ethylbenzene (Section 5.2.1); methanol (Section 4.7.1); styrene (Section 3.9); terephthalates (Section 2.3); ethylene oxide (Section 2.4) vinyl acetate (Section 2.15.7), and many others: in a word, heterogeneous catalysis is huge.

Some of these reactions use finely divided metals (often supported on an inexpensive "inert" material, often an oxide or a clay mineral); while others use metal compounds (*e.g.*, molybdenum sulfide for hydrodesulfurization (Section 3.2.3), tunsten oxide (for metathesis, Section 6.3), or titanium or chromium salts (for olefin polymerization, Section 7.3).

The conversions of the substrate by the reagents occur at specific positions (active sites) on the heterogeneous catalyst and thus many investigations of such sites have been undertaken, and have led to the study of surface science.

Catalytic activity is generally measured in terms of turnovers. Thus the Turnover Number, TON, is the number of times the catalytic reaction occurs per active site, while the Turnover Frequency (TOF) is the TON per unit of time. The TON is easily understood for defined reactions where the number of active sites is known: homogeneous catalyses that occur at clearly understood metal centres, enzymatic catalysts or some heterogeneous catalysts whose active sites are easily determined. If the number of active sites on a heterogeneous catalyst is not easily determined, then that number is often replaced by the total surface area of the catalyst. In industry TOF is sometimes measured as grams of product produced per gram catalyst per hour, as this gives a useful measure of the catalyst cost.

A2.2   Active sites

The link between homogeneous and heterogeneous catalysis is seen in the concept of the active site. In the case of the homogeneous catalyst, reactions generally occur at single metal centres situated within a coordination complex. For a heterogeneous catalyst reactions must occur at the surface in order that reactants are easily able to reach it and the products are easily able to diffuse away. However there may be many metal centres on a surface; indeed a monolithic metal catalyst, for example a single crystal, has only metal atoms on the surface. It thus becomes necessary to distinguish between those sites that are catalytically active and those that are not.

A2.3   Surface Studies

The concept has thus grown that reagents preferentially adsorb at surface discontinuities such as peaks, fissures, terraces, and kinks, and that the subsequent reactions occur there. Such sites correspond to metal atoms in inorganic chemistry which have fewer nearest neighbours, that is, sites with low coordination numbers and hence a higher tendency to bind other molecules. In other words, they have higher reactivity than those with many near neighbours, which have higher coordination numbers and hence are expected to have lower reactivity towards incoming molecules.

There is a problem with such a simple-minded analysis, as there is with the similar situation in a homogeneous system: a very strongly adsorbed molecule will not necessarily have high activity for a further chemical reaction and may therefore be inactive in a catalytic cycle; it may even act as a poison towards catalysis. To facilitate reaction a molecule must adsorb at a metal site but the adsorption must not be so strong as to prevent other reagent molecules from arriving to react with it.

Various types of electron microscopy (TEM, transmission electron microscopy; SEM, scanning electron microscopy) and atomic force microscopy (AFM) as well as a host of other spectroscopic techniques (XANES, Auger, LEED, EELS, PES, surface IR, UV, solid state NMR, *etc*) give information about the surface structures of solids and also about the state of molecules adsorbed at the surface.

In some cases such information is readily interpreted and understood. For example, carbon monoxide adsorbed on a metal will show strong carbonyl absorptions ($v(CO)$) in the infra-red region similar to those found in molecular metal carbonyl complexes (See A1.1.3). The exact positions of the absorption, the number of bands and their shapes, can be correlated with the type of binding (to one, or more, metal atoms) and even to the strength of the bonds.

Although there is a lot of information available about surfaces and surface species there is unfortunately no direct translation between such information and a more detailed understanding of the actual heterogeneous catalysis processes. There are several reasons for this: surface studies normally have to be carried out in ultra-high vacuum and at low temperatures, a far cry from the conditions of a catalytic reaction which may typically only occur at temperatures of greater than 100°C and pressures of several bar. The residence times of the species that participate in catalyses are normally very short and their concentration is very low, thus it has only quite recently become possible to derive any meaningful information from the surface studies about the catalyses. Furthermore, although many species are detected on surfaces, most are *spectator* species, which do not participate in, and are unaffected by, the desired catalyses.

A2.4   Classes of Heterogeneous Catalysts

A2.4.1   Metal Catalysts

A single crystal of a metal is made up of a regular 3-dimensional array of atoms packed in one of three ways: hexagonal close packed (*hcp*), face-centred cubic (*fcc*), or body centred cubic (*bcc*) (Figure 1, based on the packing of identical spheres). In the *hcp* or *fcc* structures each atom has 12 nearest neighbours (six in the same plane, three above and three below), while in the *bcc* each metal atom has 8 nearest neighbours. Such arrangements contain octahedral and tetrahedral holes into which smaller atoms can be placed which then have 6- or 4-coordination respectively. Similar packing characterizes some metal oxides, where the oxide ions form *hcp*, *fcp* or *bcc* arrays, with the (smaller) metal ions in the interstitial holes.

Such organizations continue to the surfaces which can be viewed as the cross-sections of the 3-dimensional arrays along defined planes of cleavage. The

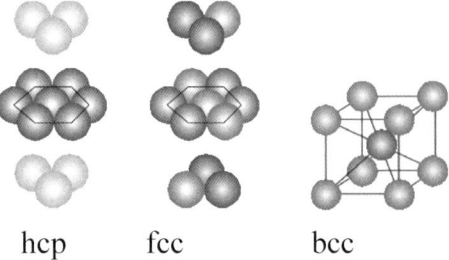

hcp   fcc   bcc

Figure 1 *Schematic representation of exploded ccp, hcp and bcc packing arrangements of identical spheres.*

regular arrays of surface atoms form *terraces*, where each stom has 9 nearest neighbours. The terraces end in *steps* where the surface atoms have 7 nearest neighbours. Other discontinuities (*kinks*, with still lower coordination numbers) can also occur in the surfaces. It should however be noted that atoms are mobile and thus metal surfaces easily undergo change in morphology. Such changes can occur on heating and also during catalysis: not only the reactants are affected by the catalytic reaction but also the ''reagent'' substrate metal (or oxide). This makes an accurate description of a surface difficult. One major problem in heterogeneous catalysis is therefore to have a reasonably consistent and catalytically long-lasting surface structure. Metals dispersed on supports (Section A2.7) will of course have proportionately larger surface areas with more discontinuities than single crystal arrangements. It is highly significant that the more finely dispersed metals generally have much higher catalytic activities, indicating that more active surface sites are present.

A2.4.2 Metal Oxide Catalysts

Various kinds of oxide materials, including single oxides, mixed oxides, molybdates, heteropoly-ions, clays, and zeolites, are used in catalysis: they can be amorphous or crystalline, acid or basic. Furthermore the oxides can be the actual catalysts or they can act as supports on which the active catalysts have been deposited. Silica and alumina are commonly used to support both metals and other metal oxide species. Amorphous silica/alumina is a solid acid catalyst, it is also used as a support for metals, when bifunctional (acid and metal) catalysis is required, *e.g.*, in the cracking of hydrocarbons. Other acid catalysts are those obtained by the deposition of a soluble acid on an inert support, such as phosphoric acid on silica (SPA, used in the alkylation of benzene to cumene, Section 5.2.3). They show similar properties to those of the soluble parent acids, while allowing easier handling and fixed bed operation in commercial units.

An important class of mixed oxides is constituted by *zeolites*. Zeolites were first defined to comprise only microporous crystalline aluminosilicates (microporous, pore diameter < 20 Å, mesoporous, 20–500 Å, macroporous, > 500 Å). However today other microporous crystalline materials are included, such as metal silicates (obtained by the isomorphous substitution of lattice Si^{4+} by a transition metal, *e.g.*, Ti^{4+}, Fe^{3+}), aluminophosphates (ALPO), silicoalumino-phosphates (SAPO) and their isomorphously metal substituted derivatives (MAPO and MAPSO). More than 150 framework types of zeolites are known, but only *ca.* 10 have found application in catalysis, the lion's share being taken by zeolite Y (Figure 2) the cornerstone of FCC (Fluid Catalytic Cracking).

Zeolites are built up of linked SiO_4 and AlO_4 (or MO_4, with M = Ga, Ti, Fe, V, ...) tetrahedra that share one O atom. Their different spatial arrangements give rise to the variety of known zeolites. These are generally classified as small pore (pore openings formed by 8 T-atom tetrahedra, with diameters in the range 3.5–4.5 Å), medium pore (10 tetrahedra, diameters 4.5–6.0 Å), and large pore (12 tetrahedra, diameters 6.0–7.5 Å). Cavities and pore intersections provide a somewhat larger porosity (diameters up to 9–11 Å). Extra porosity can be introduced artificially by chemical and physical post treatments of the zeolite crystallites. For instance 20–200 Å mesoporosity is produced by the high temperature steaming of Y zeolite, leading to stabilized Y and ultra stable Y (USY) that are largely used in the FCC and hydrocracking processes, in which very bulky molecules are processed.

The substitution of Si^{4+} by Al^{3+} produces an excess of negative charges on the surface that should be neutralized by cations in the channels. Acidic materials are obtained when the cation is a proton, H^+. Therefore the acidity of a zeolite is related to lattice Al but is not always directly proportional to its content. In practice the acid strength of individual Al sites grows with the increase of Si/Al ratio, up to a maximum for Si/Al of 9–10. Above this range a direct proportionality is found between acid catalytic activity and lattice Al content. The acidic strength of high silica zeolites has been compared to that of 98% H_2SO_4.

Isomorphous substitution with a suitable transition metal, notably Ti^{4+}, produces an oxidation catalyst. The larger ionic radius, coupled with the preference of Ti^{4+} for octahedral coordination, produces strain in the silica lattice, favouring the splitting of SiO-Ti bonds by a protic molecule, *e.g.*, H_2O

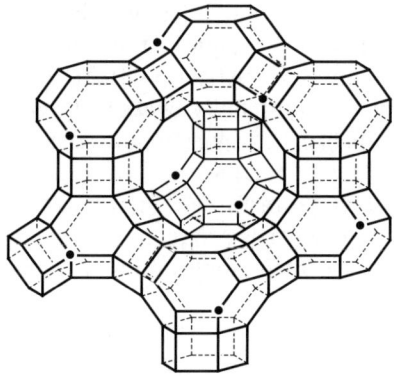

Figure 2 *Schematic representation of the structure of zeolite Y; the black dots represent the Al sites.*

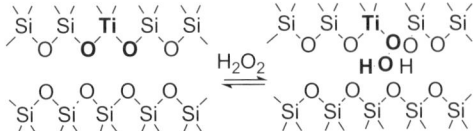

Figure 3   *Schematic representation of the formation of Ti-OOH species by the reaction of hydrogen peroxide with lattice SiO-Ti bond in TS-1.*

or H_2O_2. In the latter case a Ti-OOH oxidant species is obtained (Figure 3), very active for the epoxidation of olefins, the hydroxylation of alkanes, and the oxidation of a variety of O-, N- and S-compounds.

Both acidic and oxidation properties are obtained when Fe^{3+} is inserted in the zeolite lattice. The deposition of discrete metal particles (e.g., Pd, Pt) on the surface leads to bifunctional catalysts with both acidic and hydrogenation properties. These are frequently used to slow down the deposition of carbonaceous residues that eventually deactivate the catalyst by blocking the pores ("pore plugging").

Clays are layered crystalline materials grouped in two classes, cationic and anionic, according to the nature of ions in the interlayer volume. Most common are cationic clays composed of negatively charged aluminosilicate layers, with compensating interstitial cations. Similar to zeolites, the charges originate from the substitution of lattice Si^{4+} by Al^{3+} and other trivalent and bivalent ions. Both Lewis and Brønsted acid sites are present in clays, arising from the exposed Al^{3+} ions inserted in the tetrahedral sheets, and from terminal OH groups, respectively. Their acid strength however is less than that of the zeolites. Redox sites may also be present as a consequence of Fe^{3+} substitution for Si^{4+} or as a result of ion exchange with transition metals (Cu^{2+}, Fe^{3+}, ...).

The interlayer distance can vary and depends on the size and nature of adsorbates and on thermal treatments. Swelling by the adsorption of polar molecules can proceed up to the loss of crystallinity and full separation of sheets. This property can be exploited for the introduction of pillars in the structure and produces materials in which a two dimensional planar network of channels is introduced. This is accomplished by ion exchange of the clays with polycharged polyoxocations, *e.g.*, $[Al_{13}O_4(OH)_{24}(H_2O)_{12}]^{7+}$, followed by thermal treatment to form chemical links between layers and pillars. A major difference to zeolites is the broader range of pore sizes. Transition metals can be incorporated in the pillars to function as redox sites. Pillared clays can be designed for specific catalytic transformation of bulky reactants (see below) for both fine and commodity chemicals production. Because of their mesoporosity and good thermal stability, their applications in refinery processes are also being investigated.

A number of commercial oxidation catalysts are based on V and Mo oxides. They are multicomponent materials, such as the mixed oxides Mo-Fe-O (oxidation of methanol to formaldehyde), V-P-O (oxidation of butane to maleic anhydride), Bi-Mo-O (propylene to acrylonitrile), Bi-Fe-Mo-O

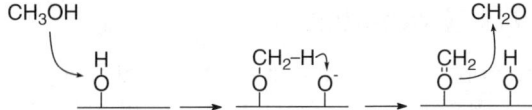

Figure 4 *Schematic representation of the oxidation of methanol to formaldehyde.*

(isobutene to methacrolein, propylene to acrolein), Mo-V-O (oxidation of acrolein to acrylic acid), and Mo-V-P heteropolyacid compounds (oxidation of methacrolein to methacrylic acid). The compositions claimed in patents often include several other elements, functioning as promoters and stabilizers. Excess P and promoters are added to stabilize the crystalline $(VO)_2P_2O_7$ active phase for the oxidation of *n*-butane to maleic anhydride. The partial exchange of H^+ by Cs^+ in heteropolycompounds is probably designed to enhance thermal stability and to modulate surface acidity.

Mechanisms proposed for reactions envisage complex reaction pathways. Thus in the oxidation of methacrolein catalyzed by $H_3PMo_{12}O_{40}$, the chemisorption of the aldehyde precedes the redox step and is thought to occur through the acid-promoted formation of a gem-diol molybdate ester. The subsequent oxidation to methacrylic acid involves Mo(VI) undergoing reduction to Mo(V). Molecular oxygen only serves to reoxidize the reduced polyanion.

$$RCHO \underset{}{\overset{H+}{\rightleftharpoons}} RCH(OM)_2 \overset{Redox}{\longrightarrow} RCOOM \rightarrow RCOOH$$

$$(M = Mo \text{ in } H_3PMo_{12}O_{40})$$

The mechanism for the oxidation of methanol to formaldehyde on iron molybdate catalysts (illustrated in Figure 4) envisages the H-abstraction and electron transfer of surface methoxy species; desorption of the products and reoxidation of metal sites complete the cycle.

A2.4.3   Metal Sulfides

MoS_2 and WS_2 modified with Ni and Co have major uses in the hydrotreating of various refinery streams. The composition chosen is a function of the feed and of the specific treatment needed: hydrodesulfurization (HDS; see Chapter 3.2.4), hydrodenitrogenation, demetallization, and reduction of unsaturation.

Although the hydrogenation activity of metal sulfides is lower by several orders of magnitude than that of metal catalysts, sulfides allow operations under conditions that are impractical for metals. They are generally used as highly dispersed materials on a high surface area support, such as γ-alumina, in fixed bed operation. Most important is catalyst design to minimize deactivation due to the deposition of metals (V, Ni) in the feed and of coke at the mouths of the pores. Metal sulfides can also be used as finely dispersed phases in continuous slurry reactors to reduce the mass transport limitations of heavy oils.

A2.5   Catalyst Promoters

Promoters are often needed in addition to the actual catalyst itself. In general promoters increase the rate of a desired reaction; this can occur because of a general increase in reaction rates or because of an increase in selectivity towards one product by comparison to others. Most promoters tend to be specific to a particular reaction and catalyst but some promoters can accelerate different reactions. For example potassium (usually added as K_2O) promotes the silver catalyzed ethylene oxidation to ethylene oxide (Section 2.4), and also promotes carbon monoxide hydrogenation over Fe catalysts (Section 4.8).

A2.6   Catalyst Poisoning/Deactivation

The activity of a catalyst often decreases with time. This can arise from several different effects: the active sites can be poisoned by impurities in the feed, fouled by carbonaceous by-products, or segregated by pore blockage due to coke deposition. Alternatively deactivation of the surface can occur by sintering leading to changes in the surface structures and loss of active sites, vaporization of catalytic oxide species (e.g., for Mo oxides), and other transformations of the active phase. Common poisons include sulfur and sulfides, but phosphorus and nitrogen compounds, olefins (especially dienes and polyenes), and carbon monoxide can also act as poisons. Some deactivations, such as coking, can be counteracted by heating the catalyst to a high temperature in the presence of air, when the coke is "burnt off".

To minimize sulfur poisoning refinery streams are hydrotreated on less active metal sulfide catalysts instead of noble metals. On MoS_2 based catalysts sulfur is also removed from feeds destined to naphtha reforming or to second stage hydrocracking units. However poisoning is not always an inconvenience to be avoided, since controlled poisoning of a surface can improve selectivity, as in the oxidation of ethylene to ethylene oxide on silver catalysts (Section 2.4). The addition of ppm chlorine compounds in the feed reduces combustion reactions, thus increasing selectivity. Excess chlorine, however, acts as a poison and reduces the activity.

A2.7   Supported catalysts

In many cases of catalysis by metals the actual catalyst particles are supported. This capitalizes on the observation that only a small proportion of the metal atoms in the surface are catalytically active; further a larger surface area is achieved for a given quantity of metal if it is dispersed on a high surface carrier. This also means that the catalyst is made in such a way that the support (frequently a metal oxide, though a wide variety of other matrices have been used) comprises the bulk of the material. This also has an obvious economic benefit since the actual (expensive) catalyst is "diluted" by a much cheaper material. Some support materials are inert – they play no role in the catalyses.

However frequently the support material does have a very important function. This is particularly so when the support acts as a (Brønsted) acid or a base. An example is in the catalytic cracking, alkylation, and isomerization of hydrocarbons (Section 5.2.6) The role of the transition metal is in oxidation or hydrogen transfer reactions while the support, for example acidic oxides such as aluminosilicates, act to protonate, rearrange and dehydrate organic species.

Supported catalysts need to possess certain properties at both the macroscopic and the microscopic level, with regard to shape and size, mechanical resistance, surface area, pore volume and pore size distribution. The size, shape, and mechanical resistance should be such that pressure drops along the reactor are negligible, allowing the reactants to flow across the catalyst bed homogeneously, without following specific channels, and catalyst pellets are resistant to attrition and are not crushed into powders under operating conditions. The support should also be chemically and physically stable to process conditions and inert with respect to the supported active species. Surface area and porosity are of the greatest importance for catalytic performance since they are strictly related to catalyst dispersion, and therefore to catalytic activity, and to mass transfer to and from the catalytic sites. It is generally critical that the reaction itself (involving adsorption/reaction/desorption), rather than the travel of the molecules to the active site, should be rate limiting. However too high a surface area and porosity is not always a desirable property: it can have an adverse effect on the selectivity of processes in which chemically unstable products are formed (e.g., in ethylene epoxidation), as they can be retained allowing side-reactions to occur on the catalyst. Sintering occurs when the particles of the active phase migrate and aggregate, reducing the dispersion and hence the number of active sites; this is often an effect of temperature.

Catalysts are manufactured by various methods (such as precipitation, extrusion and spray drying) in the form of cylinders, rings, multi-lobed extrudates and other shapes. They range in size from a few millimetres to several centimetres; small spheres are used in fluidized bed reactors. Active phases can be dispersed on the pre-shaped support by several methods such as by impregnation of a solution of the active components. Alternatively the catalysts can be made by the extrusion of mixtures of solid components: the support, active phase, and binder. For some reactions that are diffusion limited, the catalytically active species are not uniformly distributed; instead they are deposited on the outer shell of the catalyst particle (egg-shell catalysts), since those inside the particle cannot be involved in the reaction.

A2.8   Other Types of Non-homogeneous Catalysts

In addition to the classical heterogeneous catalysts such as zeolites, or metals on supports, there are now a host of other catalysts which are not truly homogeneous and where the catalyst, the reactants and the products are in different phases. Such catalysts can offer the advantages of high activity and selectivity associated with the more usual homogeneous catalysts, as well as the ease of product isolation and of catalyst recovery associated with hetero-geneous catalysts. Examples include, polymer supported catalysts (metals associated with natural or synthetic polymers), ionic liquids, supercritical solvents, and other two-phase systems such as aqueous-organic emulsions. Only the last of these is yet used commercially on any large scale but there is a great deal of interest in the field and it is likely that further applications will shortly be made.

A2.9   Special Topics in Heterogeneous Catalysis

A2.9.1   Microporosity and Shape Selectivity

Active sites in zeolites are located in a regular network of channels of uniform dimensions, whose shapes and sizes are characteristic of the individual zeolite. The diameter of the pore ports are of similar magnitude to the molecular dimensions of smaller organic compounds. Shape selectivity is the first and best known consequence of this. It can be defined as the capacity of zeolites to discriminate reactants, products and intermediates on the basis of their sizes and shapes (see also p 77).

Figure 5 schematically illustrates the concepts of *reactant shape selectivity*. Only linear paraffins that are able to diffuse and are adsorbed inside the pores can undergo a chemical transformation, *e.g.*, acid catalyzed cracking. The property is exploited in some chemical processes, such as the dewaxing of lubes and middle distillates, through the selective cracking or isomerization of the linear paraffin fraction.

Figure 6 illustrates *product shape selectivity*, *i.e.*, the capacity of favouring the formation, among all possible products, of those that diffuse faster out of the pores. While the entire range of products can be present inside zeolite channels and cavities, the effluent stream is mainly composed of the less hindered

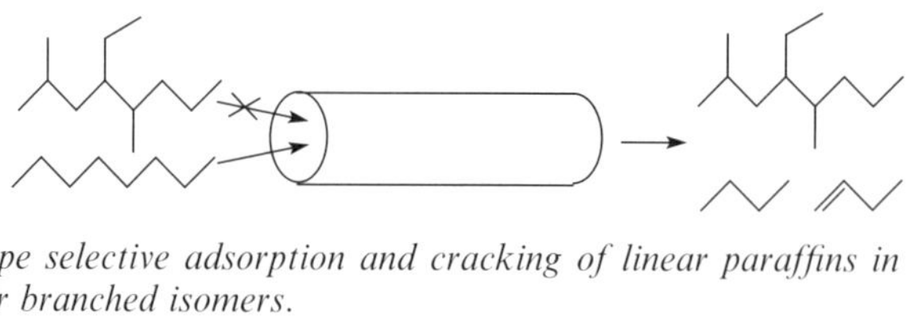

Figure 5   *Shape selective adsorption and cracking of linear paraffins in a mixture with their branched isomers.*

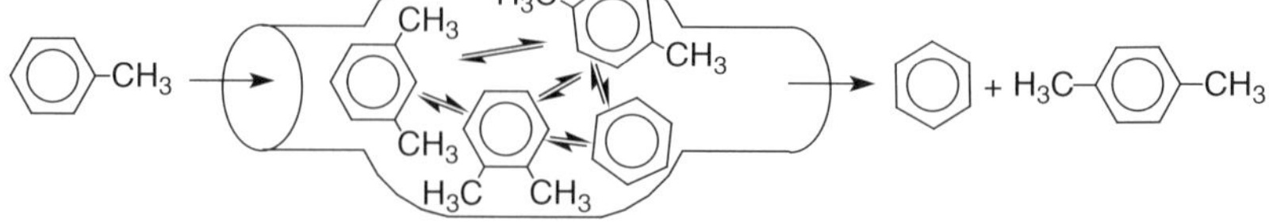

Figure 6   *Shape selective trans-alkylation of toluene to benzene and p-xylene.*

benzene and *p*-xylene because of their smaller cross sections. Again this property is at the heart of important commercial processes such as the production of *p*-xylene by the isomerization of a stream of xylene isomers and the trans-alkylation of toluene.

Transition state shape selectivity is invoked when a reaction path involves a bulky transition state that is not compatible with the size of the pores (Figure 7).

The absence of shape selectivity in both homogeneous and conventional heterogeneous catalysts leads to very different results in the above reactions.

A2.9.2 Adsorption and Desorption on Solid Supports and Catalysts

The term *adsorption* is the complex phenomenon of the interaction of an organic or inorganic molecule with a surface. *Physisorption* and *chemisorption* are generally distinguished according to the nature of forces involved.

Physisorption results from the interaction of a sorbent and a sorbate mostly through Van der Waals forces. Generally such an interaction is relatively weak, but there are notable exceptions, *e.g.*, the physisorption of hydrocarbons in zeolites which have heats of adsorption up to several tens kJ/mol.

Chemisorption on the other hand, describes a surface process in which chemical bonds are formed between surface atoms and a molecule, that either retains its structural integrity (molecular adsorption) or undergoes fragmentation (dissociative adsorption). Typical examples are the sorption of ammonia on strong acid sites, with the formation of ammonium ions (Figure 8), and of hydrogen on a metal surface forming M-H species.

The term desorption comprises the corresponding inverse surface processes. It is generally carried out thermally and in vacuo, with more drastic conditions being needed for the release of chemisorbed compounds.

Figure 7 *Shape selective trans alkylation of m-xylene to 1,2,4-trimethylbenzene.*

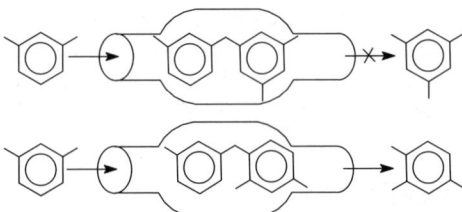

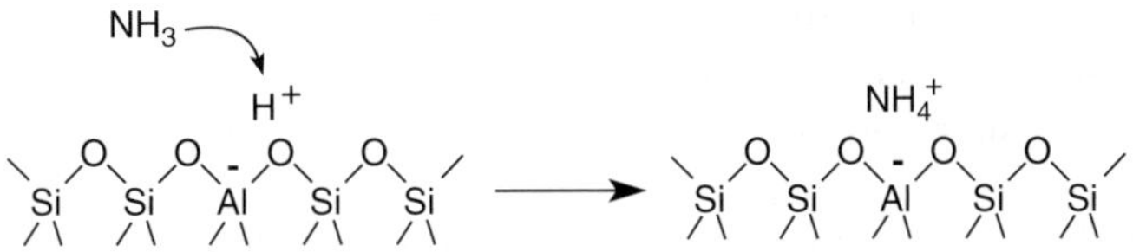

Figure 8 *Chemisorption of ammonia at a Brønsted acid site.*

A2.9.3   Petroleum Refining

One of the most important uses for catalysts is in oil refineries in which distillate fractions and most products are subjected to catalytic conversion or upgrading. In particular, the heavy fraction of crude oil (mainly consisting of large hydrocarbons, but also including N-, O-, S-, and even metal-containing heterocycles) is broken down to make smaller and cleaner hydrocarbons suitable for use as fuels in internal combustion engines. The processes are hydrocracking, hydrotreating, and catalytic reforming (Chapter 3, pages 84–89). Both oxide and metal catalysts are needed.

FCC (Fluid Catalytic Cracking) catalysts account for *ca.* 45.5% of all refinery catalysts by value (together with hydroprocessing 35.5%, hydrocracking 7.5%, reforming 7.5%, others 6%). Their use allows the cracking of heavy petroleum fractions (vacuum gas oils and residues) into more valuable lower boiling products in the gasoline and middle distillate boiling range. During the cracking, excess carbon is ejected from hydrogen deficient compounds as coke deposits on catalyst particles. The cracking occurs in the gas phase on fluidized bed reactors from which a fraction of the catalyst is continuously withdrawn and sent to a regenerator where the coke is burnt off with air.

Stabilized zeolite-Y is the principal acidic constituent of the catalyst, responsible for most of the cracking activity. The metal catalyst is a supported Pt, added in small amounts, and which functions as a CO oxidation catalyst, allowing cleaner gaseous emissions from the regenerator.

A2.9.3a   Hydrocracking

Hydrocracking combines the molecular weight reduction of the crude oil with an increase of the hydrogen content to give a useful products range (gasoline, middle distillate), in contrast to carbon loss in FCC. Hydrocracking requires bifunctional catalysts, containing catalytically active hydrogenation and cracking sites, and are generally composed of metal sulfides supported on an acidic material, such as silica-alumina or stabilized Y zeolite. More active Pd can be used, in place of metal sulfides in catalysts destined for second stage hydrocracking reactors, after most of the sulfur in the feed has been removed.

A2.9.3b   Hydrotreating

Hydrotreating summarizes various refinery operations in which hydrogen is used to improve the properties of refinery streams through the hydrodesulfurization (Section 3.2.1), hydrodenitrogenation, demetallization and hydrogenation (of aromatic or unstable products). Deoxygenation also takes place. Sulfur is removed as H_2S, nitrogen as NH_3 and oxygen as H_2O. Metal sulfides supported on alumina are generally used as catalysts (section A2.4.3).

Hydrotreating operations render refinery products suitable for the final use, e.g., in diesel oils and gasoline blends. Another use is in the hydrodemetallization of vacuum gas oils and residues destined to FCC, this protects the lattice of Y zeolite from destruction and to reduce the formation of coke and methane, which would be caused by the deposition of V and Ni on catalyst particles. Hydrotreating also removes sulfur from feeds, to allow the use of noble metal catalysts in hydrocracking and reforming processes.

A2.9.3c   Catalytic Reforming

Catalytic reforming increases the octane rating of gasoline fractions, through the aromatization of part of the C_{6+} paraffin content, particularly cycloparaffins. This process also produces most of the hydrogen needed in refinery processes as well as aromatic compounds (BTX) for petrochemical uses. Bifunctional catalysts, containing Pt/Re or Pt/Ir metals supported on a mildly acidic chlorided alumina, are used for dehydrogenation of cycloparaffins and the branching of linear ones.

The *isomerization* of C_5 and C_6 paraffins to increase chain branching, is another way to increase the octane rating of gasoline fractions. It is an acid catalyzed reaction, carried out commercially using chlorided alumina and acid mordenite, both containing supported Pt metal (Section 5.6). The hydrogenating activity of the platinum helps minimize side reactions leading to coke deposition. The use of large pore mordenite circumvents the restrictions of shape selectivity on the formation of branched C_5 and C_6 paraffins.

References

J. M. Thomas and W. J. Thomas, *Principles and practice of heterogeneous catalysis*, VCH, Weinheim, 1997.

B. C. Gates, *Catalytic chemistry* J. Wiley-Interscience, New York, 1992.

G.A. Somorjai, *Introduction to surface chemistry and catalysis*, J. Wiley-Interscience, New York, 1994.

R A Van Santen, J A Moulijn, P W N M. Van Leeuwen and B A Averill, *Catalysis an integrated approach*, NIOK/Elsevier, 1999.

J. Hagen, *Industrial Catalysis*, Wiley-VCH, Weinheim, 1999.

G. Centi, F. Cavani and F. Trifiró, *Selective Oxidation by Heterogeneous Catalysis*, Kluwer Academic/Plenum Publishers, New York, 2001.

N.Y. Chen, W.E. Garwood and F. G. Dwyer, *Shape Selective Catalysis in Industrial Applications*, Marcel Dekker, New York and Basel, 1989.

P.B. Venuto, *Microporous Mater.*, 1994, **2**, 297.

J.A. Rabo and M.V. Schoonover, *Appl. Catal. A: General*, 2001, **222**, 261.

M.G. Clerici and P. Ingallina, *J. Catal.*, 1993, **140**, 73.

M.G. Clerici, *Top. Catal.*, 2001, **15**, 257.

G. Langhendries, D.E. De Vos, G.V. Baron and P.A. Jacobs, *J. Catal.*, 1999, **187**, 453.

E.G. Derouane, *J. Mol. Catal. A: Chem*, 1998, **134**, 29.

M. Misono, *Catal. Rev. – Sci. Eng.*, 1987, **29 (2&3)**, 269.

I. Chorkendorff and J. W. Niemantsverdriet, *Concepts of Modern Catalysis and Kinetics*, Wiley-VCH, Weinheim, 2003.

R. L. Augustine, *Heterogeneous Catalysis for the Synthetic Chemist*, Marcel Dekker, New York, 1996.

R. A. Sheldon and H. van Bekkum Eds., *Fine Chemicals through Heterogeneous Catalysis*, Wiley-VCH, Weinheim, 2001.

A complete list of zeolites is provided by the *Atlas of Zeolites*, published by the International Zeolite Association.

Subject Index